Principles of
POLYMER SCIENCE

Principles of
POLYMER SCIENCE

Second Edition

P. Bahadur ☐ **N.V. Sastry**

Alpha Science International Ltd.
Oxford, U.K.

P. Bahadur
Department of Chemistry
South Gujarat University, Surat
Gujarat, India

N.V. Sastry
Department of Chemistry
Sardar Patel University, Vallabh Vidyanagar
Gujarat, India

Copyright © 2002, 2005
Second Edition 2005
Reprint 2006

Alpha Science International Ltd
7200 The Quorum, Oxford Business Park North
Garsington Road, Oxford OX4 2JZ, U.K.

ISBN 1-84265-246-X

Printed in India

Preface

One of the most exciting areas in chemistry, chemical engineering and materials science is the preparation, characterisation and application of polymers. Polymers (or macromolecules) constitute a group of materials made of long chains built up by covalently bonded molecules which include plastics, fibres, rubbers, adhesives and surface coating materials. Polymers have become the most versatile commodity in today's life style. We start our day with toothbrush, wear synthetic polymer clothing, move on rubberised wheels of the conveyance and go to bed on foamed urethane mattresses with cosy acrylic blankets.

Polymers belong to a relatively new branch of science (as the concept of macromolecules was first proposed by German physical chemist Hermann Staudinger in 1920, much criticised at that time but won him Nobel Chemistry prize in 1953) although these substances did exist since the first life began on the earth (proteins and nucleic acids which are the essential components of the cell, are polymers). There has been tremendous development in polymers in past 70 years or so and as on today it is impossible to think of world without polymeric materials that we use in our day-to-day life. The use of synthetic polymers is increasing rapidly year by year and in many applications they are replacing conventional materials like metals, ceramics, wood, natural fibers such as cotton, wool etc. In fact, science of polymers has its roots in physics, chemistry, biology, engineering and technology and medicine and has grown to a full fledged discipline of science. In polymer chemistry alone all aspects (viz. analytical, organic, physical, inorganic and biological) have been extensively developed. The commercial and economic value of polymers is also considerable because nowadays commodity and specialty polymers have been produced and several industries are flourishing by dealing with them.

Looking into the advances in polymer science in general and polymer chemistry in particular, it is almost impossible to cover all aspects of polymers in a single book.

This book familiarises the readers (who have little or no prior knowledge of the subject) to this fascinating world of polymers. The level will range from highly introductory treatment to a bit advanced aspects covering all facets of polymer science. The book is not intended to provide an in-depth coverage of many aspects of polymers such as books included in the further reading list, but rahter an overview. One can never do full justice in writing a book on a subject such as polymer science that has roots everywhere and it may not be surprising if we have failed in including some important topic or did not do justice by writing some other with sufficient details. Enough care has been taken to eliminate typographical errors and make the text available for enjoyable reading. Readers who have suggestions for improving the book are cordially invited to send them to us. Polymer chemistry is now an important subject in almost all undergraduate/ postgraduate curricula in science and engineering. The book, aimed for undergraduate and postgraduate students of chemistry and material science, would also be useful to engineering students and chemists working in industries.

To cater the needs of students coming from multidisciplinary background, an introductory course in polymer science should provide information not only on chemistry especially physical chemistry of polymers but also on other aspects such as basic information on important application of synthetic polymers as fibers, thermoplastics, elastomers and technological principles involved in their production. Mention of not only organic polymers but also natural, inorganic and specialty polymers should also be made. This book has been formulated on these lines. The book is divided into four main aspects of polymer science namely synthesis, analysis and characterisation, important solid and solution state properties of polymers, technology of polymers and much required emphasis on natural, inorganic and specialty polymers. These parts are preceded by general introduction which aims at defining and explaining salient and unique features such as chain constitution, inter-molecular forces, configurational and conformational forms in polymer chains and milestones in the development of modern polymer science. Additional supplementary material such as some common laboratory experiments, brief bio-sketch and contributions of polymer science pioneers is given in the appendix part.

The book has been written in simple English and contains neat illustrations and easily understandable tables. Several questions of different type including numerical problems, concept based problems, multiple choice questions, etc. along with their answers have been given in the end. In all this we have made an ambitious effort in explaining several facets of polymers and we would be highly satisfied if the book helps the needs of students and teachers alike.

We are grateful to Mr. Saurabh Soni and John George for their skilful assistance in the art work. The handling of the manuscript at different stages till its publication by Dr. Mayank Dalal, Dr. Nirmesh Jain and Dr. Alex George is also gratefully acknowledged. One of the authors PB* thanks his wife Anita for her continued patience, understanding, endless love and faith throughout the enormous task of completion of the book. We can offer here only an inadequate acknowledgement of our appreciation to Shri N.K. Mehra, M/s Narosa Publishing House, for his enthusiasm and great patience.

P. BAHADUR
N.V. SASTRY

Contents

1

General Introduction

1.1 Introduction

A **polymer** may be defined as a substance built up of a number of repeating chemical units held together by chemical bonds. A high polymer is one in which the number of repeating units is in excess of about 1000. This number is termed as **"Degree of Polymerisation (DP)"**. The molecular weight of a polymer is often given by the product of the molecular weight of the repeating units and DP. The compound or compounds, used in the preparation of polymer are called **monomers.** The repeating units constituting the polymer molecule are called constitutional repeat units (or **CRU**). Fig. 1.1 illustrates the monomer, DP and CRU for polyvinyl chloride.

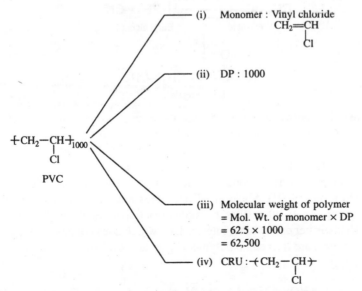

(i) Monomer : Vinyl chloride
$CH_2{=}CH$
$|$
Cl

(ii) DP : 1000

$+CH_2{-}CH+_{1000}$
$|$
Cl

PVC

(iii) Molecular weight of polymer
= Mol. Wt. of monomer × DP
= 62.5 × 1000
= 62,500

(iv) CRU : $+CH_2{-}CH+$
$|$
Cl

Fig. 1.1   The monomer, DP, molecular weight and CRU in polyvinyl chloride (PVC)

It is important to note here that the term polymer was first used by Berzelius in 1827, but the concept of polymer as molecule with high molecular weight (at least few tenthousands) was introduced by the German Scientist Hermann Staudinger almost a century later (in 1920) which was at that time severely criticised. A macromolecule is the term interchangeably used for polymers, more often of biological origin. For Example insulin, a protein-like hormone, is made up of several different amino acids (51 in total) and can be called as a macromolecule but not strictly as a high polymer. Not all the substances can be polymerised. It is necessary that a chemical substance can act as monomer only if it has at least bi-functionality (which could be due to one double bond or two reactive functional groups).

As stated before, the bi-functionality may result due to the presence of at least a double bond (like in olefins) or two reactive functional groups (two -OH groups in ethylene glycol).

$$H_2C{=}CH_2 \longrightarrow {\leftarrow}CH_2{-}CH_2{\rightarrow}_{\overline{n}}$$

Ethylene   Polyethylene

$$- HO{-}CH_2{-}CH_2{-}OH \longrightarrow HO{\leftarrow}CH_2{-}CH_2{-}O{\rightarrow}_{\overline{n}}H$$

Ethylene glycol   Polyethylene glycol

Let us consider following examples to show that bi-functionality in a monomer is a must for the growth of a polymer chain through the reaction of functional groups successively. For two compounds each possessing only one functional group (as in the case of acetic acid and ethyl alcohol), the reaction is complete in one step leading to the formation of ethyl acetate.

$$CH_3{-}\overset{\overset{\textstyle O}{\|}}{C}{-}O\mathrm{H} + HO{-}CH_2{-}CH_3$$

Acetic acid   Ethyl alcohol

$$CH_3{-}\overset{\overset{\textstyle O}{\|}}{C}{-}O{-}CH_2{-}CH_3$$

Ethyl acetate

For a reaction between two compounds one possessing a monofunctional group such as a carboxylic group in acetic acid and the other with bifunctional groups of two hydroxyl groups in ethylene glycol, it can be easily seen that the reaction completes in two steps leading to a monoester (I) in the first step. In the second step the pre-formed monoester reacts further with one molecule of acetic acid to yield a diester (II). Since the diester does not have any more reactive functional groups on its either sides, no further reaction takes place. Thus it can be seen that when diester or any other intermediate which does not have any reactive functional groups, are formed no polymer chains are resulted.

Ist step

$$HO{-}CH_2{-}CH_2{-}OH + H{-}O{-}\overset{\overset{\textstyle O}{\|}}{C}{-}CH_3$$

Ethylene glycol   Acetic acid

$$HO{-}CH_2{-}CH_2{-}O{-}\overset{\overset{\textstyle O}{\|}}{C}{-}CH_3$$

Monoester
[1]

[I] combines with another molecule of CH_3COOH in the second step to form a diester [II];

IInd step

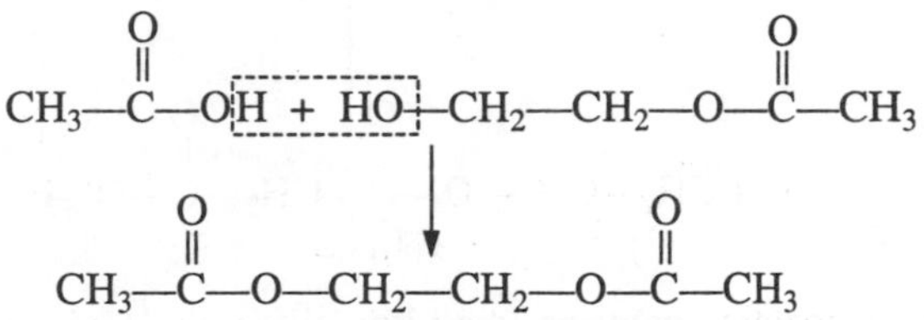

$$CH_3-\overset{\overset{O}{\|}}{C}-O\underline{H + HO}-CH_2-CH_2-O-\overset{\overset{O}{\|}}{C}-CH_3$$

$$\downarrow$$

$$CH_3-\overset{\overset{O}{\|}}{C}-O-CH_2-CH_2-O-\overset{\overset{O}{\|}}{C}-CH_3$$

Diester

[II]

Now if we consider a reaction between two compounds each containing two reactive functional groups (e.g. ethylene glycol and malonic acid), it can be inferred that the addition of each of the monomer to another continues till the concentration of one of the species is depleted. Then one can have high molecular weight polymers formed from these bifunctional monomers.

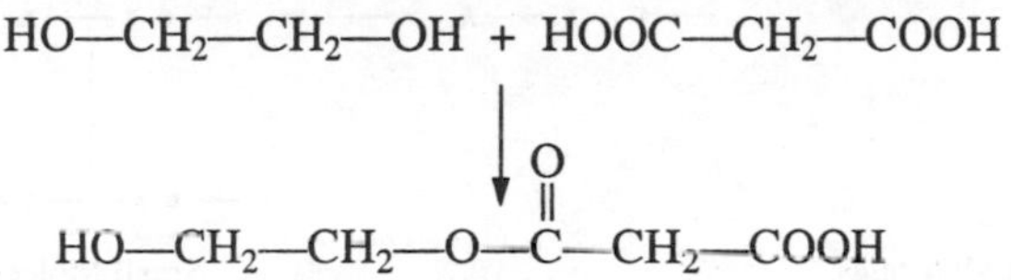

$$HO-CH_2-CH_2-OH + HOOC-CH_2-COOH$$

$$\downarrow$$

$$HO-CH_2-CH_2-O-\overset{\overset{O}{\|}}{C}-CH_2-COOH$$

Monoester

Monoester can combine with either glycol or malonic acid at the corresponding functional group end to facilitate further growth of the chain;

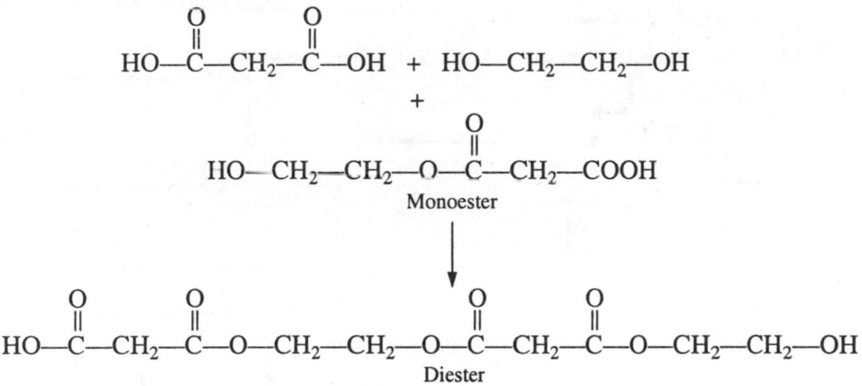

$$HO-\overset{\overset{O}{\|}}{C}-CH_2-\overset{\overset{O}{\|}}{C}-OH + HO-CH_2-CH_2-OH$$

$$+$$

$$HO-CH_2-CH_2-O-\overset{\overset{O}{\|}}{C}-CH_2-COOH$$

Monoester

$$\downarrow$$

$$HO-\overset{\overset{O}{\|}}{C}-CH_2-\overset{\overset{O}{\|}}{C}-O-CH_2-CH_2-O-\overset{\overset{O}{\|}}{C}-CH_2-\overset{\overset{O}{\|}}{C}-O-CH_2-CH_2-OH$$

Diester

Thus if n moles of a bifunctional species such as malonic acid is reacted with the same number of moles of ethylene glycol which has two reactive hydroxyl groups on both the ends, the esterification reaction occurs step wise forming mono, di-, tri-, tetra-, multi- and polyesters (see the following scheme).

$$
nHO-\underset{\parallel}{\overset{O}{C}}-CH_2-\underset{\parallel}{\overset{O}{C}}-OH \;+\; nHO-CH_2-CH_2-OH
$$

$$
HO+CH_2-CH_2-O-\underset{\parallel}{\overset{O}{C}}-CH_2-\underset{\parallel}{\overset{O}{C}}-O)_n H
$$

Polyester

Thus, the reaction between two bifunctional molecules leads to a linear high polymer. Further, if we add some trifunctional group, it may form either branched polymeric chain or a three dimensional network structure.

The reactions between different molecules is schematically shown in Fig. 1.2

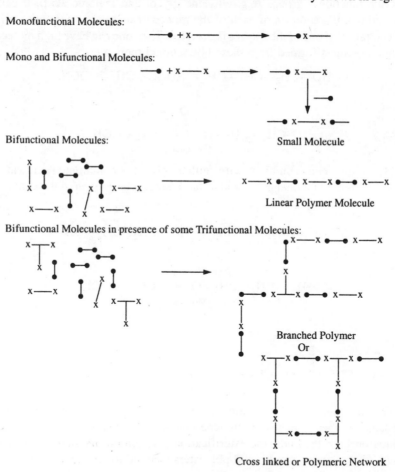

Fig. 1.2   **Reactions between molecules with different functionality**

1.2   Classification of Polymers

There are several ways to classify polymers. The classification is based on several considerations. The source of polymers i.e. natural or synthetic, the type of polymerisation process used in the synthesis, nature and type of chain and solid state behavour of polymer chains etc. The different classification schemes can be outlined as shown in Fig. 1.3.

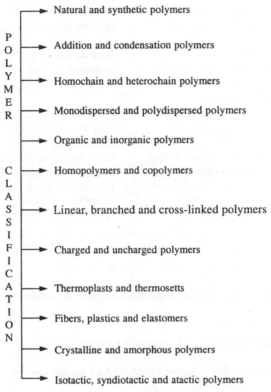

Fig. 1.3 Classification of polymers

Most linear polymers take on new shapes by the application of heat and pressure. They are thus called **'thermoplastics'** whereas the cross-linked polymers cannot be made to flow or melt irreversibly and are said to be **'thermosetting'** resins. The differences between these two types are listed in Table 1 (a). A polymer may contain monomers of identical or different chemical structure. Polymers made up of only one type of monomeric units are called **'homopolymers'** whereas those polymeric compounds which are built up of two different types of monomer units in their chain are called **'copolymers'** (or mixed polymers). Polymers with three different types of monomeric units are some times called **'terpolymers'**. The advantage of such polymeric structures is that a single polymer molecule can have the properties of both the entities, which can be selected suiting to the end use of application. Such possibility hardly exists in simple molecules.

Copolymers are further classified as alternating copolymers or statistically regular (where both different repeating monomeric units are joined side by side) and random copolymers or statistically irregular (where there is no regularity in

Table 1(a) Comparison between thermoplastics and thermosetts

Thermoplastics	Thermosetts
1. They are often formed by addition polymerisation leading to long linear chain polymers with no cross-links	They are often formed by condensation polymerisation and three dimensional network structures are formed
2. They soften readily on heating because secondary forces between the individual chains can break easily by heat or pressure	They do not soften on heating, On prolonged heating, however charring of polymers is caused
3. They can be reshaped and reused	They can not be reshaped and reused
4. They are usually soft, weak and less brittle	They are usually hard, strong and more brittle
5. These can be reclaimed from wastes	They can not be reclaimed from wastes
6. They are usually soluble in suitable solvents	Due to strong bonds and intra- and interchain cross-links, they are insoluble in almost all organic solvents

the joining of units of two different monomers). Copolymers can also be of block or graft types. In Fig. 1.4 are schematically shown structures of different homo-, co- and ter-polymers made using the monomers A, B and C.

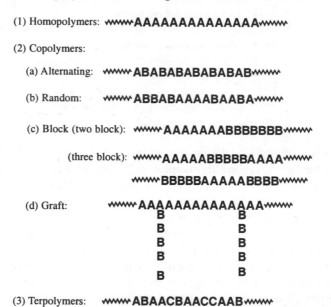

(1) Homopolymers: wwwwAAAAAAAAAAAAAAwwww

(2) Copolymers:

   (a) Alternating: wwww ABABABABABABAB wwww

   (b) Random: wwww ABBABAAAABAABA wwww

   (c) Block (two block): wwww AAAAAAABBBBBBB wwww

      (three block): wwww AAAAABBBBBAAAA wwww

          wwww BBBBBAAAAABBBB wwww

   (d) Graft: wwww AAAAAAAAAAAAAAAA wwww
                B   B
                B   B
                B   B
                B   B
                B   B

(3) Terpolymers: wwww ABAACBAACCAAB wwww

Fig. 1.4 Molecular architecture of homo-, co- and terpolymers

Polymers can be classified as linear, branched or cross-linked polymers. High density polyethylene (HDPE) is a linear polymer, while low density polyethylene (LDPE) is a branched polymer. Natural rubber has two configurational forms. The *cis* form has less density and is available as latex. The other form is the *trans* form and is brittle and hard (gutta percha). Further the natural rubber on vulcanisation (or mastication with sulphur) develops crosslinking and becomes processable. The various topologies of the polymer chains are schematically presented in Fig. 1.5.

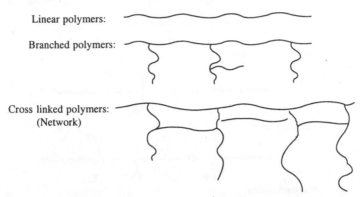

Linear polymers:

Branched polymers:

Cross linked polymers:
(Network)

Fig. 1.5 Schematic representation of linear, branched and cross-linked polymers

According to the structure of their main chain, all polymers can be broadly classified as homochain and heterochain polymers. In case of homochain polymers the main chain is made up of linkages between the atoms of same element e.g. carbon in organic polymers.

$$-\overset{|}{\underset{|}{C}}-\overset{|}{\underset{|}{C}}-\overset{|}{\underset{|}{C}}-\overset{|}{\underset{|}{C}}-\overset{|}{\underset{|}{C}}-\overset{|}{\underset{|}{C}}-\overset{|}{\underset{|}{C}}-$$

In heterochain polymers, the main chain consists of linkages between different atoms e.g. between carbon and oxygen in polyester, polyethers, and between carbon, nitrogen and oxygen in polyurethanes etc.

$$-\overset{|}{\underset{|}{C}}-\overset{|}{\underset{|}{C}}-O-\overset{|}{\underset{|}{C}}-\overset{|}{\underset{|}{C}}-O-\overset{|}{\underset{|}{C}}-$$

Polymers are also classified as organic, elementoorganic and inorganic polymers. Organic polymers have chains consisting of C—C linkages and have apart from carbon atoms, hydrogen, oxygen, nitrogen, sulfur and halogen atoms in the side chains. Elementoorganic (or hetero organic) polymers include (i) macromolecules whose chains are composed of carbon as well as heteroatoms (except N, S, O and halogen atoms) and (ii) inorganic chains in which side groups contain carbon atoms directly linked to chain. Inorganic polymers are polymers containing no carbon atoms but have Si—Si, Si—O, N—PX, P—O and B—O linkage e.g. polysilanes, polysiloxanes, polyphosphazenes, polyphosphoric acid or polyphosphates and polyboron oxides. Inorganic polymers have been studied to a little extent so far and it is difficult to provide a classification. However, inorganic polymers possess superior thermal, electrical and mechanical properties over the organic polymers. Silicones, polyphosphazenes and several organometallic and coordination polymers are from this class and worth mentioning (these are described in detail in chapter 7). Some common inorganic polymers are shown in Fig. 1.6.

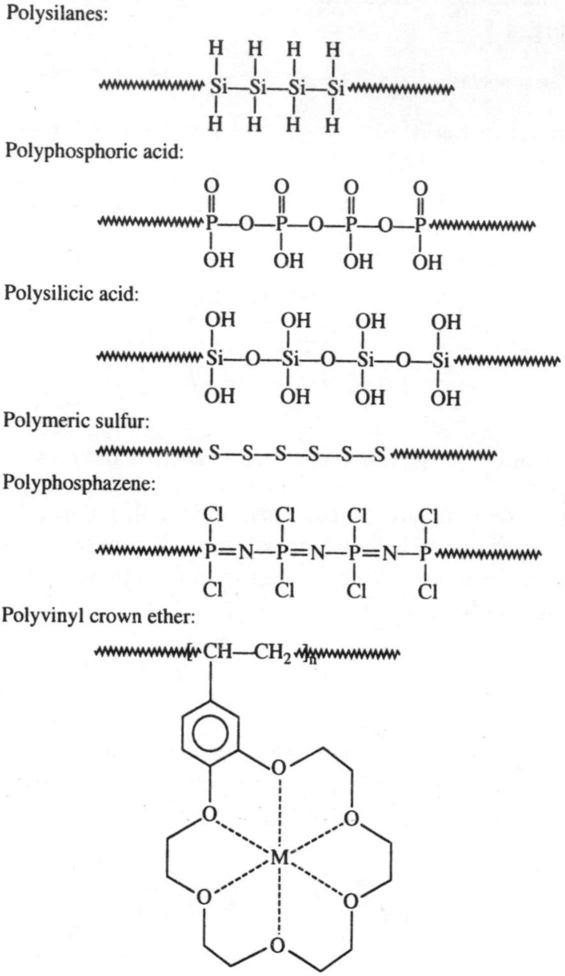

Polysilanes:

Polyphosphoric acid:

Polysilicic acid:

Polymeric sulfur:

Polyphosphazene:

Polyvinyl crown ether:

Fig. 1.6   Some common inorganic polymers

Polymers in which the side branches are present in every unit, being joined by different chemical groups with the main polymer chain are known as comb-like polymers. Branched polymers which resemble a star by their structure are known as star like polymers. Collinear double chain polymers are known as ladder polymers. Model structures for various branched polymers are shown in Fig. 1.7.

Polymers can also be classified as (i) **natural** or (ii) **synthetic.** The common natural polymers include polysaccharides (starch, cellulose, gums etc), proteins (gelatin, albumin, enzymes, insulin), polyisoprenes (natural rubber, gutta percha) and nucleic acids (RNA and DNA). Natural polymers are sometimes also called 'Biopolymers' or 'Biological macromolecules'.

Polymers can also be classified further as fibers, plastics, resins and rubbers

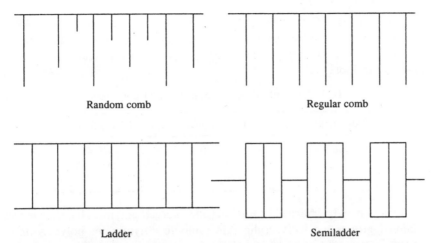

Random comb   Regular comb

Ladder   Semiladder

Fig. 1.7 Possible model structures for branched and cross-linked polymers

based on the nature and extent of secondary valence forces and mobility among the constitutional repeat units. Important examples belonging to each class are shown in Table 1(b).

Table 1(b) Some important polymeric materials useful as plastics, fibers, rubbers and resins

Form	Type	Polymeric material
Fibres	Natural	Cotton, wool, silk and asbestos
	Semisynthetic (Cellulosic)	Viscose rayon, acetate rayon and cuprammonium rayon
	Synthetic	Polyamides, polyesters, acrylics, polyurethanes, polyolefins, polypropylene, polyvinyl chloride, polyvinyl alcohol
Plastics	Cellulosic plastics	Celluloid, cellophane
	vinyl plastics	Polyolefins-(polyethylene, polypropylene, polytetrafloroethylene (teflon)), polystyrene, polyvinyl chloride, polyvinyl acetate, polyvinylpyrrolidone,
	Polyacryloids	polyacrylic acid and polymethylmethacrylate
Rubber	Natural	polyisoprene (*cis*-1, 4)
	Synthetic	(polybutadiene, polychloroprene and polyisoprene), butadiene copolymers (e.g. NBR, SBR or Buna N and Buna S), butyl rubber, polyurethanes, polysulfides and polysiloxanes (silicone rubber)
Resins	Water Soluble	Modified starches, cellulose derivatives, polyvinyl alcohol, polyvinyl pyrrolidone, polyacrylic acid and polyacrylamide
	Water Insoluble	Phenol plasts and amino plasts, polyurethane resins, silicone resins, epoxy resins and unsaturated polyesters

Plastics, fibers and elastomers possess different physical properties. While fibers are crystalline materials, elastomers are amorphous. Plastics lie between these two extreme cases in physical properties viz. modulus and elongation as shown in the Table 1(c).

Table 1(c)   Physical characteristics of polymers

Polymers	Modulus, dyne cm^{-2}	Elongation, %
Elastomers	$10^6 - 10^7$	1000
Plastics	$10^8 - 10^9$	100 – 200
Fibers	$10^{10} - 10^{11}$	10 – 30

Polymers may be **charged** or **uncharged.** Charged polymers have some free functional groups e.g. polyacrylic acid (anionic polymer) or polyethylimine (cationic polymer). Charged (or ionic) polymers that are soluble in water are called as "**polyelectrolytes**". Water insoluble charged polymers are often called "**ionomers**".

Polymers may also be classified as **amorphous** or **crystalline** depending upon their morphological behaviour. These two types of polymers behave differently as crystallinity influences the properties such as hardness, stiffness and elasticity, thus making them useful as plastic, rubber, fiber or resin. LDPE is less crystalline while HDPE is highly crystalline polymer. However, it is important to note that solid polymers when crystallised hardly attain hundred percent crystallinity. Thus solid state of polymers is often characterised by both amorphous and crystalline domains. The two polyethylenes have different properties as shown in Table 1(d).

Table 1(d)   Properties of LDPE and HDPE – a comparison

LDPE	HDPE
Branched (% crystallinity < 50)	Linear (% crystallinity > 90)
Density, 0.92 g cm^{-3}	Density, 0.96 g cm^{-3}
Softening point, 360 K	Softening point, 400 K
Tensile strength, 85–136 atm	Tensile strength, 205 – 315 atm

Linear low density polyethylene (LLDPE) is now manufactured and has less branching. The linearity provides the polymer strength and branching gives toughness.

1.3   Nomenclature of Polymers

Polymers are prepared by addition polymerisation of an unsaturated monomer or by step polymerisation of monomer (or monomer pair) containing at least two functional groups. The nomenclature of addition polymers is relatively simple. Condensation polymers, for example, terylene and nylons are named according to the repeat unit.

Addition polymers are usually named by prefixing **poly-** before the monomer from which the polymer is derived. For example polyvinyl chloride and polyethylene are synthesised from vinyl chloride and ethylene monomers, respectively.

Copolymers are named by inserting -**co**- (or -**alt**-, -**rand**-, -**block**-, or -**graft**- if the position of the two monomers is known) as prefix. For example; poly (styrene-*co*-butadiene) is a copolymer of styrene and butadiene and poly (styrene-*block*-butadiene) is a diblock copolymer of styrene and butadiene. It is also a common practice to designate a diblock copolymer as styrene-butadiene copolymer and a triblock copolymer as styrene-butadiene-styrene block copolymer. An AB block copolymer is also designated as poly (A-*b*-B) and an ABA copolymer as poly (A-*b*-B-*b*-A). The block sequence arrangement of a diblock copolymer is also represented by a long dash such as A—*block*—B and the corresponding copolymer as poly **A**—*block*—poly B.

A wide diversity exists in naming polymers viz. based on source, structure, industrial names, trivial names and even IUPAC names. The nomenclature of few polymers has been illustrated in Table 1(e). However no single nomenclature is universally adopted.

IUPAC names of some other common polymers are polystyrene (poly-(1-phenylethene)), polyphenylene oxide (poly(oxyl, 4 phenylene)) and polyvinyl chloride (poly (1-chloroethene)). Similarly IUPAC names for various monomers are; adipic acid (1, 6 hexandioic acid), acrylonitrile (propennitrile), bisphenol A (2, 2, bis-4-hydroxyphenylpropane), butadiene (1, 3-butadiene), caprolactam (2-oxohexamethyleneimine), chloroprene (2-chloro-1, 3-butadiene), divinyl-benzene (1, 4-diethenylbenzene), ethylene glycol (1, 2,-ethane diol), ethylene oxide (1, 2 epoxyethane), hexamethylene diamine (1, 6-hexanediamine), isoprene (2-methyl 1, 3-butadiene), melamine (2, 4, 6-triamino-s-triazine), methacrylic acid (2-methylpropenoic acid), styrene (ethenyl benzene), terphthalic acid (benzene-1, 4-dicarboxylic acid) and vinyl chloride (chloroethene). Also, these IUPAC names given in the parenthesis are seldom used in the polymer literature.

1.4 Isomerism in Polymer Chains

Various types of isomerism are possible in case of polymers. These can be grouped as constitutional, orientational, geometrical or stereoisomerism as shown in Fig. 1.8.

A good example of constitutional isomerism is that of the monomer with the CRU as C_2H_4O. Three different polymers viz. polyethylene oxide, polyvinyl alcohol and polyacetaldehyde are possible, each one has quite different property. It is not always that a polymer chain possesses regularity; any polymer which lacks such an ordered arrangement is called irregular. Polydienes, for example 1, 2- and 1, 4-polybutadienes show chain irregularity. In case of polyisoprenes (1, 2-; 3, 4- and 1, 4-) isomers are possible. Chain irregularity may be of different kinds e.g. polymerisation of vinyl polymers gives rise to 'head to tail' (major) as well as 'head to head' or 'tail to tail' (minor) products as shown in the Fig. 1.8.

The stereo isomerism in polymer chains leads to tacticity (spatial order of arrangement). Based on tacticity, the macromolecular substances are divided into three categories as shown in Fig. 1.9.

Table 1(e)   Different nomenclature schemes for polymers

Repeat unit#	Name based on chemical structure	IUPAC name	Commercial or Trivial name
1. $-CH_2-CH_2-$	Polyethylene	Polymethylene	Polyox
2. $-CH_2-CH_2-O-$	Polyethylene oxide or Polyethylene glycol	Polyoxyethylene	Carbowax
3. $-CH_2-CH-$ OH	Polyvinyl alcohol	–	Vinylon
4. $+CH-CH_2\}_n\{CH_2-CH-CH_2-CH_2\}_m$ C_6H_5 CH_3	Styrene-hydrogenated isoprene block copolymer	–	
5. $-NH-C-(CH_2)_5-$ O	Poly (ε-caprolactam)	Poly[imino(1-oxy hexamethylene]	Nylon, 6
6. $-O-CH_2-CH_2-O-C-\bigcirc-C-$ O O	Polyethyleneterephthalate	Poly (oxyethyleneterephthaloyl)	Mylar (film) Terylene or Dacron Fiber)

\# 1- source $CH_2 = CH_2$ or Br $(CH_2)_{12}$ Br + 2Na, 2 — two sources e.g. ethylene oxide or ethylene glycol, 3-source does not exist and obtained by hydrolysis of polyvinylacetate, 4-source is hydrogenation of styrene-isoprene copolymer, 5-from caprolactam or ε-aminocaproic acid and 6 - from ethylene glycol and terephthalic acid

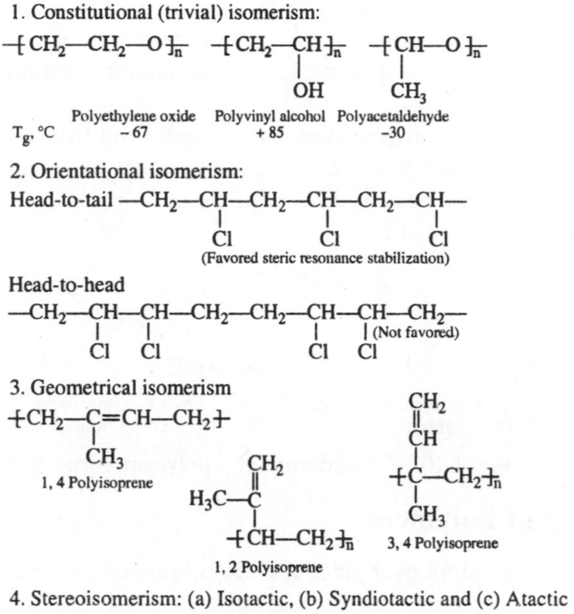

1. Constitutional (trivial) isomerism:

$$-\!\!\left[CH_2\!-\!CH_2\!-\!O\right]_{\!n}\quad -\!\!\left[CH_2\!-\!CH\right]_{\!n}\quad -\!\!\left[CH\!-\!O\right]_{\!n}$$
$$\qquad\qquad\qquad\qquad\quad\ \ \stackrel{|}{OH}\qquad\ \ \stackrel{|}{CH_3}$$

	Polyethylene oxide	Polyvinyl alcohol	Polyacetaldehyde
T_g, °C	– 67	+ 85	–30

2. Orientational isomerism:

Head-to-tail —CH$_2$—CH—CH$_2$—CH—CH$_2$—CH—
$\qquad\qquad\qquad\quad\ \stackrel{|}{Cl}\qquad\ \ \stackrel{|}{Cl}\qquad\ \ \stackrel{|}{Cl}$
(Favored steric resonance stabilization)

Head-to-head

—CH$_2$—CH—CH—CH$_2$—CH$_2$—CH—CH—CH$_2$—
$\qquad\quad\ \stackrel{|}{Cl}\ \ \stackrel{|}{Cl}\qquad\qquad\ \ \stackrel{|}{Cl}\ \ \stackrel{|}{Cl}$ (Not favored)

3. Geometrical isomerism

$$+CH_2\!-\!\underset{\underset{CH_3}{|}}{C}\!=\!CH\!-\!CH_2+$$

1, 4 Polyisoprene

$$H_3C\!-\!\underset{\underset{+CH-CH_2+_n}{|}}{\overset{\overset{CH_2}{\|}}{C}}$$

1, 2 Polyisoprene

$$\underset{\underset{CH_3}{|}}{\overset{\overset{CH_2}{\overset{\|}{\underset{|}{CH}}}}{+C\!-\!CH_2+_n}}$$

3, 4 Polyisoprene

4. Stereoisomerism: (a) Isotactic, (b) Syndiotactic and (c) Atactic

Fig. 1.8   Isomerism in polymers

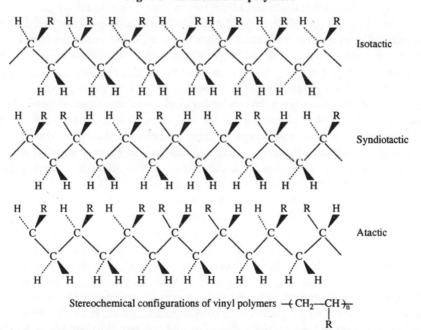

Isotactic

Syndiotactic

Atactic

Stereochemical configurations of vinyl polymers —(CH$_2$—CH)$_{\overline{n}}$
$\qquad\qquad\qquad\qquad\qquad\qquad\qquad\qquad\qquad\qquad\ \stackrel{|}{R}$

Fig. 1.9   Isotactic, syndiotactic and atactic polymers

Isotactic polymers — these polymers consist of chain segments, which display a regular repetition of monomer units with tertiary C-atoms of the same steric configuration; syndiotactic polymers — the chains of these polymers consist of regular series of monomer units in which every second C-atom of the chain possesses opposed steric configuration and atactic polymers — the polymer chain lacks regularity in the distribution of steric configuration of monomer units.

The *cis-trans* isomerism in polymers can be easily seen from the structures of two natural polymers (natural rubber and gutta-percha). As shown in Fig. 1.10, both are polyisoprenes but possess different physical behaviour e.g. the former is elastic whereas the later is hard.

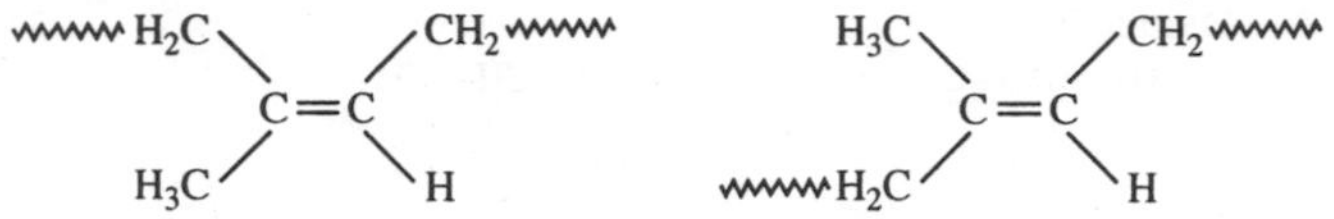

cis-polyisoprene   *trans*-polyisoprene
Natural rubber   Gutta-percha

Fig. 1.10   *Cis* and *trans* 1, 4 polyisoprenes

1.5   History of Polymers

Polymers existed in nature ever since the life originated on earth as the prime bio-molecules viz. proteins, nucleic acids and polysaccharides, which are all naturally occurring polymers (biopolymers or biological macromolecules). Staudinger's concept of macromolecules (severely criticised initially) got support from the research works of Hermann Mark and William Carothers. The development of molecular weight determination methods proved in no uncertain terms that molecules with longer chains and bigger size do exist. Natural rubber (which is a polymer of isoprene) was known to man for more than 500 years. Christopher Columbus on his voyage to explore the world, found Brazilians playing with a ball-like material made from the sap (or latex) of the rubber trees. however, first successful application of natural rubber was possible only after 1840 when Charles Goodyear cross-linked natural rubber with elemental sulphur (the process is known as **vulcanisation**). Unvulcanised rubber was gummy material which creeped of its own weight. Different amounts of sulphur added to natural rubber made it more and more hard, tough and elastic material. Baekeland made the first synthetic polymer by the condensation of phenol and formaldehyde (now known as PF resin or Bakelite). The accidental discovery of cellulose nitrate by Schonbein and its plasticization with camphor to produce celluloid by Hyatt was done much before the concept of macromolecules was put forward by Staudinger in 1920. In the decade of 1920s several new polymers like plasticized PVC, urea-formaldehyde resins, thiokol rubber and alkyd resins were made. Polyethylene, polystyrene, polymethyl methacrylate, neoprene rubber, nylon 6, 6, melamine resin and polyesters were soon commercialised. In 1940, Plunkett at DuPont company in USA accidentally discovered polytetrafluoroethylene (PTFE, teflon) which is the most inert polymer commonly known for its use as a coating material for kitchen ware particularly in non-sticking frying pans. However, a very important development in the history of synthetic polymers is the Ziegler-Natta polymerisation. Karl Ziegler (from erstwhile East Germany) was able to produce linear polyethylene (now called HDPE) under ordinary conditions of pressure and temperature. The demands of modern day life for high performing and speciality polymers has thrown a challenge to polymer chemists, physicists and technologists to come up with new polymers and also do modifications on existing polymers.

In earlier days, very drastic conditions were needed for the polymerisation of ethylene, which yielded branched polymer (LDPE). Both HDPE and LDPE are now important polymers with varied applications. Using Ziegler's co-ordination catalysts, Italian scientist Giulio Natta polymerised propylene with ordered arrangement of pendant methyl groups. This gave highly crystalline polypropylene, again a polymer with a great demand in today's market. Some important dates in the polymer history are outlined in Table 1 (f). Since 80s so many monomers were polymerised. Copolymerisation using two or more monomers was achieved. The discovery of thermoplastic elastomers, thermally stable polymers, conducting polymers and liquid crystalline polymers can be considered as important landmarks in the polymer history. Polymer chemistry has shown the fastest development in

Table 1(f)   Important events in the polymer history

Period	Event
1840–1920	Vulcanisation of natural rubber by Charles Goodyear, Schonbein made cellulose nitrate, Menard and Hyatt synthesised collodion and celluloid respectively. Chardonnet prepared regenerated cellulose fibers, Graham classified substances as crystalloids and colloids, oligomers of ethylene glycol and amino acids were made, Baekeland prepared the first synthetic polymer bakelite
1920	Staudinger gave the concept of macromolecules, alkyd resins, plasticised PVC, U-F and thiokol resins were synthesised
1930	Synthesis and commercialisation of melamine resins, polymethyl methacrylate (Perspex, ICI; lucite, DuPont), neoprene, polystyrene, polyethylene and nylon 6, 6 (by Carothers at DuPont)
1940	PET and unsaturated polyesters (terylene, ICI; dacron, DuPont) and teflon (by Plunkett at DuPont) were patented
1950	Synthesis and development of ABS, spandex (polyurethane elastomers), HDPE (by Ziegler), stereoregular PP (by Natta) and polycarbonates were reported
1960	Ionomeric thermoplastic elastomers, aramides (kevlar), flame retardant polymers and carbon fibers (cyclised, dehydrogenated PAN) were developed
1970	Thermotropic liquid crystalline polymers were invented
1980	Polyetherether ketone (PEEK) was made
1990	Development of polymer composites, high performance polymers, design and synthesis of biodegradable polymeric vehicles for sustained release of pharmaceuticals and fertiliser formulations, design of hydro- and micro gels based on water soluble polymers, development of conducting polymers, polymer solid electrolyte, electro-luminescent polymers and synthesis of multifunctional star and dendrimers

past half century and has emerged as a full-fledged science having its roots in physics, biology, engineering and technology. Detailed mathematical treatments have been provided to explain the mechanisms/kinetics of polymerisation and. for the solid and solution properties of polymers.

1.6   Industrial Scenario

Polymers have now become indispensable materials for us and it is difficult today to think of daily life without them. Virtually modern synthetic polymers have replaced the use of metals in many cases. Polymers are in fact engineering materials and are fast replacing metals in every application. Some of the advantages and disadvantages in using polymers in place of metals are listed as follows. The advantages of polymers are (i) they are light weight, (ii) they have good thermal/ electrical insulation capacity, (iii) they are resistant to corrosion effects and are chemically inert, (iv) polymers have easy workability and their fabrication costs are low, (v) they have good strength, dimensional stability and toughness, (vi) they are transparent in appearance, have good dyeability and possess decorative surface effects and (vii) polymers absorb the mechanical shocks and show resistance to abrasion effects. The major disadvantages however are (i) polymers are high cost materials, (ii) they are easily combustible, (iii) they have poor *ductility*, (iv) they deform under applied load, (v) they have low thermal stability and (vi) polymers embrittle at low temperatures.

One of the most severe disadvantages that has been of recent concern in using plastic/polymeric materials is associated with the problem of waste disposal. Several thousands of polymers are now available for use. These are utilised as thermoplasts (non-fiber), thermoplasts (fibers), thermosetts (cross-linked materials used as resins and surface coating materials) and elastomers for a variety of uses. The major polymers, thus classified are commercially produced in the proportion shown in Table 1 (g).

Table 1(g)   The percentage composition of commercial polymers from different classes

Type	Percentage	Main Polymers
Thermoplastics (non-fibers)	~ 65%	Polyethylene, Polyvinyl chloride, Polypropylene and Polystyrene
Thermoplastics (fibers)	~ 15%	Polyesters and Nylons
Thermosetts	~ 10%	P-F and U-F resins
Elastomers	~ 10 %	SBR, NBR and Polybutadiene

Very few polymers, however, are used in producing articles for domestic and speciality applications. The most important of these along with their production data are shown in Table 1 (h). Fig. 1.11 also shows some data for the period 1970–2000. Other commercially available polymers are often produced in smaller quantities and are used for highly specialised applications. Instead of getting a new monomer for the production of polymers, the trend is to mix two polymers (**polyblends**) or a polymer with some non-polymeric material (**composites**).

Table 1(h)   US production of polymeric materials

		(In million pounds) 1990 as base year
Thermosetts	Phenolics	2900
	Urea-F resins	1500
	Polyester resins	1300
	Epoxies	500
Thermoplastics	LDPE	9700
	PVC and its copolymers	9100
	HDPE	8100
	Polypropylene	7200
	Polystyrene	5000
Fibers	Polyesters	3200
	Nylons	2700
	Polyolefins	1800
	Acrylics	500
	Rayon	300
Elastomers	SBR	1900
	Polybutadiene	900
	Ethylene/propylene copolymers	500
	NBR	200

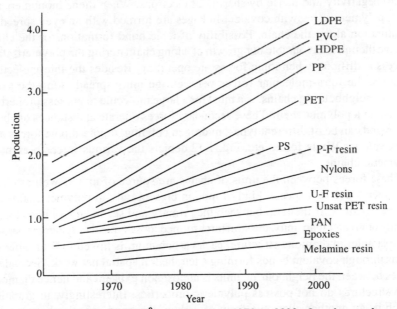

Fig. 1.11   US production (10^9 kg/year) from 1970 to 2000 of various polymers

Polyblends are just physical mixtures (heterogeneous), often designed to improve properties like processibility, mechanical strength, abrasion resistance and flame retardance. The term **engineering plastics or performance plastics** refers to a group of polymeric materials possessing the following characteristics: plasticity at some stage of processing, high load bearing capacity, high mechanical strength, rigidity, abrasion resistance dimensional and thermal stability, light weight and high performance properties which permit them to be used in the same manner as metals, alloys and ceramic materials.

The other thrust aspect of polymer industry has been to produce speciality polymers for very specialised applications. Many industries are carrying out trials for synthesising highly conducting polymers, novel conjugated polymers with light emitting and photo conducting properties. Efforts are also on to synthesise polymer tubes based on polycatenanes and polyrotoxanes. Supramolecular chemistry has also been applied in joining small chemical molecules via noncovalent bonds to develop large physical structures which can be used as polymer tubes for entrapping biomolecules. Multibranched polymers with a main core having a trifunctionality and a first generation dendrimer (with up to six to nine end groups) are also being in the process of commercialisation.

1.7   Intermolecular Forces in Polymers

The main linkage in polymer chains in general is always through the covalent bond between the terminal atoms of successive CRU units. The covalent bonds are quite strong having bond energies of about 50 to 100 kcal. The formation of covalent bonds occurs between the atoms (or elements) having similar electronegativity and hence by sharing of electrons. When diene monomers are used, polymer chains with covalent linkages are formed with an even spread of unsaturation along the chain. Possibility of ionic bond formation for the chain linkage though is favourable but growth of a long chain during the polymerisation process is difficult when ionic forces are operating. Besides the intermolecular linkages, the intra-molecular forces between the units spread along the same chain or neighbouring chains, is a must for a large molecule to possess properties unique for a polymer chain. These secondary forces operate at distances of about 3–4 Å and can be of different types; hydrogen bonding, dipole-dipole forces and weak van der Waals forces etc. Fig. 1.12 depicts the various forces present in polymeric chains.

These forces between the units of chains are source of unit as well as chain movements due to a special kind of motion present only in polymer chains i.e. internal rotation. If the ratio between the primary and secondary valence forces is unity or greater than unity, the structures formed loose typical polymer properties. For example, in diamond structure, each carbon atom is linked to four other C-atoms through covalent bonds forming a tetrahedral spatial net work. Secondary forces between the neighbouring units is almost non existent and hence diamond type structures do not possess polymeric properties. Interestingly, in graphite, which is, an another allotropic form of carbon, the bond energies between the atoms in the same plane differs considerably from that of other planes. This indicates the presence of secondary forces of different magnitudes in different planes. Thus graphite structures possess typical polymer properties. The arrangement of carbon atoms in diamond and graphite structure are shown in Fig. 1.13.

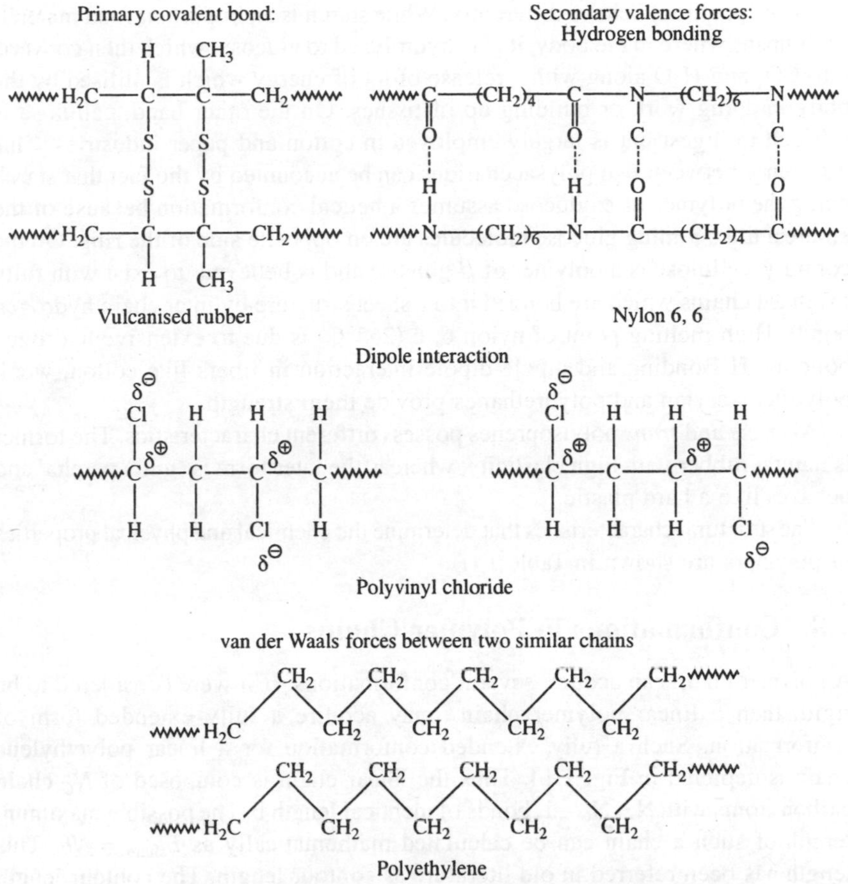

Primary covalent bond:

www H₂C—C—C—CH₂ www
(with H and CH₃ substituents, S—S bridges)

www H₂C—C—C—CH₂ www
(with H and CH₃ substituents)

Vulcanised rubber

Secondary valence forces:
Hydrogen bonding

www C—(CH₂)₄—C—N—(CH₂)₆—N www

www N—(CH₂)₆—N—C—(CH₂)₄—C www

Nylon 6, 6

Dipole interaction

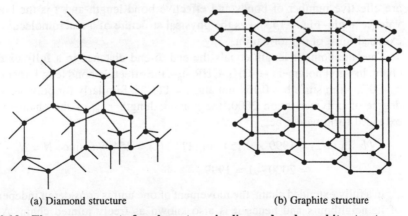

Polyvinyl chloride

van der Waals forces between two similar chains

Polyethylene

Fig. 1.12 Molecular forces in macromolecules

(a) Diamond structure   (b) Graphite structure

Fig. 1.13 The arrangement of carbon atoms in diamond and graphite structures

The arrangement of atoms or groups in space gives rise to enormous change in the properties of macromolecules e.g. starch and cellulose both being the polymers of glucose (glucose units are joined by ethereal oxygen with carbon at

1st and 4th position) behave differently. While starch is an important food material for humans where in the body, it gets hydrolyzed to glucose, which then converts into CO_2 and H_2O along with a release of lot of energy which is utilised by the body in doing work or building up of tissues. On the other hand, cellulose is difficult to digest but is largely employed in cotton and paper industries. This difference between two polysaccharides can be accounted by the fact that starch being the polymer of α-glucose assumes a helical conformation because of the ethereal links joining glucose molecules are on opposite side of the ring. On the contrary, cellulose is a polymer of β-glucose and is believed to exist with fully extended chains which are bonded into a sheet structure by inter chain hydrogen bonds. High melting point of nylon 6, 6 (265°C) is due to extensive hydrogen bonding. H-Bonding and dipole-dipole interaction in fibers like cotton, wool, polyester, acrylan and polyurethanes provide them strength.

Also, *cis* and *trans* polyisoprenes possess different characteristics. The former is natural rubber with high elasticity, whereas the later form is 'gutta percha' and behaves like a hard plastic.

The structural characteristics that determine the chemical and physical properties of polymers are shown in Table 1 (i).

1.8 Conformations in Polymer Chains

A polymer chain can acquire several conformations. If it were considered to be rigid, then a linear polymer chain may acquire a fully extended form of conformation. Such a fully extended conformation for a linear polyethylene chain is depicted in Fig. 1.14. Thus the linear chain is composed of N_C chain carbon atoms with $N = N_C - 1$, bonds of identical length b. The possible maximum length of such a chain can be calculated mathematically as $L_{chain} = Nb$. This length has been referred in old literature as contour length. The contour length r_{cont} of a chain in fact is given by $r_{cont} = N b \sin(\tau/2) = N_e b_e$, where N_e and b_e are effective number of bonds and effective bond length and τ is the bond or valence angle. Fig. 1.15 shows the physical structure of a macromolecule in an exended form of conformation.

Thus the contour length equals the end-to-end distance in a fully extended chain. In vinyl polymers $-(-CH_2-CHR-)_n-$, the effective bond length (per CRU) $b_e = 0.254$ nm since b = 0.154 nm and $\tau = 111.5°$. Similarly for polymer with a degree of polymerisation 2000, the contour length r_{cont} of the chain in fully extended form would be;

$$N b \sin(\tau/2) = 1999 \times 0.154 \sin(111.5/2) = 254 \ nm, \text{ since } N = N_c - 1$$

$$= 2000 - 1 = 1999.$$

In a fully extended chain, the movement of one unit is considered independent of its neighbours and hence it is also known as freely jointed chain. A long polymer chain often assumes a conformation close to that of a coil. The reason for this once again the presence of secondary valence forces operating through-out the chains at distances of about 3–4 Å. Thus it is not always possible to calculate the length of various conformers that a polymer chain attains, it may be

Table 1(i) Essential structural characteristics that determine the chemical and physical properties of polymers

Structural Characteristics	Description and Examples
I. *Periodically recurring primary units*	
(a) Structure Joining of similar primary units	Aliphatic, alicyclic, aromatic and heterocyclic PVC, Teflon, polystyrene, styrene-butadiene rubber, nylon 6, 6 and nylon 6
(b) Joining of different primary units	Copolymers e.g. SBR
(c) Joining of polymer segments	Block and graft copolymers
II. *Nature of the joining atoms of the primary units in the macromolecule*	C–O and C–N among ether, ester, amide and urea groups etc.
III. *Structure build-up of the macromolecule*	
(a) Linear	HDPE, polyesters and polyamides
(b) Weakly branched	LLDPE
(c) Strongly branched	Glycogen and ovalbumin
(d) Cross-linked	LDPE, Phenolplasts, cured epoxy resins, polyurethanes and polyester resins
IV. *Topo-chemical characteristics*	
(a) Geometrical isomers	*Cis-* and *trans-* forms of natural rubber
(b) Optical isomers	Polypeptides and polysaccharides
(c) Tacticity	Isotactic, syndiotactic and atactic polyolefins
(d) Helical structures	Polypeptides, isotactic polymers and starch
(e) Head to tail, head to head (or tail to tail) joining	Polystyrene and polyvinyl chloride
V. *Molecular weight*	In several thousands

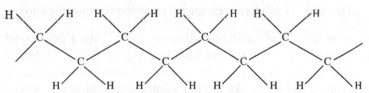

Fig. 1.14 Extended, planner, zig-zag conformation of polyethylene

useful to calculate end-to-end distance of polymer chain. The end-to-end distance is the shortest distance between chain ends in the molecule. For branched polymers, which have several ends, radius of gyration is often used to estimate the size of the coil. In solutions, the information about these parameters can be obtained experimentally by light or neutron scattering techniques and also from viscosity measurements. The experimental results can be compared with theoretical predictions obtained using realistic models of the chain molecule. Based on random flight technique and classical statistics, the conformers of polymer molecules can be better understood. The various possible conformations (see

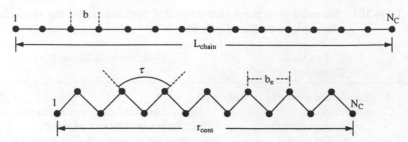

Fig. 1.15 **Chain length and contour length of a polymer chain in fully extended form of conformation N_C = 15 chain atoms, N = 14 chain bonds, N_e = 8 effective bonds and τ = bond angle**

Fig. 1.16) and the root mean square distance can be expressed as shown in Fig. 1.16.

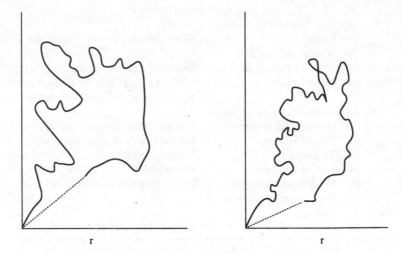

Fig. 1.16 **A polymer molecule in two different conformations**

The root mean square end-to-end distance, $(\bar{r}^2)^{1/2}$ for a polymer chain is related to number of bonds N_C and bond length b by $(\bar{r}^2)^{1/2} = N_C^{1/2}b$. The polymer chain acquires unperturbed dimensions in θ-solvents in which the polymer chain is assumed to behave like a freely jointed chain. In non-θ- solvents, the chains are described by real chains which differ from the idealistic freely joined chains. the ratio between the root mean square end-to-end distance for a polymer chain in θ- and non-θ- conditions estimates a parameter called steric or hindrance parameter, σ. For real chains, σ value is more than unity. The first of the more sophisticated models takes into account the fact that bond angles in molecule are very nearly constant. It also assumes that rotations about the skeletal bonds are unhindered. This model is known as the freely rotating chain. For tetrahedral bonding, this modification approximately doubles the characteristic ratio to 2.2. The ratio is generally increased further if it is assumed that different rotational angles have different energies. The final refinement involves taking into account the cooperativity of neighbouring rotational isomeric states. In other words, it assumes that the movement of one unit influences the movement of its neighbour. Most experimental values of this characteristic ratio are in the range of 4–10.

1.9 Polymer Waste Disposal and Remedies

A huge amount of polymer materials is used all over the world as plastics, fibers and elastomers, Fortunately, these materials do not offer serious health hazards. However, the main problem associated with the polymer waste is its disposal. Another problem associated with the exceedingly large use of polymers is the resource problem, as these are produced chiefly from petrochemicals. There are two ways that can be thought of to combat this problem. Besides cutting down the use of polymeric materials to a certain extent, recycling of used polymers and also making them biodegradable could offer solutions. Recycling means different things. For example, waste PVC can be burnt to give HCl and ethylene which by suitable chemical reactions can be converted into vinyl chloride monomer and can finally be polymerised to give back PVC. However, producing the polymer again by polymerising the monomers obtained from the decomposition of the polymer waste has to be economic. Moreover a finished polymer product is often associated with several additives viz. fillers, plasticizers, antioxidants, thermal and photostabilizers and colourants. Thus for making recyclic process economically, energetically and technically viable, it is necessary that the used polymer made from the same lot and brand should be collected separately. This is possible only when the users are educated thoroughly on this problem and necessary provision is created by the concerned public authorities for collecting the used polymer products variety wise. Also, it may be noted that severe problems can be encountered in making the recycling process industrially viable.

Recycling of glass, paper and rubber has shown promise and the processes are industrially feasible. Everyone is aware of recycled glass and recycled paper. Recycling of rubber, for example, rubber tyres used as automotive bumper guards, capped tyres, fillers, rubber toys, rubber gaskets, tyre hides, hoses and belts etc. is also almost established. The rubber obtained from waste rubber articles when recycled gives reclaimed rubber. Though the recycled rubber possesses less tensile strength, lower elasticity and wear resistance, it is a useful product as it is very cheap and combats problem of rubber wastes. Recycled polyethylene is used in making carry bags. There is much focus on polypropylene based products as it is the single largest material consumed in terms of volume due to its low cost, good strength and easy moulding. Other recycled plastics such as polystyrene and polyesters etc. are used to make cheaper items. It is important to note that the thermosett plastics and crosslinked rubbers can not be recycled. Using biodegradable polymers also offers an important way to fight against polymer waste disposal but more research needs to be done in this field.

Suggested Questions

1. What are polymers? Explain why polymer science is relatively new but fastest growing science?

2. Name five important events in the history of polymer chemistry.
3. Enlist the contribution of following scientists in the development of polymer chemistry: (a) Hermann Staudinger (b) Hermann Mark (c) P.J. Flory (d) Hyatt (e) Carothers (f) Charles Goodyear (g) Karl Ziegler (h) Guilio Natta (i) Henry Baekeland (j) Fredrick Sanger (k) Merrifield (l) P.G. deGennes.
4. Define the terms 'monomer' and 'monomer functionality'. Describe briefly how do polymers behave differently from low mol mass substances. Which important properties make them useful as material?
5. Who are the scientists associated with the following which lead them Nobel prize on polymer research: (i) Concept of macromolecules (ii) Stereoregular polymerization (iii) Solid phase polypeptide synthesis (iv) Theory of polymer solutions and (v) Conducting plastics.
6. Show the structure of repeating units in: (a) Polycaprolactam (b) Polybutylene terephthalate (c) Polyisoprene (d) Polyisobutylene (e) Polyacrylonitrile (f) Polyvinyl alcohol (g) Polyvinylidine chloride (h) Polyacrylamide (i) Teflon (j) Polyoxypropylene (k) Polyphenylene (l) Polyoxyphenylene (m) Polyoxymethylene (n) Polyvinyl acetate (o) Polyvinyl chloride (p) SBR (q) PMMA (r) Poly α-methylstyrene (s) Polyphosphoric acid
7. Write the monomers used in making: (a) Polyurethane (b) Nylon 66 (c) Epoxy resin (d) SBR (e) Polycarbonate (f) Terylene (g) Thiokol rubbers (h) Silicones (i) Glyptals (j) Melamine resin.
8. Write repeating units in: (a) Nylon (b) Nylon 6 (c) Nylon 6, 6 (d) Nylon 6, 10 (e) Nylon 11.
9. Write the structure of repeating units in following copolymers:
   (a) Poly (divinyl benzene-co-styrene) (b) Poly styrene-block-poly (ethylene oxide) (c) Poly (ethylene-co-maleic anhydride) (d) Polystyrene-block-poly (ethylene-co-maleic anhydride) (e) Poly (styrene-graft-isoprene).
10. Draw structures of the following copolymers considering A and B as monomers: (a) Random copolymer (b) Alternating copolymer (c) Two block copolymer and three block copolymer (d) Graft copolymer (e) Radial copolymer.
11. Define the following with suitable examples:
    (a) Ladder polymer (b) Flame retardant polymers (c) Organometallic polymers (d) Inorganic polymers (e) Ter-polymers (f) Block and graft polymers (g) Homopolymers (h) Telechelic polymers (i) Reactive polymers (j) Spiropolymers.
12. Write the structure of repeating units and the method of preparation of polyvinyl butyral and polyvinyl alcohol.
13. Write the structure of the following heat resistant polymers: (a) Poly xylyene (b) Polyphenylene (c) Graphite (d) Polysulphone.
14. What are the following: (a) Coumarone-indene resins (b) Galalith (c) Ebonite (d) Shellac
15. Write the IUPAC (or structure based names) for; (a) Polyacrylonitrile (b) Polystyrene (c) Polyethylene terephthalate (d) Polymethyl methacrylate (e) Polyvinyl acetate (f) Nylon 6.
16. Make an exhaustive chart classifying macromolecules in different possible ways.
17. What are the following polymers? Mention their properties and applications: (a) High impact polystyrene (HIPS) (b) Polyimides (c) Crosslinked poly (styrene sulphonate).
18. What do the following trade names stand for: (a) Kraton (b) Teflon (c) Thiokol (d) Pluronics (e) Saran (f) Plexiglas (g) Orlon (h) Nylon (i) Mylar (j) Neoprene (k) Lucite (l) Bakelite.
19. Discuss the importance of inter-molecular forces in polymers with suitable example for each case.

20. Show schematically the inter-molecular (secondary) forces present in PVC, PAN, HDPE, Nylon 6, and Terylene.
21. Write structures showing (a) head-to-tail and head-to-head arrangement and (b) isostactic, syndiotactic and atactic arrangements in polypropylene.
22. Describe giving suitable examples of the following isomerism in polymers: (a) Constitutional isomerism (b) Orientation isomerism (c) Geometrical isomerism (d) Stereo-isomerism.
23. Write the structures for the following monomers and the names of polymers they produce; (a) Adipic acid (b) DMT (c) Bisphenol A (d) Epichlorhydrin (e) TDI.
24. Suggest few remedies that can help make the synthetic polymers eco-friendly.
25. What are natural polymers and how do they differ from synthetic ones?

Suggested Further Readings

Batzer, H., and F. Lohse, *Introduction to Macromolecular Chemistry.* New York: Wiley, 1982.

Billmeyer, F.W., Jr., *Text Book of Polymer Science,* 3rd ed. New York: Wiley, 1984.

Bovey, F.A., and F.H. Winslow (eds.) *Macromolecules, An Introduction to Polymer Science.* New York: Academic Press, 1979.

Boyd, R.H., and P.J. Phillips, *The Science of Polymer Molecules.* New York: Cambridge Univ. Press, 1994.

Carraher, C.E., Jr., *Seymour/Carraher's Polymer Chemistry—An Introduction,* 4th ed. New York: Dekker, 1996.

Coleman, M.M., and P.C. Painter *Fundammentals of Polymer Science.* Lancaster, Pa.: Technomic, 1994.

Cowe, J.M.G., *Polymers: Chemistry and Physics of Materials,* 2d. ed. London: Blackie, 1991.

Elias, Hans-Georg, *An Introduction to Polymer Science.* Weinheim: VCH, 1997.

Gowarikar, V., N.V. Viswanathan, and J. Sreedhar, *Polymer Science.* New York: Wiley, 1986.

Hiemenz, P.C., *Polymer Chemistry.* New York: Dekker, 1984.

Nicholson, J.W., *The Chemistry of Polymers.* Boca Raton, Fla.: CRC Press, 1991.

Ravve, A, *Principles of Polymer Chemistry.* New York: Plenum, 1995.

Rodriguez, F., *Principles of Polymer Systems,* 4th ed. Washington, D.C.: Taylor and Francies, 1996.

Rosen, S.L. *Fundamental Principles of Polymeric Materials,* 2d ed. New York: Wiley, 1993.

Seymour, R.B., and C.E. Carraher., Jr., *Polymer Chemistry—An Introduction,* 3d ed. New York: Dekker, 1992.

Sperling, L.H., *Introduction to Physical Polymer Science,* 2d ed. New York: Wiely, 1992.

Stevens, M.P., *Polymer Chemistry,* 2nd ed. New York: Oxford Univ. Press, 1990.

Young, R.J., and P.A. Lovell, *Introduction to Polymers,* 2nd ed. New York: Chapman and Hall, 1991.

2

Synthesis of Polymers

2.1 Chain Growth Polymerisation (Addition Polymerisation)

2.1.1 Introduction

Two major types of polymerisation methods are used to convert small molecules (monomers) into polymers. These methods were originally referred to as (i) **Addition polymerisation** and (ii) **Condensation polymerisation**. The addition polymerisation is also called as chain-, chain growth-, or chain reaction polymerisation. While condensation polymerisation is also described under headings such as step-, step growth-, or step reaction polymerisation and sometimes polycondensation.

As it has been mentioned before, a monomer needs to have at least bifunctionality, which may arise from one double bond or two reactive functional groups to undergo polymerisation. Unsaturated monomers usually follow addition polymerisation while those, containing functional groups undergo polycondensation. For example, polyethylene and polypropylene are addition polymers (as the monomers undergo addition polymerisation) whereas nylons, terylene etc. are condensation polymers (as the monomer pairs undergo condensation). Each method has some characteristic and distinguishing features as shown in Table 2(a) and Fig. 2.1. Some common polymers made from each method are shown in Figs. 2.2 and 2.3.

2.1.2 Mechanism of Polymerisation (Free Radical, Cationic and Anionic)

The monomers contain at least a double bond that participates in the polymerisation reaction. The mechanism of this kind of polymerisation is similar to that of chain reaction involving the halogenation of alkanes. It consists of three steps namely initiation, propagation and termination. The initiation step comprises of the reaction of the active species generated from an initiator (or catalyst by its decomposition) and adds to the C=C bond of the monomer. The active species generated from an initiator may be a free radical, a cation, an anion or a coordination complex. These species on reaction with the monomer form new free radicals such as a carbocation, a carbanion or a coordination complex.

Table 2(a)   Distinction between addition and condensation polymerisation

Addition polymerisation (Chain-growth polymerisation)	Polycondensation (Step-growth polymerisation)
1. Monomers are unsaturated	Monomers contain two or more functional groups
2. Nothing is eliminated as a result of this type of polymerisation	Usually small molecules like H_2O, CH_3OH, HCl, etc. are eliminated
3. High molecular weight in polymer is attained at once	High molecular weight is attained only at very high conversions
4. Only monomers and polymers are present during the course of polymerisation	All possible molecular weight species viz. dimers, trimers, tetramers and multimers etc. are present
5. Only monomers add to the growing chain	All species are reactive
6. Chain grows at active centres	Stepwise intermoleculer condensation
7. Polymer molecular weight is equal to DP × molecular weight of monomer	Not so, some small molecules often get eliminated
8. Involves opening of double bond by active species like free radical or ion	Involves reaction between functional groups
9. Examples of polymers are polyolefins, polydienes, vinyl polymers and acrylic polymers	Examples are polyesters, polyamides and polycarbonates
10. Initiator (catalyst) generates active species that attack monomer (the active species can be free radical, anion or cation). The process of polymerisation is exothermic	Most reactions have high ΔE_a and hence usually heating is required. Uncatalysed or catalysed (e.g. by acids) processes are possible
11. Can be done by bulk, solution, suspension, and emulsion polymerisation techniques	Polycondensation can be achieved in melt, solution as well as at interfacial boundary between two liquids in which the respective monomers are dissolved
12. Can quickly lead to a polymer with very high molecular weight, cross-linking can be achieved by using monomers with two double bonds e.g. divinylbenzene	Slow stepwise addition process, molecular weight is < 1,00,000, highly dependent on monomer stoichiometry, little amount of tri or multi-functional monomers develops extensive cross-linking

The important characteristics of chain polymerisation are: (i) once the initiation occurs, the polymer chains form very quickly i.e. in the time scale of 10^{-1} to 10^{-6} s, (ii) the catalyst concentration needed is very low and that means during the course of polymerisation only monomers and polymer are present, (iii) the process is exothermic (see the thermodynamics of polymerisation) and (iv) high polymers with molecular weights of 10,000 to 10 million can be obtained.

The initiators for chain polymerisation may be different depending upon the nature of initiation. The initiators commonly used for generating various active species are listed below;

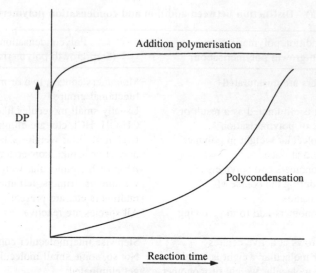

Fig. 2.1   The addition polymerisation and polycondensation

Free radical initiators	Peroxides (e.g. benzoylperoxide, BPO), azo compounds (e.g. azobisisobutyronitrile, AIBN), redox systems (e.g. persulphate + bisulphite), light or ionising radiation
Cationic initiators	Proton or Lewis acids, carbocations, oxonium ions, high energy radiations
Anionic initiators	Lewis bases, organic alkalis e.g. sodium + ammonia, butyl lithium and naphthalene + sodium
Coordination initiators	Transition metal complexes

In free radical polymerisation, the initiator generates free radicals as the active species through its homolytic decomposition. The different mechanisms through which initiator species are initially produced are shown in Fig. 2.4. The decomposition of peroxide type initiators in aqueous systems is generally accelerated in the presence of a reducing agent. This allows high rates of free radical formation at low temperatures, for example in emulsion polymerisation. A typical redox system is that of ferrous ion and hydrogen peroxide. Another widely used redox system is of persulfate with thiosulfate or bisulfite as reducing agents.

Cationic polymerisation involves initiators, which produce a cationic species that interacts with the monomer. More traditional initiators used here include Lewis acids such as BF_3, $AlCl_3$ etc. Protonic acids like sulphuric acid, perchloric acid, phosphoric acid and trichloroacetic acids do act as initiators.

However, pure Lewis acids are not effective as initiators and do need a proton containing Lewis base such as water (called co-catalyst). The Lewis base coordinates with electrophilic Lewis acid and the proton is the actual initiator.

Anionic polymerisation uses initiators, which produce anionic species. Organo alkali compounds like butyllithium, electron transfer reagents such disodium-

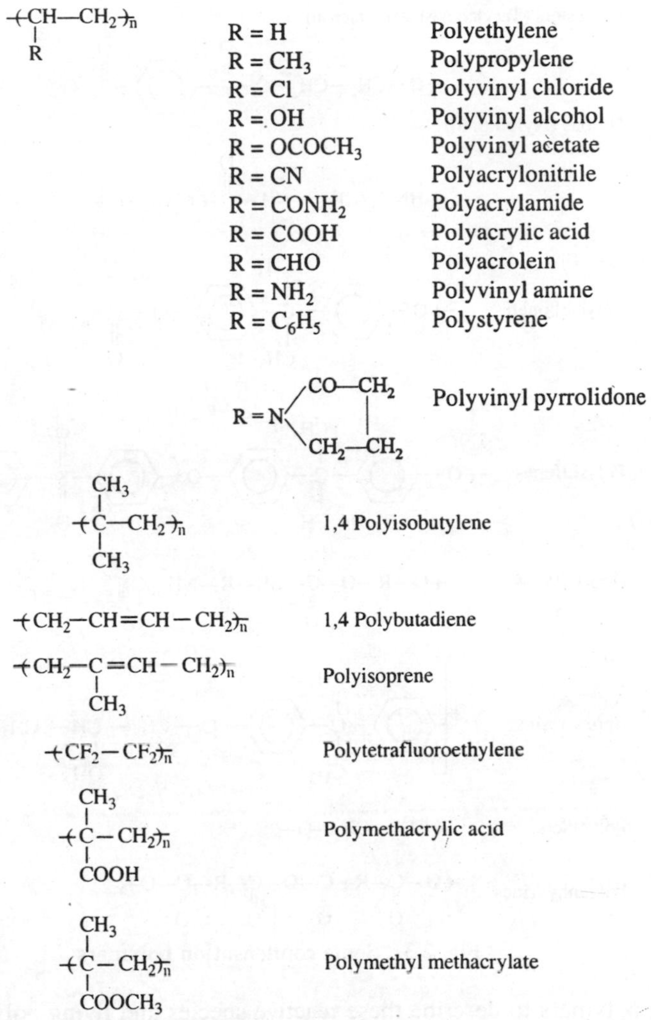

$$-(\!\!\begin{array}{c} CH-CH_2 \\ | \\ R \end{array}\!\!)_n$$

R = H	Polyethylene
R = CH$_3$	Polypropylene
R = Cl	Polyvinyl chloride
R = OH	Polyvinyl alcohol
R = OCOCH$_3$	Polyvinyl acetate
R = CN	Polyacrylonitrile
R = CONH$_2$	Polyacrylamide
R = COOH	Polyacrylic acid
R = CHO	Polyacrolein
R = NH$_2$	Polyvinyl amine
R = C$_6$H$_5$	Polystyrene

$$R=N\big\langle\begin{array}{c} CO-CH_2 \\ | \\ CH_2-CH_2 \end{array}\qquad \text{Polyvinyl pyrrolidone}$$

$$-(\!\!\begin{array}{c} CH_3 \\ | \\ C-CH_2 \\ | \\ CH_3 \end{array}\!\!)_n \qquad \text{1,4 Polyisobutylene}$$

$$-(CH_2-CH=CH-CH_2)_n \qquad \text{1,4 Polybutadiene}$$

$$-(\!\!\begin{array}{c} CH_2-C=CH-CH_2 \\ | \\ CH_3 \end{array}\!\!)_n \qquad \text{Polyisoprene}$$

$$-(CF_2-CF_2)_n \qquad \text{Polytetrafluoroethylene}$$

$$-(\!\!\begin{array}{c} CH_3 \\ | \\ C-CH_2 \\ | \\ COOH \end{array}\!\!)_n \qquad \text{Polymethacrylic acid}$$

$$-(\!\!\begin{array}{c} CH_3 \\ | \\ C-CH_2 \\ | \\ COOCH_3 \end{array}\!\!)_n \qquad \text{Polymethyl methacrylate}$$

Fig. 2.2   Some addition polymers

naphthalene complex and NaNH$_2$ are common initiators for anionic polymerisation. Vinyl monomers with electron withdrawing group e.g. acrylonitrile and styrene etc. are readily polymerised by anionic mechanism. The reaction depends on the polarity of the solvent, electronegativity of the initiator and the resonance stability of the carbanion. Monomers with strong electron withdrawing groups may need only weakly polar initiators like Grignards reagent while monomers with weak electron withdrawing groups need strongly polar initiators like butyllithium. Unlike cationic polymerisation, here the stereoregularity is observed. Soluble anionic initiators in polar solvents at low temperatures form syndiotactic polymers while in non-polar solvents they give isotactic polymers. The termination step does not take place itself in anionic polymerisation and the living macroanion has to be terminated by the addition of some reagents. Szwarc thus used the term

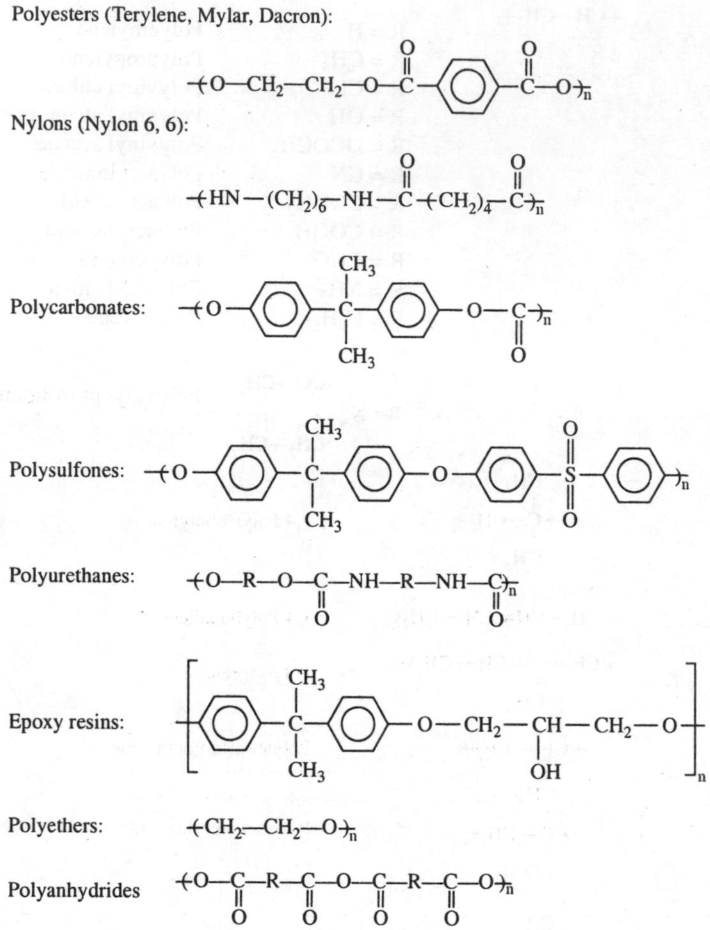

Polyesters (Terylene, Mylar, Dacron):

$$+O-CH_2-CH_2-O-\underset{O}{\overset{O}{\underset{\|}{C}}}-\left\langle\bigcirc\right\rangle-\underset{\|}{\overset{O}{C}}-O+_n$$

Nylons (Nylon 6, 6):

$$+HN-(CH_2)_6-NH-\underset{\|}{\overset{O}{C}}+CH_2)_4-\underset{\|}{\overset{O}{C}}+_n$$

Polycarbonates:

$$+O-\left\langle\bigcirc\right\rangle-\underset{CH_3}{\overset{CH_3}{\underset{|}{C}}}-\left\langle\bigcirc\right\rangle-O-\underset{O}{\overset{}{C}}+_n$$

Polysulfones:

$$+O-\left\langle\bigcirc\right\rangle-\underset{CH_3}{\overset{CH_3}{\underset{|}{C}}}-\left\langle\bigcirc\right\rangle-O-\left\langle\bigcirc\right\rangle-\underset{O}{\overset{O}{\underset{\|}{S}}}-\left\langle\bigcirc\right\rangle+_n$$

Polyurethanes:

$$+O-R-O-\underset{O}{\overset{}{C}}-NH-R-NH-\underset{O}{\overset{}{C}}+_n$$

Epoxy resins:

$$\left[\left\langle\bigcirc\right\rangle-\underset{CH_3}{\overset{CH_3}{\underset{|}{C}}}-\left\langle\bigcirc\right\rangle-O-CH_2-\underset{OH}{\overset{}{CH}}-CH_2-O\right]_n$$

Polyethers:

$$+CH_2-CH_2-O+_n$$

Polyanhydrides

$$+O-\underset{O}{\overset{}{C}}-R-\underset{O}{\overset{}{C}}-O-\underset{O}{\overset{}{C}}-R-\underset{O}{\overset{}{C}}-O+_n$$

Fig. 2.3   Some condensation polymers

living polymers to describe these reactive species and living polymerisation to describe the anionic polymerisation. The absence of self-termination in anionic polymerisation is useful in the synthesis of block copolymers.

The reactive species is a coordination complex in case of coordination polymerisation (Ziegler-Natta polymerisation is a particular case). These catalysts involve a variety of transition metal complexes. These complexes are usually based on Ti, V, or Cr type metals and organometallic compounds like $(C_2H_5)_3Al$. After the initiation, the second step is propagation, where the newly generated 'active species' adds to another monomer in the same manner as in the initiation step. This procedure is repeated over and over again until the final step of the polymerisation process 'termination', occurs.

The termination step is one when the growing chain with active species reacts with other growing chain or by the spontaneous decomposition of the active site. The termination follows either coupling or disproportionation mechanism. The detailed mechanism of free radical polymerisation is shown in Fig. 2.5.

Thermal decomposition

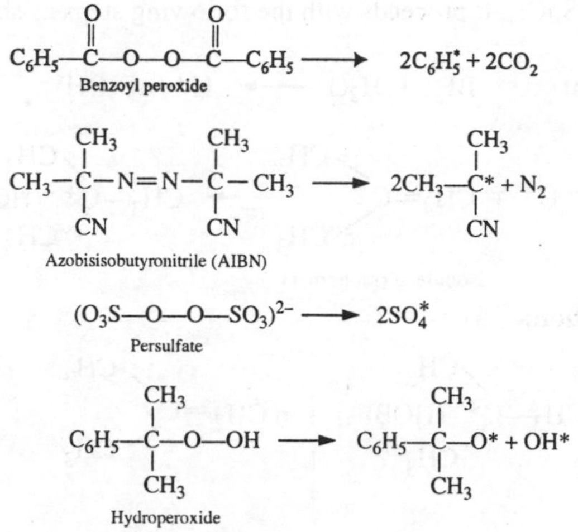

$$C_6H_5-\overset{\overset{\displaystyle O}{\|}}{C}-O-O-\overset{\overset{\displaystyle O}{\|}}{C}-C_6H_5 \longrightarrow 2C_6H_5^* + 2CO_2$$

Benzoyl peroxide

$$CH_3-\underset{\underset{\displaystyle CN}{|}}{\overset{\overset{\displaystyle CH_3}{|}}{C}}-N=N-\underset{\underset{\displaystyle CN}{|}}{\overset{\overset{\displaystyle CH_3}{|}}{C}}-CH_3 \longrightarrow 2CH_3-\underset{\underset{\displaystyle CN}{|}}{\overset{\overset{\displaystyle CH_3}{|}}{C}}^* + N_2$$

Azobisisobutyronitrile (AIBN)

$$(O_3S-O-O-SO_3)^{2-} \longrightarrow 2SO_4^*$$

Persulfate

$$C_6H_5-\underset{\underset{\displaystyle CH_3}{|}}{\overset{\overset{\displaystyle CH_3}{|}}{C}}-O-OH \longrightarrow C_6H_5-\underset{\underset{\displaystyle CH_3}{|}}{\overset{\overset{\displaystyle CH_3}{|}}{C}}-O^* + OH^*$$

Hydroperoxide

Redox systems

(i) $S_2O_8^{2-} + S_2O_3^{2-} \longrightarrow SO_4^{2-} + SO_4^*$

Persulfate Thiosulfate

(ii) $Fe^{2+} + H_2O_2 \longrightarrow Fe^{3+} + OH^* + OH^-$

Fig. 2.4   Formation of free radicals from initiators

Initiation: $CH_2{=}CH_2 + {}^*R \longrightarrow {}^*CH_2{-}CH_2{-}R$

Propagation:

$CH_2{=}CH_2 + {}^*CH_2{-}CH_2{-}R \longrightarrow {}^*CH_2{-}CH_2{-}CH_2{-}CH_2{-}R$

$CH_2{=}CH_2 + {}^*CH_2{-}CH_2{-}(CH_2)_2{-}R \longrightarrow {}^*CH_2{-}CH_2{-}(CH_2)_{n+2}R$

Chain transfer: $CH_2{=}CH_2 + {}^*CH_2{-}CH_2{-}(CH_2)_{\overline{n}}R$

$$\downarrow$$

$${}^*CH_2{-}CH_3 + {}^*CH_2{=}CH{-}(CH_2)_{\overline{n}}R$$

Termination:

(a) Combination: $R{-}(CH_2)_{\overline{n}}CH_2{-}CH_2^* + {}^*CH_2{-}CH_2{-}(CH_2)_{\overline{n}}R$

$$\downarrow$$

$$R{-}(CH_2)_{\overline{n}}CH_2{-}CH_2{-}CH_2{-}CH_2{-}(CH_2)_{\overline{n}}R$$

(b) Disproportionation:

$$R{-}(CH_2)_{\overline{n}}CH_2{-}CH_2^* + {}^*CH_2{-}CH_2{-}(CH_2)_{\overline{n}}R$$

$$\downarrow$$

$$CH_2{=}CH{-}(CH_2)_{\overline{n}}R + CH_3{-}CH_2{-}(CH_2)_n{-}R$$

Fig. 2.5   Mechanism of free radical polymerisation

The cationic polymerisation takes place in presence of catalysts such as protonic acids viz. HCl, HClO$_4$, H$_2$SO$_4$, CCl$_3$COOH and H$_3$PO$_4$ and Lewis acids such as BF$_3$, AlCl$_3$ and SnCl$_4$. It proceeds with the following steps as shown in Fig. 2.6.

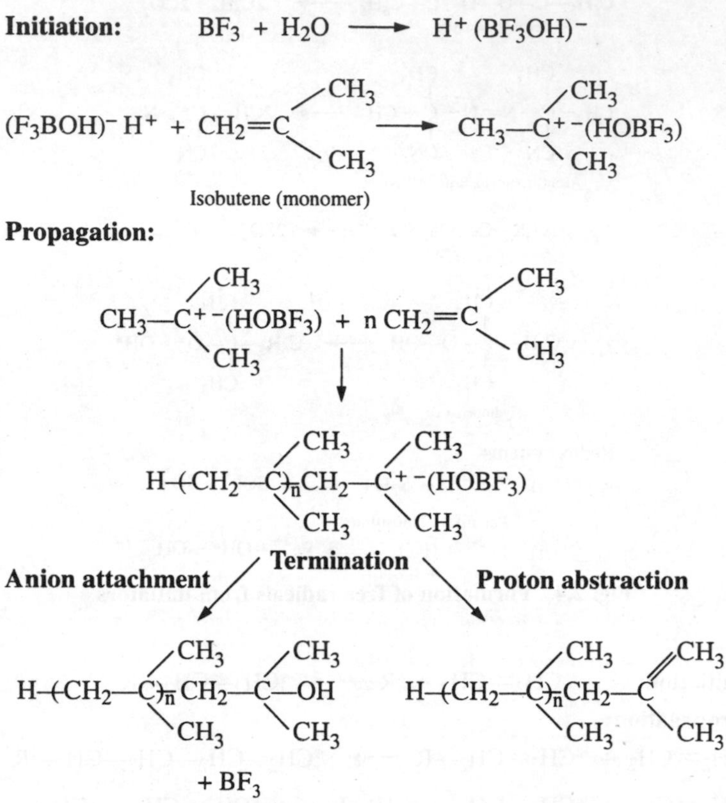

Initiation: BF$_3$ + H$_2$O $\longrightarrow$ H$^+$ (BF$_3$OH)$^-$

Isobutene (monomer)

Propagation:

Anion attachment   **Termination**   **Proton abstraction**

+ BF$_3$

Fig. 2.6   Mechanism of cationic polymerisation

The salient features of cationic polymerisation are: (i) the required activation energy is small and the polymerisation may occur at low temperatures, (ii) the similarly charged macrocations and also carbocations would repel, thus the final termination process cannot take place by combination, (iii) the mechanism depends on the nature of the solvent, electrophilicity of the monomer and the nucleophilicity of the counterion generated and (iv) the polymerisation order in the olefin monomers follows; isobutene > propene > ethene.

Anionic polymerisation has following specific features: (i) monomers with electron withdrawing groups e.g. – CN and – COOH are readily polymerised, (ii) termination is to be deliberately achieved as it does not take place by itself. Hence it is also known as living polymerisation, (iii) it can be used in synthesising tailor made block copolymers and (iv) soluble anionic initiators at low temperatures in polar solvents yield syndiotactic polymers and in non-polar solvents produce isotactic polymers. The various steps i.e. initiation, propagation and termination involved in anionic polymerisation are depicted in Fig. 2.7.

Initiation:

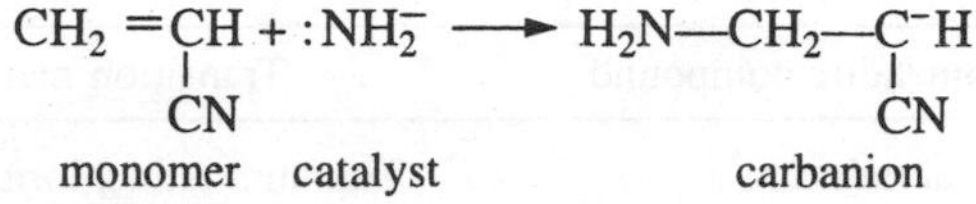

$$CH_2 =\!\!CH + :NH_2^- \longrightarrow H_2N\text{—}CH_2\text{—}\overset{|}{C}^-H$$

CN	CN
monomer catalyst	carbanion

Propagation:

$$H_2N\text{—}CH_2\text{—}\underset{CN}{\overset{|}{C}}^-H + n\,CH_2 =\!\!\underset{CN}{\overset{|}{C}}H$$

$$\downarrow$$

$$H_2N\!\!\left[\!CH_2\text{—}\underset{CN}{\overset{|}{CH}}\!\right]_{\!n}\!CH_2\text{—}\underset{CN}{\overset{|}{C}}^-H$$

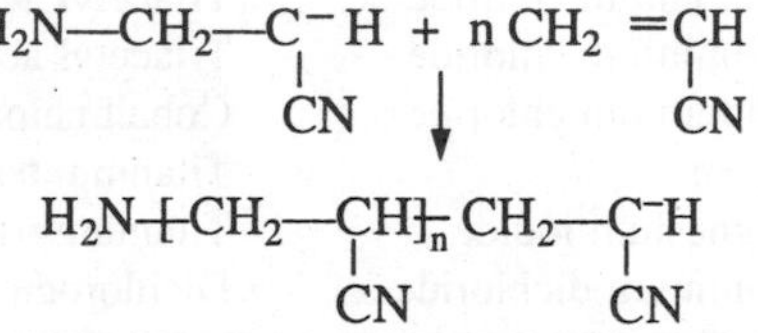

Termination: Does not take place by itself and the living chains are made to terminate e.g. by adding ammonia.

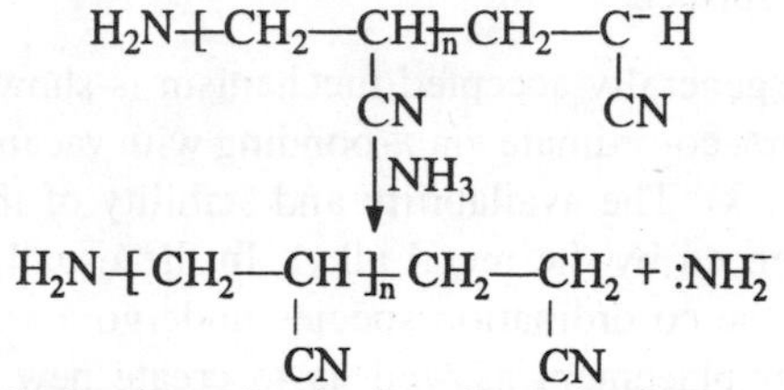

$$H_2N\!\!\left[\!CH_2\text{—}\underset{CN}{\overset{|}{CH}}\!\right]_{\!n}\!CH_2\text{—}\underset{CN}{\overset{|}{C}}^-H$$

$$\downarrow NH_3$$

$$H_2N\!\!\left[\!CH_2\text{—}\underset{CN}{\overset{|}{CH}}\!\right]_{\!n}\!CH_2\text{—}\underset{CN}{\overset{|}{CH_2}} + :NH_2^-$$

Fig. 2.7 Mechanism of anionic polymerisation

2.1.3 Coordination Polymerisation

Before 1950, the only available commercial polymer, polyethylene, was a branched polymer and was manufactured at extremely high pressure. Karl Ziegler in the early 1950s, prepared linear polyethylene at low pressure and ambient temperatures using $TiCl_4/(C_2H_5)_3Al$ mixture as a catalyst. Using such complex catalysts, Giulio Natta in 1955 produced crystalline polypropylene. Soon the industrial usefulness of the process was known and the manufacture of linear PE, isotactic PP, ethylene-propylene copolymers, *cis*-1, 4 polydienes was started. Ziegler and Natta for their outstanding work on using coordination catalysts for the synthesis of polymers shared Nobel prize in Chemistry for the year 1963. A number of coordination catalysts can be used to polymerise olefins. The polymerisation as such is referred to as coordination polymerisation. In general, a Ziegler-Natta catalyst may be described as a transition metal compound coming from group IV to VIII and organometallic compound of a metal from group I to III of the periodic table. The transition metal compound is customarily called a catalyst and the organometallic compound a cocatalyst. The most commonly used catalyst system is $R_3Al + TiCl_4$. The typical catalyst components are given in Table 2(b).

While the technique is highly developed, it is still not perfectly clear to what mechanism exactly the polymerisation follows. The characteristic features of this process are:

(i) the polymerisation of non-polar monomers is easy and is stereo-selective. The extent of stereo-selectivity depends on the amount of exposure of the active site,

(ii) the reaction has the characteristics of anionic polymerisation and

Table 2(b)   Ziegler Natta catalyst components

Organometallic compound	Transition metal salt
Triethylaluminium	Titanium tetrachloride
Diethylaluminium chloride	Vanadium trichloride
Diethylaluminium chloride	Triacetyl acetone vanadium
Diethylaluminium chloride	Triacetyl acetone chromium
Diethylaluminium chloride	Cobalt chloride-pyridine complex
Butyllithium	Titanium tetrachloride
Butyl magnesium iodide	Titanium trichloride
Ethyl aluminium dichloride	Dichlorodicyclopentadienyl titanium.

(iii)   the catalyst is usually heterogeneous and the reaction takes place on a solid surface.

The most generally accepted mechanism is shown in Fig. 2.8. The olefin is believed to first co-ordinate via π-bonding with vacant d-orbitals on the transition metal (complex). The availability and stability of these co-ordination sites are highly influenced by the metal alkyl. In the usually considered monometallic mechanism, the co-ordination species undergo a *cis*-rearrangement to provide stereo-regular placement as well as to create new vacancies in the transition metal structure, which co-ordinates with the next monomer unit (the propagation step).

The monomer gets inserted between Ti atom and the terminal C atom in the growing chain. The polymerisation is thus also called **insertion polymerisation**. The extent of stereo-regularity depends on the amount of exposure of the active sites. At least 98% isotactic polypropylene is formed by the polymerisation at ambient temperature. Low temperatures favour syndiotactic polymer.

The Al/Ti ratio (in AlR_3 and $TiCl_4$) controls the *cis-* or *trans*-structure of polydienes. For example if ratio Al/Ti > 1, 95%, *cis*-polyisoprene is formed whereas for the ratio, Al/Ti < 1, 95% *trans*-polyisoprene is formed. The molecular weight of the polymer is regulated to some degree by chain transfer by monomer or with the co-catalyst. Hydrogen is often added in the commercial process for causing the chain transfer.

A comparison in terms of a schematic representation of propagation steps involved in free radical, cationic, anionic and co-ordination polymerisations is given in Fig. 2.9.

2.1.4   Ring Opening Polymerisation

Some polymers can be prepared by reactions involving ring opening. Few typical examples of monomers for ring-scission polymerisation are lactones, cyclic ethers, lactams, cyclic anhydrides and N- carboxyanhydrides. Fig. 2.10 illustrates some useful polymers made by ring opening polymerisation.

The polymerisation of such cyclic compounds can follow the kinetics and mechanism resembling to both chain polymerisation and polycondensation. For example, the monomer is added to form a chain (chain polymerisation), the polymer growth is through step wise addition and thus molecular weight of the

Initiation

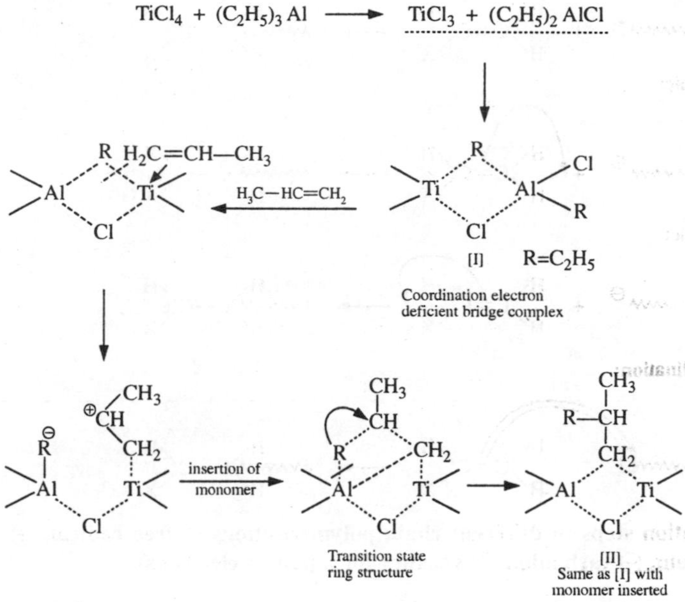

$$TiCl_4 + (C_2H_5)_3\,Al \longrightarrow TiCl_3 + (C_2H_5)_2\,AlCl$$

[I]   R=C$_2$H$_5$

Coordination electron
deficient bridge complex

insertion of
monomer

Transition state
ring structure

[II]
Same as [I] with
monomer inserted

Propagation

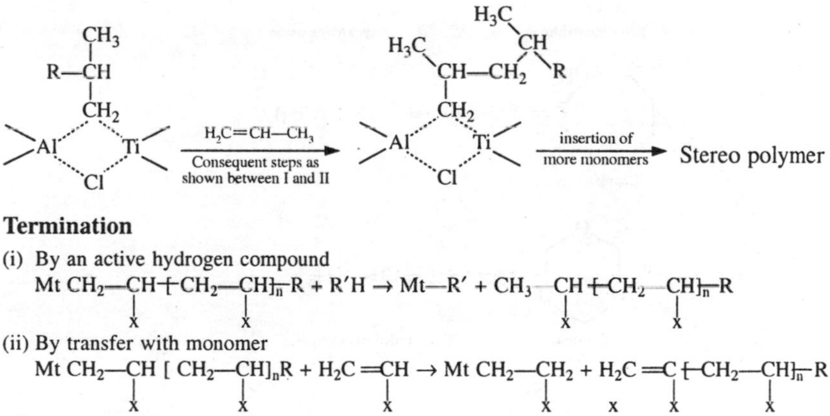

$H_2C=CH-CH_3$
Consequent steps as
shown between I and II

insertion of
more monomers → Stereo polymer

Termination

(i) By an active hydrogen compound

$$Mt\ CH_2-CH+CH_2-CH]_{\overline{n}}R + R'H \rightarrow Mt-R' + CH_3\ CH+CH_2\ CH]_{\overline{n}}R$$
$$\qquad\quad | \qquad\qquad\quad | \qquad\qquad\qquad\qquad\qquad\qquad\ | \qquad\qquad\ |$$
$$\qquad\quad X \qquad\qquad\quad X \qquad\qquad\qquad\qquad\qquad\qquad\ X \qquad\qquad\ X$$

(ii) By transfer with monomer

$$Mt\ CH_2-CH\ [\ CH_2-CH]_n R + H_2C=CH \rightarrow Mt\ CH_2-CH_2 + H_2C=C+CH_2-CH]_{\overline{n}}R$$
$$\qquad\quad | \qquad\qquad\quad | \qquad\qquad\ | \qquad\qquad\qquad\qquad\qquad\qquad | \qquad\quad | \qquad\ |$$
$$\qquad\quad X \qquad\qquad\quad X \qquad\qquad\ X \qquad\qquad\qquad\qquad\qquad\qquad X \qquad\quad X \qquad\ X$$

(iii) By spontaneous internal transfer

$$Mt\ CH_2-CH+CH_2-CH]_{\overline{n}}R \rightarrow Mt-H + H_2C=C+CH_2-CH]_{\overline{n}}R$$
$$\qquad\quad | \qquad\qquad\quad | \qquad\qquad\qquad\qquad\qquad\qquad | \qquad\qquad\ |$$
$$\qquad\quad X \qquad\qquad\quad X \qquad\qquad\qquad\qquad\qquad\qquad X \qquad\qquad\ X$$

Mt = Ti, Mo, Cr, V, Ni or Rh

Fig. 2.8 Bimetallic mechanism for Ziegler-Natta polymerisation

polymer increases during the course of reaction (polycondensation). Thus monomers can be polymerised by different methods. The choice of a particular method for given polymer synthesis depends upon several factors viz. ease of method, desired molecular characteristics such as total molecular weight, molecular weight distribution and type of chain etc.

The polymerisation of ethylene oxide leading to a high mol. wt. polyethylene oxide (PEO) can be accomplished by different ways.

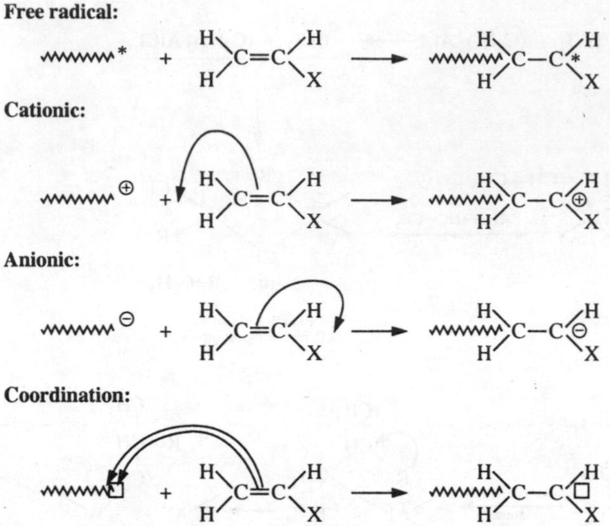

Free radical:

Cationic:

Anionic:

Coordination:

Fig. 2.9 **Propagation steps in different chain polymerisations (* free radical, $\oplus$ carbonium, $\ominus$ carbanion, $\square$ vacancy for a pair of electrons)**

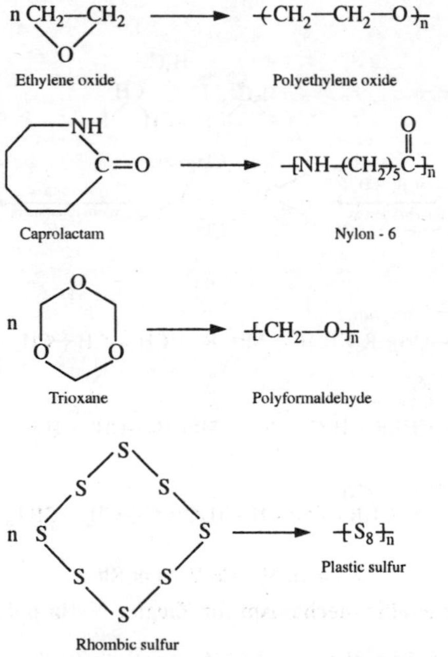

Ethylene oxide   Polyethylene oxide

Caprolactam   Nylon - 6

Trioxane   Polyformaldehyde

Rhombic sulfur   Plastic sulfur

Fig. 2.10 **Ring opening polymerisation**

The utility of different polymerisation methods is shown in Table 2(c).

Lewis acids provide low mol. wt. polymers while anionic polymerisation yields high mol. wt. PEO. The three steps of chain polymerisation namely initiation, propagation and termination are shown in Fig. 2.11.

Basically, polymerisation kinetics suggests that the rate of polymerisation

Table 2(c)   Preferred methods of polymerisation of monomers

Monomers	Polymerisation method			
	Free radical	Anionic	Cationic	Co-ordination
Ethylene	+		+	+
Propylene			+	+
Styrene	+	+	+	+
Isobutylene			+	
1, 3 Butadiene	+	+		+
Isoprene	+	+		+
Vinyl pyrrolidone	+		+	
Methyl methacrylate	+	+		+
Acrylonitrile	+	+		

Initiation:

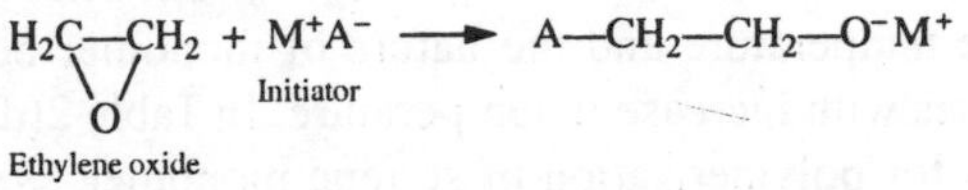

$$H_2C\text{---}CH_2 + M^+A^- \xrightarrow{\text{Initiator}} A\text{---}CH_2\text{---}CH_2\text{---}O^-M^+$$

Ethylene oxide

Propagation:

$$A\text{---}CH_2\text{---}CH_2\text{---}O^-M^+ + H_2C\text{---}CH_2$$

$$\downarrow$$

$$A\text{---}CH_2\text{---}CH_2\text{---}O\text{---}CH_2\text{---}CH_2\text{---}O^-M^+$$

$$A\text{---}(CH_2\text{---}CH_2\text{---}O)_{\overline{n}-1}CH_2\text{---}CH_2\text{---}O^-M^+ + H_2C\text{---}CH_2$$

$$\downarrow$$

$$A\text{---}(CH_2\text{---}CH_2\text{---}O)_{\overline{n}}CH_2\text{---}CH_2\text{---}O^-M^+$$

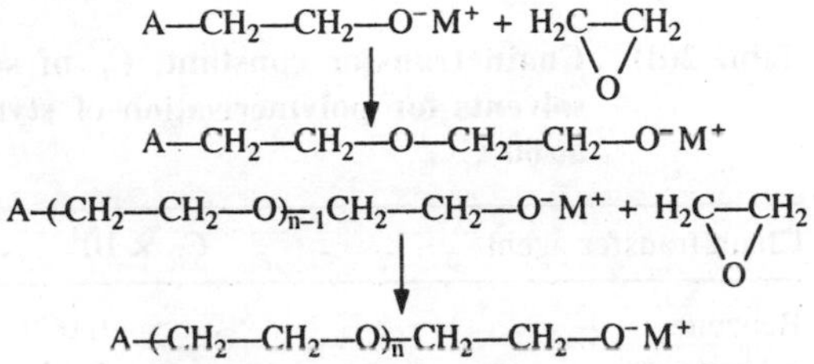

Termination: Many of these Polymerisation reactions have the characteristics of living polymerisation and thus have no termination step

Fig. 2.11   Mechanism of ring opening polymerisation

will depend on several factors and will also be proportional to the first power of the monomer concentration and square root of the initiator concentration. Thus, doubling the initiator concentration means that the rate increases only by a factor of 1.4. This has been experimentally confirmed and the detailed kinetics is described later.

The Gel Effect (Tromsdorf effect, auto-acceleration)

In some polymerisation processes, after a certain per cent conversion, the polymerisation appears to accelerate rather rapidly. This is because of the increase in viscosity, often leading to gels, which decreases the termination rate constant. Since the rate of polymerisation is a function of propagation and termination rate constants, an acceleration effect is observed. This sudden increase in viscosity particularly in bulk polymerisation some times leads to explosion.

The Chain Transfer

Often the growing chain terminates before it fully grows by interaction with a small molecule say RH in such a way that a portion of the small molecule terminates the active chain and at the same time produces a new radical R. The chain transfer is a common feature of solution polymerisation. The chain transfer reaction can be shown as follows.

$$\text{wwwM* + RH} \xrightarrow{K_{tr}} \text{wwwMH + R*}$$

The chain transfer can be done by the monomer itself (though difficult as the removal of hydrogen atom or other atom from the monomer is not easy), through solvent, from initiator (though insignificant as the concentration of the initiator is always very low) or from polymer. It can also be intentionally done by adding substances called *chain transfer agents* or *chain modifiers*. The chain transfer during the polymerisation leads to low mol. wt. polymers (sometimes desirable and at some other times undesirable). The capability of chain transfer agents can be expressed as the ratio of the rate constant for chain transfer reaction, K_{tr} to the rate constant for the propagation reaction, K_p, which is called the chain transfer constant, C_{tr} which is thus defined as $C_{tr} = K_{tr}/K_p$. The chain transfer constant depends upon the temperature and the nature of monomer being polymerised. Usually, it increases with increase in temperature. In Table 2(d) are shown chain transfer constants for polymerisation of styrene monomer.

Table 2(d)   Chain transfer constant, C_{tr} of some solvents for polymerisation of styrene at 60°C

Chain transfer agent	$C_{tr} \times 10^4$
Benzene	0.029
n-Heptane	0.42
m-Cresol	11.0
CCl$_4$	90.0
CBr$_4$	22,000
n-Butylmercaptan	2,10,000

The extent to which the molecular weight or degree of polymerisation decreases by different chain transfer agents can be examined by measuring degree of polymerisation or mol. wt. as a function of the concentration of chain transfer agent. Under the conditions when the chain transfer is due to the chain transfer agent only i.e. no transfer from the initiator, monomer, or the polymer, the Mayo equation holds;

$$1/DP = 1/DP_0 + C_{tr} \cdot [\text{transfer agent}]/[\text{monomer}] \quad (2.1)$$

To determine the degree of polymerisation for a particular monomer in the presence of varying concentrations of the chain transfer agent, a plot between $1/DP$ and [transfer agent]/[monomer] can be constructed (see Fig. 2.12).

A linearty in the behaviour can be seen through Mayo equation. The slope of the straight line gives chain transfer constant. The values of chain transfer constants are given in the parenthesis. It is clear that benzene is not effective as chain transfer agent whereas butylmercaptan is highly effective.

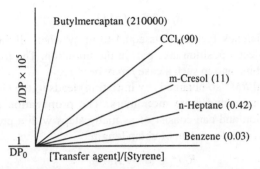

Fig. 2.12   **Effect of chain transfer agents in the polymerisation of styrene**

2.1.5   Kinetics of Free Radical Addition (Chain) Polymerisation

As it has been discussed in the mechanism part that the polymerisation is a chain reaction that involves three steps viz initiation, propagation and termination. In the first part of the process, free radical is generated from the initiator, which combines with the monomer generating new free radical. The initiator undergoes thermal, photo- or high-energy radiation decomposition and produces free radicals (free radicals can also be generated by chemical reaction). The rate of decomposition of initiator follows first order kinetics and depends upon temperature and the solvent. The rate constant for decomposition of initiator varies with temperature in accordance with the well known Arrhenius equation, $K_d = Ae^{-E_a/RT}$. The rate constant and energy of activation for the decomposition of two initiators are given below:

Initiator	Solvent	Temp., °C	K_d, s^{-1}	E_a, k cal mol^{-1}
BPO	benzene	30	4.8×10^{-8}	27.8
AIBN	benzene	40	5.4×10^{-7}	30.7

The decomposition of the initiator (I) can be explained as follows:

Considering one initiator molecule decomposing to give two free radicals, the rate of decomposition reaction can be shown as:

$$I \xrightarrow{K_d} 2R* \tag{2.2}$$

and
$$R_d = - d[I]/dt = K_d[I] \tag{2.3}$$

where I is initiator, $R*$ is free radical generated, K_d is the rate (or decay) constant and R_d is the rate of decomposition. The free radicals so generated attack onto the monomer, M, and produce a new free radical (initiation step):

$$R* + M \rightarrow RM* \tag{2.4}$$

and
$$R_i = - d[M]/dt = d[RM*]/dt = K_i[R*][M] \tag{2.5}$$

where K_i is the initiation rate constant and R_i is the rate of initiation. The rate of initiation is rate controlling step and is related to the efficiency of production of two free radicals from each initiator molecule. Thus,

$$R_i = 2K_d \cdot f \cdot [I] \tag{2.6}$$

where f is the efficiency factor and is equal to unity when all the free radicals generated by the decomposition take part in the initiation. The free radicals can sometimes recombine and in such a case f may be < 1.

The free radical $RM*$ so obtained (by initiation) leads to the chain growth by combining with more and more monomers. The propagation step is thus a bimolecular reaction and can consist of as many steps with a propagation rate constant K_p, which is independent of chain length:

$$RM* + M \rightarrow RMM* \tag{2.7}$$

$$RMM* + M \rightarrow RMMM* \tag{2.7a}$$

$$R—M* + M \rightarrow R—MM* \tag{2.7b}$$

So, the rate of propagation, R_p will be

$$R_p = - d[M]/dt = K_p[R—M*][M] \tag{2.8}$$

or

$$[R—M*] = R_p/K_p[M] \tag{2.8a}$$

In the third step, where the termination occurs, e.g. by coupling

$$R—M* + *M—R \rightarrow R—MM—R \tag{2.9}$$

termination rate constant R_t is given by

$$R_t = - d[M*—R]/dt \tag{2.10}$$

$$= 2K_t[M*—R]^2$$

where K_t is the termination rate constant.

Using steady state approximation i.e. considering the rate of initiation equal to rate of termination, it can be written that

$$R_i = R_t \tag{2.11}$$

or

$$2K_d f[I] = 2K_t[M*—R]^2 \tag{2.11a}$$

or

$$[M*—R] = \{K_d \cdot f \cdot [I]/K_t\}^{1/2} \tag{2.11b}$$

Substituting the $[R—M*]$ value in the equation of propagation rate, propagation step can thus be written as

$$R_p = K_p/K_t^{1/2} (fK_d)^{1/2}[M][I]^{1/2} \tag{2.12}$$

$$= K[M][I]^{1/2}$$

where

$$K = [K_p^2 K_d f/K_t]^{1/2}$$

This equation shows the propagation as the first order with respect to monomer concentration and half order with respect to initiator concentration.

Limitations: The kinetic expression is good for head to tail polymerisation, when termination is by coupling and when auto-acceleration is absent. The degree of polymerisation *DP* can be written as

$$DP = R_p/R_i = \{K_p[M](K_d \cdot f \cdot [I]/K_t)^{1/2}\}/2K_d[I] \qquad (2.13)$$

$$= K_p[M]/2(K_d \cdot K_t \cdot f \cdot [I]^{1/2})$$

or $$= ([M]/[I]^{1/2}) * (K_p/2)(K_d \cdot K_f \cdot f)^{1/2}$$

$$= ([M]/[I]^{1/2}) * K''$$

where $$K'' = K_p/(2K_d \cdot K_t \cdot f)^{1/2} \qquad (2.13a)$$

Thus, the degree of polymerisation or the molecular weight is directly proportional to $[M]$ or $(1/[I]^{1/2})$. The characteristic features of free radical addition polymerisation thus can be summarised as

(i) $R_p \propto [M]$ and $[I]^{1/2}$
(ii) $R_t \propto [I]$
(iii) $DP \propto [M]$ and $[I]^{-1/2}$
(iv) The monomer concentration steadily decreases during the course of polymerisation and becomes zero in the end.

Increase in temperature increases the $[I]$ and thus the rate of reactions, but decreases the average mol. wt.. No polymerisation will take place above a certain temperature called ceiling temperature e.g. this temperature for styrene monomer is 310°C. The kinetic chain length v is defined as the average number of monomer molecules consumed by each free radical generated. Thus

$$v = R_p/R_i; = R_p/R_t \qquad (2.14)$$

so $$v = K_p \cdot (K_d^{1/2}/K_t^{1/2})(f \cdot [I]^{1/2}[M] \cdot (1/K_d f[I]) \qquad (2.14a)$$

$$= K_p[M]/(2(K_d K_t f[I]^{1/2}) \qquad (2.14b)$$

Thus degree of polymerisation $DP = 2v$ (for termination by coupling) and $DP = v$ (for termination by disproportion). Table 2(e) lists some of the kinetic parameters for free radical polymerisation.

Table 2(e)   **Typical kinetic parameters for free radical polymerisation**

Rate constant	Activation energy/k cal mol^{-1}
$K_d/10^{-3}$ s^{-1}	30 to 50
$K_i/10^3$ liter mol^{-1} s^{-1}	5 to 7
$K_p/10^3$ liter mol^{-1} s^{-1}	4 to 10
$K_t/10^7$ liter mol^{-1} s^{-1}	0 to 6

The chain growth polymerisation, proceeds via contraction of volume and thus the kinetics can be conveniently examined dilatometrically (a method that measures the changes in specific volumes). The increase in molecular weight during the course of polymerisation can be monitored viscometrically or turbidimetrically.

2.1.6   Kinetics of Cationic Polymerisation

As discussed in the mechanism part before, the cationic polymerisation begins when the initiator releases a proton that combines with the monomer in the initiation step on the polymerisation. Considering BF_3 as the catalyst and H_2O as the cocatalyst, the different rate equations can be written as:

Initiation: $M + H^+BF_3OH^- \rightarrow HM^+ + BF_3OH^-$

$$R_i = K_i[C^-][M], \text{ where } C = \text{catalyst or cocatalyst}$$

Propagation: $R_p = K_p[M][M^+]$

The rate depends on the dielectric constant of the solvent. The solvents with high dielectric constant speed up the rate as this promotes the separation of carbocation and counterion. The rate is assumed essentially same for all the propagation steps.

Termination: $R_t = K_t[M^+]$

This first order process simply involves the dissociation of the macrocation-counterion complex to BF_3, H_2O and dead polymer molecule.

Assuming steady state, i.e. $R_i = R_t$ one can get

$$K_i[C][M] = K_t[M^+] \quad \text{and} \quad [M^+] = K_i[C][M]/K_t \tag{2.15}$$

The overall rate of polymerisation is thus expressed as

$$R_p = K_p[M][M^+] = (K_p K_i/K_t)[C][M^2] = \text{constant } [C][M^2] \tag{2.16}$$

$$DP = R_p/R_t = K_p[M][M^+]/K_t[M^+] = \text{constant } [M] \tag{2.17}$$

2.1.7   Kinetics of Anionic Polymerisation

Here the propagation species are carbanions as described in the mechanism part. Considering the polymerisation of acrylonitrile by KNH_2

$$: \overset{\ominus}{N}H_2 + CH_2{=}CH{-}CN \rightarrow H_2N{-}CH_2{-}\overset{\ominus}{C}H{-}CN$$

the initiation step can be written as $R_i = K_i[C][M]$ where $[C] = [:NH_2]$

for the propagation step $R_p = K_p[M][M^-]$

for the termination step $R_t = K_t[NH_3][M]$

assuming steady state, it can be written as

$$R_i = [R_p \cdot M^-] = (K_i/K_t) \cdot [C][M]/[NH_3], \tag{2.18}$$

thus $R_p = \text{constant } [C][M]^2/[NH_3]$ (2.19)

and $DP = (\text{constant}) \{[M]/[NH_3]\}$. It can now be seen that the rate of propagation and DP are inversely related to the concentration of ammonia.

2.1.8   Phase Systems in Polymerisation (Techniques of Polymerisation)

Based on the different methods of preparation, polymerisation can be classified as:

1. Bulk (or mass) polymerisation
2. Solution polymerisation
3. Precipitation polymerisation
4. Suspension polymerisation
5. Emulsion polymerisation

1. Bulk Polymerisation

The polymerisation reaction is carried out within the monomer itself. The reaction is catalysed by additives (like initiator, transfer agents etc.) or under the influence of heat or light. Since polymerisation is a highly exothermic process, the process is difficult to carry out and the polymer obtained is generally of non-uniform molecular mass distribution. If the polymer is insoluble in its monomer, it is obtained as a powder of porous solid. Since the recipe contains primarily the monomers, the polymer formed is usually pure. Bulk polymerisation on large scale is carried out with vinyl chloride, vinyl acetate and acrylic esters.

2. Solution Polymerisation

When both the monomer and the polymer produced are soluble in a suitable solvent, the polymerisation reaction is carried out with the monomer in solution. Such a polymerisation is referred to as 'solution polymerisation' and has the advantage of easy dissipation of heat evolved as a result of exothermic polymerisation reaction. The polymers so formed have low degree of branching and relatively uniform molecular weight, but the disadvantage is that solid polymeric product can only be isolated in pure form from the solution with great difficulty or some times not at all. It is because the solvent is occluded and firmly traps the polymer. For this reason solution polymerisation is mainly applied when solutions of polymers are required (for ready-made use) for technical applications e.g. as lacquers, adhesives etc. The products obtained by this method are usually relatively low molecular weight because of the possibility of chain transfer to solvents.

3. Precipitation Polymerisation

In such a polymerisation reaction, the monomer is soluble in the solvent and the polymer precipitates out as a result of polymerisation (polymer being insoluble in solvent unlike solution polymerisation). The so precipitated polymer can be separated in the form of a gel or powder by centifugation or simple filtration. Further there is no problem in heat dissipation and the degree of polymerisation is also high. Polyethylene, polyvinyl esters, polyacrylic esters are obtained commercially using hydrocarbons as solvents. Polyacrylonitrile is prepared using water as solvent.

4. Suspension Polymerisation (or Pearl Polymerisation)

In suspension polymerisation, the monomer containing initiator, modifier etc. is dispersed in a solvent (generally water) by vigorous stirring. For a stable suspension of monomer in the solvent some stabilisers are added. These may be protective colloids (e.g. carboxymethyl cellulose, polyvinyl alcohol), finely divided inorganic solids (e.g. hydroxides of various metals) or surfactants. The polymerisation takes place within the dispersed (or suspended) monomer particles and the agglomeration of which is prevented by stabiliser. After the reaction is completed, the polymer is separated by centrifugation or filtration, washed and dried.

5. Emulsion Polymerisation

In emulsion polymerisation, the liquid monomer is dispersed in an insoluble liquid, which in turn gives an emulsion. For most of the polymerisation reactions under this head, the dispersion medium is water. So the monomer-in-water emulsion (containing catalyst and stabilised by emulsifying agents e.g. surface active agents) is then used as the polymerising mixture. The events that occur in emulsion polymerisation are schematically shown in Fig. 2.13.

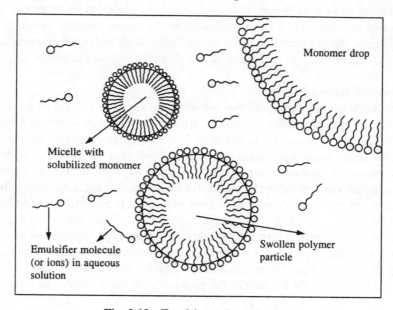

Fig. 2.13   Emulsion polymerisation

A typical recipe for emulsion polymerisation is monomer, water (monomer is immiscible with water), emulsifier (a surfactant at concentration above critical micelle concentration) and a water soluble initiator. The monomer, though insoluble, has some limiting solubility in water. Thus it remains as (i) the large dispersed droplets stabilised by surfactant adsorption, (ii) solubilized in the hydrophobic core of the micelle and (iii) in the molecularly dissolved state in water. The latter is initiated by the water soluble initiator e.g. benzoyl peroxide or a redox system. Polymerisation thus takes place within the surfactant micelles (size about 10 nm) which grow in size at the expense of dispersed monomer droplets (size about 1000 nm). In the end a latex (or a polymer colloid) with dispersed polymer (size about 100 nm) is left.

Once the polymerisation is complete, the polymer can be isolated either by evaporation of water in a spray drier, or by the coagulation of the polymer latex and subsequent filtration and drying. The polymers of varying degree or purity are obtained depending upon the finishing treatment; the spraydried product contains emulsifier, catalyst, modifier and other foreign substances; the separated product can be washed before drying. The degree of polymerisation is usually high since there exists only a few possibilities of chain termination of a growing polymer chain within the micelles. Table 2(f) shows the comparison between different polymerisation techniques.

Table 2(f) Comparison of various polymerisation techniques

Method	Advantages	Disadvantages
Homogeneous		
Bulk (batch)	Low impurity level, casting possible	Thermal control difficult
Bulk (continuous)	Improved thermal control	Isolation is difficult and requires devolatilization
Solution	Improved thermal control	Difficult to remove solvent, solvent recovery costly and chain transfer may limit molecular weight
Heterogeneous		
Suspension	Low viscosity, simple polymer isolation, easy thermal control and may yield particles of desirable size	Highly sensitive to agitation rate, control of particle size is difficult, possible contamination by suspending agent and washing, drying, and compaction are necessary
Emulsion	Low viscosity, Good thermal control, latex may be directly usable, 100% conversion with high MW at high rates and small particle size can be obtained.	Emulsifier, surfactants and coagulants must be removed and high residual impurity level may affect certain polymer properties, high cost, washing, drying and compacting may be necessary.

2.1.9 Industrial Polymerisation

The polymerisation of most of the monomers by using different methods can be carried out at laboratory scale using simple reaction vessels. These vessels are usually thick walled standard glass mono, two- and three-necked flasks, and long tubes etc. At the laboratory scale the monomer concentrations or quantity taken is very low and hence the elimination of effects due to heat generation, viscosity build up and heterogenities that are developed during polymerisation process, is achieved by continuous and controlled stirring. But at Industrial scale production, the quantity of monomer feed and other reagents are very high. Thus high viscosities, low diffusivity and large heats produced because of polymerisation process, are to be considered while designing reactors for large scale polymer production. The polymer engineer, takes into account the flow rates, temperatures and composition, while designing the geometry and other requirements of a reactor.

Reactors

Several types of reactors have been used in polymer industries. Among them three types of reactors namely batch reactor, tubular-flow reactors, stirred-tank reactor and cascade reactors are most commonly employed. A brief description of these reactors is as follows.

Batch Reactor

These reactors are simplest in design. They are designed such that the reactor withstands maximum temperature and pressure that can be built up. These reactors are highly useful, when the polymerisation is conducted batch wise. The reactants containing monomer, initiator, stabiliser and other chemicals are mixed by agitation and polymerisation is initiated by applying heat from the external sides of the reactor. The initial heat generated at early stages of the polymerisation facilitates the further continuation of the process. The control of heat is achieved by allowing the refluxing of solvent media. The major disadvantage of batch reactors is that, termination of a process needs to be done either by quenching the contents or transferring or pouring the product to other containers before the viscosity of the system becomes too high.

Tubular-Flow Reactors

The objective of this type of reactors is to provide a progressive flow of the monomer into the reactor and separating the polymer and unreacted monomers. This allows the maintenance of uniform distribution of thermal exposure and residence times. The operation of such a process is not easy. Usually, radial mixing and flow redistribution are achieved by number of stationary techniques such as coiling the tube into a spiral and utilisation of temperature and pressure gradients that are developed along the tube. Another way of achieving radial mixing is to use extruders and casting slips. A typical stationary tubular-flow reactor is shown in Fig. 2.14. This type of reactor is commonly used for polymerising olefins at high pressures.

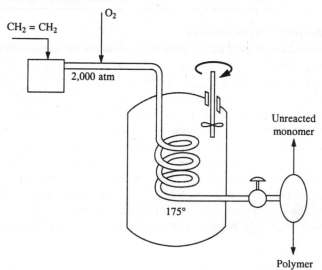

Fig. 2.14   Tubular-flow reactor for ethylene polymerisation

Stirred-Tank Reactor

This type of reactor is sturdy and also known as autoclave. They are designed mainly for easy and faster dissipation of heat, by initiating the reaction and controlling the heat by appropriate controlled cooling. The reactor's design is such that the monomer feed and other agents are mixed throughout the volume of the reactor by setting up the agitation system reasonably constant. This minimises the temperature and composition variation in the reaction media. The yield of polymerisation is affected when there is a gel formation. Similarly the nature of mixing affects the molecular weight distribution. It has been shown that when the life time of a growing chain is long compared to residence time, as normally observed in step growth polymerisation, the molecular weight distribution achieved is very broad, even if perfect mixing is ensured. On the other hand when chain life is short, the distribution is narrowest with perfect mixing. This is especially true in case of polyaddition reactions employing free radicals.

2.1.10 Thermodynamic Aspects of Polymerisation

Whether a monomer will polymerize or not depends upon polarizability of the double bond. The symmetrical monomer molecules polymerise with difficulty (e.g. ethylene). Unsymmetrical monomer molecules polymerise easily (e.g. propylene, isobutylene and styrene). Monomers with polar substituents polymerise (e.g. acrylonitrile, acrylic acid) readily. A non-polar monomer that polymerises with great difficulty or even does not polymerise at all may be copolymerised with a polar monomer. The polymerisation process can be explained thermodynamically. A process may occur spontaneously at constant temperature and pressure if the free energy change is negative i.e. the free energy of polymer is less than the free energy of monomer. The free energy change is given by the relation, $\Delta G = \Delta H - T \Delta S$.

The formation of macromolecules from randomly arranged monomers results in their ordering i.e. the entropy of a polymer is less than the entropy of monomers. For most of the chain polymerisations, ΔS falls in a narrow range of -25 to -30 cal mol^{-1} K^{-1} (thus does not favor polymerisation).

Polymerisation of monomers with double bonds involves breaking up of the double bond and formation of two single bonds e.g. in ethylene

$$CH_2 = CH_2 \rightarrow -[CH_2 - CH_2]_n-$$

Now since C=C bond energy is 145.5 k cal mol^{-1} and C—C bond energy is 84 k cal mol^{-1}; the enthalpy of polymerisation can be calculated as $145.5 - 2(84) = -22.5$ k cal mol^{-1}. ΔH and ΔS along with the calculated values of free energy values for several monomers are shown in Table 2(g).

An increase in polymerisation temperature increases $T\Delta S$ term whereas ΔH does not change significantly. This decreases the magnitude of ΔG and at a certain temperature a condition is achieved when $\Delta H = T\Delta S$, i.e. $\Delta G = 0$. This means that above this temperature no polymerisation can occur. The temperature at which $\Delta G = 0$ is called **ceiling temperature** T_c. Table 2(h) lists the ceiling temperatures for several commonly used monomers.

For cyclic monomers undergoing ring opening polymerisation, the ease of

Table 2(g)   Enthalpy and entropy and free energy of polymerisation of some monomers at 25°C

Monomer	$-\Delta H$ k cal mol^{-1}	$-\Delta S$ cal mol^{-1} k^{-1}	$-\Delta G$ k cal mol^{-1}
Ethylene	22.7	24.0	15.5
Propylene	20.5	27.8	12.2
Styrene	16.7	25.0	9.2
α-Methylstyrene	8.4	24.8	1.0
Butadiene	17.4	20.5	11.3
Acrylonitrile	18.4	26.0	10.6
Methyl methacrylate	13.5	28.0	5.2
Vinyl acetate	21.0	26.2	13.2
Tetrafluoroethylene	37.2	26.8	29.2

Table 2(h)   Ceiling temperature for common liquid monomers

Monomer	T_c, °C
Styrene	310
α-Methylstyrene	61
Methyl methacrylate	220

polymerisation can be explained on thermodynamic basis considering strains in the rings. The six membered ring systems have least strain and hence hardly undergo ring-opening polymerisation. Whereas the ring systems having other than hexagonal symmetry have more strain and can be easily opened up. Thus for non-strained rings, the ring opening has no effect on the change of bonds and thus $\Delta H = 0$. Ring opening is accompanied with an increase in disorder i.e. entropy and for a ring opening, the term $T\Delta S = 2.5$ k cal mol^{-1}. Since $\Delta H = 0$, $\Delta G = 0$ for non-strained rings, polymerisation is not thermodynamically favored. However, for strained rings, ring opening is accompanied by energy generation, thus ΔH is negative and consequently $\Delta G < 0$. This means that the polymerisation of cyclic monomers with strained ring is thermodynamically favored. Table 2(i) lists the thermodynamic parameters for the different ring systems.

Table 2(i)   The thermodynamic parameters for the different ring systems

No. of carbon atoms in the ring system	ΔH/k cal mol^{-1}	ΔG/k cal mol^{-1}
3	− 27.0	− 22.1
4	− 25.1	− 21.5
5	− 5.2	− 2.2
6	+ 0.7	+ 1.4
7	− 5.2	− 4.9

2.1.11 Copolymerisation

The process of polymerising two or more than two monomers together is called copolymerisation. The term copolymer is better referred to as addition polymers. The copolymers often possess the properties shown by the homopolymers made from the constituent monomers of the copolymer. Instead of searching new monomers for the production of polymers with desired properties, it is often desired to achieve this by copolymerisation using the well known inexpensive monomers like styrene, ethylene, propylene, butadiene, vinyl chloride and other common vinyl and acrylic monomers etc. Some very useful and commercially important copolymers are SBR and NBR—the copolymers of butadiene with styrene or acrylonitrile. These copolymers are ideal elastomers and possess properties superior to polybutadiene rubber. Block copolymers of styrene and butadiene are excellent thermoplastic elastomers.

High impact polystyrene (HIPS) is a similar copolymeric plastic. Copolymer of butadiene with styrene and acrylonitrile together (ABS) is a useful plastic. Polyvinyl chloride has high T_g and less solubility. This rigid PVC is not useful when flexibility in the end product is desired. It is for this reason vinyl chloride monomer is copolymerised with small amount of vinyl acetate. Copolymers of vinyl chloride with vinylidene chloride (Saran—film or filament) and with acrylonitrile (vinyon, dynel fiber) are well known.

Ethylene copolymerised with propylene (EPM) and with propylene and diene like butadiene (EPDM) are plastics resistant to oxygen, ozone and heat. Copolymers of ethylene with methacrylic acid (ionomers) are good transparent adhesives. However, the properties of the copolymer depend upon the relative amount of the two monomers present in the copolymer and its type. The kinetics of copolymerisation is well understood and on its basis one can predict the composition of a copolymer and its nature when the two different monomers are copolymerised in different mole ratios in the feed. The kinetics of copolymerisation is described below:

When two monomers say M_1 and M_2 are simultaneously polymerised, both will form free radicals RM_1^* and RM_2^* as a result of the initiation reaction between the initiator free radical R*. Now there are four propagation steps as described below which will determine the reactivity of a monomer free radical with the same monomer or the other.

(i) $RM_1^* + M_1 \rightarrow RM_1M_1^*$ [propagation reaction (PR) type 11]

(ii) $RM_1^* + M_2 \rightarrow RM_1M_2^*$ [PR type 12]

(iii) $RM_2^* + M_1 \rightarrow RM_2M_1^*$ [PR type 21]

(iv) $RM_2^* + M_2 \rightarrow RM_2M_2^*$ [PR type 22]

Assuming that all the four types of addition take place, the growing chains ending with M_1^* or M_2^* will have the possibility of all the four types of propagation. Now, the rates of these four propagation reactions are as follows,

$$M_1^* + M_1 \rightarrow M_1M_1^* \qquad R_{11} = k_{11}[M_1^*][M_1]$$

$$M_1^* + M_2 \rightarrow M_1M_2^* \qquad R_{12} = k_{12}[M_1^*][M_2]$$

$$M_2^* + M_1 \rightarrow M_2 M_1^* \qquad R_{21} = k_{21}[M_2^*][M_2]$$

$$M_2^* + M_2 \rightarrow M_2 M_2^* \qquad R_{22} = k_{22}[M_2^*][M_2]$$

The basic assumption is that the reactivity of any growing chain depends only on the end monomer unit carrying the free-radical site and not on the number of type of monomer units already added to the chain. The rate at which the monomers M_1 and M_2 are consumed during the course of propagation can be expressed as follows:

$$-d[M_1]/dt = k_{11}[M_1^*][M_1] + k_{21}[M_2^*][M_1] \tag{2.20}$$

$$-d[M_2]/dt = k_{12}[M_1^*][M_2] + k_{22}[M_2^*][M_2] \tag{2.21}$$

Now, assuming a steady state, wherein the rate of particular chain end (say, M_1) disappearing is equal to the rate of formation of the same chain end, we can write

$$k_{12}[M_1^*][M_2] = k_{21}[M_2^*][M_1] \tag{2.22}$$

$$\frac{[M_1]}{[M_2]} = \frac{k_{12}[M_1^*]}{k_{21}[M_2^*]} \tag{2.23}$$

A combination of all the foregoing equations gives the 'copolymer equation',

$$\frac{d[M_1]}{d[M_2]} = \left\{ \frac{[M_1](k_{11}/k_{12})[M_1] + [M_2]}{[M_2][M_1] + (k_{22}/k_{21})[M_2]} \right\} \tag{2.24}$$

$$= \frac{[M_1]}{[M_2]} \left\{ \frac{r_1[M_1] + [M_2]}{r_2[M_2] + [M_1]} \right\} \tag{2.24a}$$

The terms k_{11}/k_{12} (denoted by r_1) and k_{22}/k_{21} (denoted by r_2) appearing in this equation are two important terms, known as the reactivity ratios for any given pair of monomers M_1 and M_2. These ratios (r_1 and r_2) indicate whether a growing chain carrying a free radical on a particular monomer unit would prefer to add its own monomer species or the comonomer species. In other words, the composition of the copolymer formed at any given instance is dependent not only on the concentration of the monomer species present in the system at that instance but also on their reactivity ratios. The point to be noted is that r_1 and r_2 for any given pair of monomers are dependent purely on the nature of other parameters such as the solvent, initiator and chain transfer agent. The latter, however, will have a pronounced influence on the molecular weight and the molecular weight distribution of the polymer formed. The chemical composition of the copolymer formed depends exclusively on the monomer concentration and their reactivity ratios by the equation, which can be rewritten in the following form:

$$\frac{[m_1]}{[m_2]} = \frac{[M_1]}{[M_2]} \left\{ \frac{r_1[M_1] + [M_2]}{r_2[M_2] + [M_1]} \right\} \tag{2.25}$$

where $[m_1]/[m_2]$ gives the ratio of the monomers M_1 and M_2 entered into the copolymer formed. If we express the monomer components $[m_1]$ and $[m_2]$ in terms of mole fraction instead of molar concentration, the mole fraction of monomer M_1 in the copolymer can be given by

$$\frac{[m_1]}{[m_1] + [m_2]} = \frac{d[M_1]}{d[M_1] + d[M_2]} = n_1 = 1 - n_2 \qquad (2.26)$$

where n_1 and n_2 are the mole fractions of monomers M_1 and M_2, respectively, in the monomer formed. Similarly, the mole fractions of monomer M_1 in the monomer feed can be given by

$$\frac{[M_1]}{[M_1] + [M_2]} = N_1 = 1 - N_2 \qquad (2.27)$$

where, N_1 and N_2 are the mole fractions of the monomers M_1 and M_2, respectively in the monomer feed.

Now, the 'copolymer equation' can be expressed as,

$$n_1 = \frac{r_1(N_1)^2 + N_1 N_2}{r_1(N_1)^2 + 2N_1 N_2 + r_2(N_2)^2} \qquad (2.28)$$

By knowing r_1 and r_2 and also the monomer feed ratio, one can easily predict the instantaneous molar composition of the copolymer formed. There are experimental methods by which the reactivity ratios can be determined.

Determination of Reactivity Ratios
In order to predict composition of the monomers in the copolymer formed, the values of r_1 and r_2 for a given pair of monomers are required. Experimental determination of r_1 and r_2 can be achieved by copolymerising the two monomers at different mole ratios to low conversion (upto less than 10%) and then analyzing the copolymer formed. Low conversion minimises the effect of change in monomer mole ratio as the copolymerisation progresses.

Either of the following two procedures can be employed to get the reactivity ratio. In *Mayo-Lewis scheme* the copolymer equation can be transformed into the following simple form;

$$r_2 = r_1(A^2/B) + [(A/B) - A] \qquad (2.29)$$

where $A = [M_1]/[M_2]$ and $B = [m_1]/[m_2]$

Since $[M_1]$ and $[M_2]$ are known (as taken during polymerisation at varying concentration) so that several values of A and $[m_1]$ and $[m_2]$ are determined for the copolymer formed to give B for corresponding values of A. Now using the above equation for each set of A and B, giving different values to r_1 arbitrarily, the corresponding values of r_2 are calculated. Thus plotting the r_2 on y-axis and r_1 on x-axis, a linear plot will be obtained (satisfying above straight line equation). For different sets of A and B values, different lines will be obtained; each would intersect at one fixed point. The coordinates of this point will be the actual r_1 and r_2 values for that monomer pair. It is illustrated in Fig. 2.15.

The other procedure is *Fineman-Ross method*, (see Fig. 2.16) which uses the copolymer equation of the form,

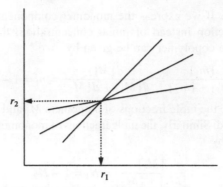

Fig. 2.15   Mayo-Lewis method to obtain r_1 and r_2

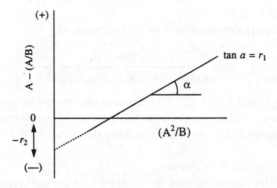

Fig. 2.16   Fineman-Ross method to obtain r_1 and r_2

$$(A - (A/B)) = -r_2 + r_1(A^2/B) \tag{2.30}$$

where $A = [M_1]/[M_2]$ and $B = [m_1]/[m_2]$ as considered above. Thus a plot between $A - (A/B)$ versus (A^2/B) shown below will be linear with intercept equal to $-r_2$ and slope r_1.

The *Q-e scheme* proposed by *Alfrey and Price*, established a semi quantitative relationship predicting the reactivity behaviour of one monomer to another standard monomer. In this method they considered the resonance stabilisation and polarisation characteristics of a given monomer. Assume that a monomer M_2 adds onto a radical M_1^* and the rate constant for the propagation reaction after the addition can be written as,

$$k_{12} = P_1Q_1 \exp(-e_1e_2) \tag{2.31}$$

where P_1 is a factor that characterises the state of the radical M_1^* at the growing chain and Q_2 is the resonance stability in the structure of monomer M_2 i.e. reactivity measure for M_2, e_1 and e_2 refer to the polarisation characteristics of monomer M_1 and M_2 as well as radicals M_1^* and M_2^*. The rate of propagation or addition of a M_1 to its non radical i.e. k_{11} can be defined as

$$k_{11} = P_1Q_1 \exp(-e_1e_1) \tag{2.32}$$

By definition $r_1 = k_{11}/k_{12}$ and $r_2 = k_{22}/k_{21}$ or

$$r_1 = \frac{P_1 Q_1 \exp(-e_1 e_1)}{P_1 Q_2 \exp(-e_1 e_2)} \qquad (2.33)$$

thus, $r_1 = Q_1/Q_2 \exp[-e_1(e_1 - e_2)]$ and $r_2 = Q_2/Q_1 \exp[-e_2(e_2 - e_1)]$.

The values Q_i and e_i have been assigned to several monomers based on styrene, which is considered as reference or standard because it has been experimentally proved that styrene undergoes copolymerisation with most of the other monomers. Thus, r_1 and r_2 values for a given monomer M_1 and styrene (M_2) have been determined by the methods previously described. Knowing the values of r_1 and r_2 for a given pair of monomer and styrene and assuming standard values of $Q = 1$ and $e = -0.8$ for styrene, Q and e values for the monomers have been obtained by using above given relations. Thus by knowing Q-e values for different monomer pairs, r_1 and r_2 values can be calculated by using the above mentioned relations. The general conclusions drawn from the Q-e scheme are, (i) monomers do not copolymerise when the values of Q for individual monomer differ largely, (ii) when the e values for a given pair of monomer differ widely, alternating copolymers are formed, (iii) the ideal or azeotropic copolymerisation is achieved with monomer pairs, when the Q values and e values are roughly equal or identical. The Q-e values of various monomers are listed in Table 2(j).

Table 2(j) Q-e values for monomers

Monomers	Q	e
Acrylic acid	1.15	0.77
Butadiene	2.39	−1.05
p-Chlorostyrene	1.03	−0.33
Ethyl acrylate	0.52	0.22
Ethylene	0.015	−0.02
Maleic anhydride	0.23	2.25
Methyl methacrylate	0.74	0.40
Propylene	0.002	−0.78
Styrene	1.0	−0.80
Vinyl acetate	0.026	−0.22
2-Vinylpyridine	1.30	−0.50

In some simple cases of reactivity ratios, we discuss here how can the copolymer composition be predicted based on monomer reactivity ratios.

1. When $r_1 = r_2 = 0$. Here reactions 11 and 22 would not occur and therefore M_1^* will react with M_2 to form M_2^* which in turn would react with M_1. Thus an alternating copolymer would form irrespective of the monomer ratio taken in the feed.

2. When $r_1 = r_2 = 1$. Here all the four propagation steps are equally possible. Thus the copolymer formed will have the same ratio as the monomers in the feed. Thus the copolymer composition can be adjusted by changing the monomer ratio in the feed. It can thus be said as an ideal polymerisation. Also, since the copolymer composition is uniform through out the process and it is called azeotropic copolymerisation.

3. When $r_1 > 1$ and $r_2 < 1$. In such a case a look into the four propagation steps will clearly show that the copolymer will be richer in M_1. Furthermore, if the two reactivity ratios differ much, it would be difficult to incorporate M_2 in the growing chain and the copolymer will essentially be homopolymer of M_1. For the reverse case i.e. when $r_1 < 1$ and $r_2 > 1$, the copolymer would be rich in M_2.

4. When $r_1 > 1$ and $r_2 > 1$. In such cases the mixture of two homopolymers can be expected.

The reactivity ratio for some monomers in free radical polymerisation are listed in Table 2(k).

Table 2(k)   Reactivity ratios for typical monomers employed in free radical polymerisation

Monomer-1	Monomer-2	r_1	r_{12}	$r_1 r_2$
Arcylamide	Acrylic acid	1.38	0.36	0.5
Acrylic acid	Styrene	0.25	0.15	0.04
Acrylonitrile	Butadiene	0.25	0.33	0.08
Acrylonitrile	Styrene	0.04	0.41	0.16
Acrylonitrile	Vinyl chloride	3.28	0.02	0.07
Butadiene	Methyl methacrylate	0.70	0.32	0.22
Butadiene	Styrene	1.39	0.78	1.08
Maleic anhydride	Methyl acrylate	0	2.5	0
Maleic anhydride	Styrene	0	0.02	0
Maleic anhydride	Vinyl acetate	0.003	0.055	0.0002
Methyl methacrylate	Styrene	0.50	0.50	0.25
Methyl methacrylate	Vinyl acetate	20	0.015	0.30
Methyl methacrylate	Vinyl chloride	12.5	0	0
α-Ethylstyrene	Maleic anhydride	0.038	0.08	0.003
α-Methylstyrene	Styrene	0.38	2.3	0.87
Styrene	p-Chlorostyrene	0.74	1.025	0.76
Styrene	Vinyl acetate	55	0.01	0.55
Styrene	Vinyl chloride	17	0.02	0.34
Vinyl acetate	Vinyl chloride	0.23	1.68	0.39
Vinyl chloride	Vinylidene chloride	0.3	3.2	0.96

2.2   Step-growth Polymerisation (or Polycondensation)

Polycondensation may be described as a simple condensation reaction carried out repeatedly. For example, polyesterification to get terylene is same as a simple esterification of an alcohol with carboxylic acid, the equilibrium of which can be shifted to the ester side by the removal of the by-product water either by simple distillation or as an azeotrope. However, a polycondensation process differs from the simple condensation in that the high molecular weight polymer produced in the former increases the viscosity of the reaction mixture if the polymer is soluble in it and if not, the polymer precipitates. Thus, powerful stirrers and a strong polymerisation vessel are needed to accommodate viscous masses.

Polycondensation is a result of reaction between functional groups in bifunctional monomers e.g. the formation of polyesters by the reaction of diols with dibasic acids. The low molecular weight intermediates called 'oligomers' are first formed. An oligomer for example in polyesterification can have two terminal–OH groups, two terminal–COOH groups, or one of each. The hydroxyl groups can react further with dibasic acid and the carboxylic groups with diol. Alternatively, two oligomers can condense. The repetition of such reactions yield final polymers. The polycondensation reaction can be stopped at any time and low molecular weight polyesters (hydroxyl or carboxyl group terminated) are isolated. This is in strong contrast to chain polymerisation in which the reaction can not be stopped to get intermediates.

Usually, small molecules like water, alcohol, hydrogen chloride, ammonia, carbon dioxide are liberated during the course of polycondensation reaction, but this is not always so. Formation of polyurethane by the reaction between diisocynate and diol, the formation of epoxy polymers by the reaction of epichlorohydrin with diol, the formation of nylon-6 from ε-caprolactam are the examples of step growth polymers which do not yield small molecules (the formation of nylon-6 from caprolactam using initiator may well be described as addition polymerisation process).

In polycondensation, the monomer concentration disappears early in the reaction because of ready formation of low molecular weight oligomers. The molecular weight of a given polymer chain increases successively throughout the course of reaction, and thus long reaction times build up high molecular weight product. Thus, again in contrast to chain polymerisation, in polycondensation process, there remains neither monomer nor fully grown polymer after an early stage of reaction. Instead there is a wide distribution of slow growing oligomers.

2.2.1 Mechanism of Polycondensation

Polycondensation reaction follows essentially the similar mechanism as proposed in general reactions of functional groups in low molecular weight organic compounds. For example, nylons, the polyamides do form via a simple SN^2 type Lewis acid-base reaction, involving the attack of Lewis base, for example, a nucleophilic amine on the electron deficient site of electrophile carbonyl group followed by a loss of proton. In the process, elimination of small molecules occur. Similarly one can envisage identical mechanism for polyesterification.

The following scheme depicts the mechanism of nylon and polyester formations.

Mechanism of nylon formation

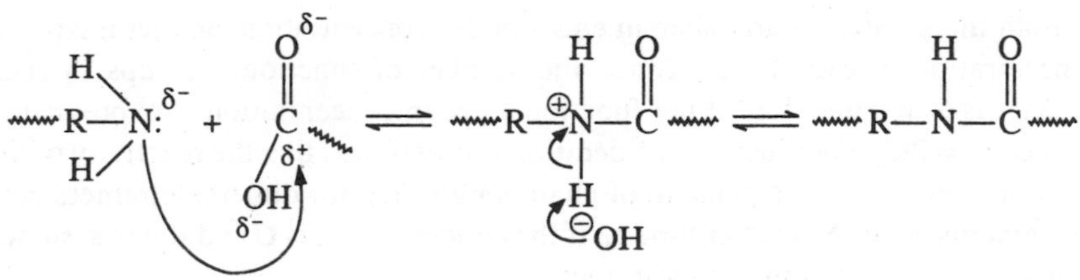

Mechanism of polyester formation

2.2.2   Phase Techniques in Polycondensation

The polycondensation involves the reactions of the functional groups and therefore for the reaction to occur the reactant needs to be in fluid phase. This can be achieved either by melting the reacting monomers (melt polycondensation) or by dissolving them in a suitable inert common solvent (solution polycondensation). A third technique is called interfacial polycondensation. This polycondensation technique uses the solutions of the two monomers prepared separately in two solvents immiscible with each other. When these two solutions are brought in contact, the polycondensation reaction begins at the interface separating the liquids. The three polycondensation techniques are described in the Table 2(I).

2.2.3   Kinetics of Polycondensation

Consider a reaction of two substances each containing two functional groups (e.g., ethylene glycol and terephthalic acid) to give a polymer,

$$X - A - X + Y - B - Y \rightarrow X - (A - B) - Y$$

Both the monomers are taken in equal molar concentration, and let the initial concentration of each be C_0. Since the number of functional groups in each reactant is 2 and therefore if the functional group concentration is expressed as n_0, then $n_0 = 2C_0$. For kinetic consideration it is assumed that the reactivity of the functional groups is independent of chain length (i.e. for dimers, tetramers etc.) and remains same. Now after time t_0, if the concentration is C and if p is assumed as the extent of reaction at time t, then

Table 2(l)   Comparison of requirements of different polycondensation techniques

Requirement	Melt	Solution	Interfacial
Temperature	High	Usually up to a temperature below b.pt. of solvent media or liquid monomers and up to m.pt. of solid monomers	
Stability to heat	Necessary	Unnecessary	Unnecessary
Kinetics	Equilibrium, stepwise	Equilibrium, stepwise	Non-equilibrium, polyaddition on a macroscopic level
Reaction time	1 hr. to several days	Several minutes to 1 hr.	Several minutes to 1 hr.
Yield	Necessarily high	Low to high	Low to high
Stoichiometric equivalence	Necessary	Less necessary	Often necessary
Purity of reactants	Necessary	Less necessary	Less necessary
Equipment	Specialized often sealed	Simple, open	Simple, open
Pressure	High, low	Atmospheric	Atmospheric

$$C = C_0(1 - p) \qquad (2.34)$$

$$n = n_0(1 - p) \qquad (2.35)$$

where n is the moles of functional groups left after time t.

The average DP of polycondensation polymer formed at time t can be written as, DP = [initial number of functional groups]/[number of functional groups after time t]

$$= n_0/n = [C_0]/[C] \qquad (2.36)$$

$$= [C_0]/[C_0](1 - p)$$

$$= 1/(1 - p)$$

Thus the degree of polymerisation is dependent on the extent of conversion p and increases as p increases. This equation is known as *Carothers* equation. The DP can be seen to vary with p in the following manner.

p	0.50	0.90	0.95	0.99	0.999	1.0
DP	2	10	20	100	1000	∞

The polycondensation reactions may be non-catalysed or catalysed. For non-catalysed reactions, e.g. of ethylene glycol and terephthalic acid, the acid itself acts as a catalyst (as well as one reactant). Hence, the polyesterification reaction rate will be proportional to the square of the concentration of the carboxylic acid group and to the first power of the hydroxyl group, i.e.

$$d[C]/dt = K[X - A - X]^2 \cdot [Y - B - Y] \tag{2.37}$$

$$= K[C]^2 \cdot [C]$$

as the concentration of acid and alcohol are same i.e. $[C]$,

$$- d[C]/dt = K[C]^3 \tag{2.38}$$

or $\qquad\qquad - d[C]/[C]^3 = Kdt \tag{2.38a}$

Integration of the above equation yields

$$2Kt = 1/[C]^2 - 1/[C_0] \tag{2.39}$$

as $\qquad\qquad [C] = [C_0](1 - p)$

$$2Kt = 1/[C_0]^2(1 - p)^2 - 1/[C_0] \tag{2.40}$$

or $\qquad 2K[C_0]^2 t + 1 = 1/(1 - p)^2 \tag{2.40a}$

This equation shows linear behaviour between $1/(1 - p)^2$ and t.

Also, since $DP_n = 1/(1 - p)$ one can get

$$DP_n^2 = 2K[C_0]^2 t + 1$$

Thus, DP_n appears as the second power and time as the first power. This means that the polymer molecular weight increases slowly with time. For acid catalysed polycondensation (e.g. by mineral acids), the following equation can be written,

$$- d[C]/dt = K[X - A - X][Y - B - Y] \tag{2.42}$$

$$= K[C][C]$$

$$= K[C]^2$$

(the reaction is first order with respect to each reactant)

Upon integration, one gets

$$Kt = 1/[C] - 1/[C_0] \tag{2.43}$$

or $\qquad\qquad Kt = 1/[C_0](1 - p) - 1/[C_0] \tag{2.43a}$

or $\qquad\qquad Kt[C_0] = 1/(1 - p) - 1$

and $\qquad\qquad DP_n = K[C_0]t + 1 \tag{2.44}$

These equations indicate a linear relationship between t and $1/(1 - p)$ and $1/(1 - p)^2$. The kinetics of polycondensation can be studied by titrating the unreacted monomer using a suitable reagent at different intervals of time. Alternatively, the polymer formed at these time intervals can be precipitated/dried/weighed. Typical plots of variation of DP with time for catalysed and uncatalysed polycondensations are shown in Fig. 2.17.

It is important to note that the rate constants for reactions of monofunctional substance is essentially the same as that for di-functional compounds and, therefore, these can be used for polycondensation reactions. Also, a similar increase in rate

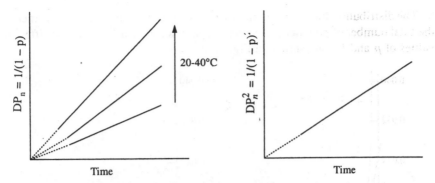

Fig. 2.17 Schematic plot of *DP* versus time for the reaction between ethylene glycol and terephthalic acid (a) acid catalysed and (b) non-catalysed polycondensation

constants (obtained from Arrhenius plots) for condensation of mono- and difunctional compounds are comparable.

Control of Polymer Molecular Weight
As it has been discussed before that the polycondensation is a quite rapid process which proceeds to almost completion, a very high mols. wt. polymer will thus result. The mechanical properties of a polymer remains almost constant after a critical mol. wt. and the melt viscosity increases exponentially. The very high melt viscosity of high mol. wt. polymers offers problems in its processing and, therefore, polymers with very high mol. wt. are not often desired. Figure 2.18 describes the general variation in the mechanical properties and melt viscosity with molecular weight.

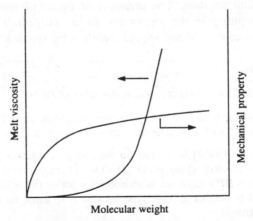

Fig. 2.18 Variation of melt viscosity and mechanical properties with molecular weight of polymers

The mol. wt. can however, be controlled by adjusting the concentrations of reactants (one having a slightly low mole composition as compared to the other) or by using a small amount of a monofunctional reactant.

The distribution functions, number fractions (i.e. number fraction of n-mer in the total number of polymer molecules), N_n and weight fraction, W_n for different values of p and DP are shown in Fig. 2.19.

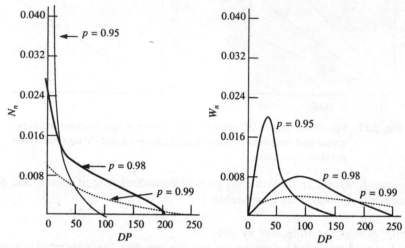

Fig. 2.19   **Distribution functions (a) number fraction, N_n and (b) weight fraction, W_n as a function of DP at different conversions during polycondensation**

The Criterion of Monomer Purity

Most vinyl monomers often contain small amounts of substances called inhibitors, which are intentionally added to prevent self polymerisation of the monomer during storage or transportation. Such substances must be removed prior to polymerisation. It can be achieved either by distilling the monomer or washing it with dilute alkali solution. The presence of traces of monofunctional or polyfunctional impurity in the monomers during polycondensation process drastically influences the DP and should therefore be avoided.

Numerical Exercises

Ex. 1. If $\overline{M}_n$ for LDPE is 1,400,000, what is the value of DP?

Ex. 2. At the end of the polymerisation of p-hydroxybenzoic acid, IR analysis shows 0.2 mole% unreacted COOH. Calculate the molecular weight and categorise it.

Solution. 99.8% of the COOH is reacted, so the extent of reaction $p = 0.998$: $DP = 1/(1-p) = 1/0.998 = 500$. Repeat unit is $-OC_6H_4CO-$ and hence its mol. wt. $= 180$. Hence mol. wt. of polymer $= DP \times$ (mol. wt. of monomer) $= 500 \times 180 = 90,000$.
Mol. wt. calculated is number average type because it is based on the estimation of reacted functional groups.

Ex. 3. The monomer HO $(CH_2)_{14}$COOH is used to prepare a polymer of number average molecular mass of 24,000. Neglecting the effect of end groups, calculate the percentage coversion to polymer necessary to reach this MW.

Solution. The repeat unit is [$(CH_2)_{14}$–COO]$_x$, monomer unit mol. wt. $= C_{15}H_{28}O_2 = 240$ g mol^{-1}, $DP =$ mol. wt. of polymer/mol. wt. of monomer unit $= 24000/240 = 100$, $DP = 100 = 1/(1-p)$ and hence extent of reaction p $= 0.990$

Ex. 4. What is the functionality of a mixture consisting of 0.7 mol. of ethylene glycol, 0.05 mol of ethanol and 0.25 mol of glycerol.

(Ans. 2.2)

Ex. 5. What is the $\overline{M}_n$ of a polymer prepared by the self polycondensation of HO—⟨○⟩—COOH, if the 99.5% of the functional groups react.
Repeat unit = (O—⟨○⟩—CO)$_n$ $p = 0.995$, $DP = 1/(1-p) = 1/(1-0.995) = 200$, the molecular weight of repeat unit is 120 and hence the mol. wt. of polymer = $120 \times DP = 120 \times 200 = 24,000$ g mol^{-1}

Ex. 6. The fractional conversion in an ester interchange reaction is 0.99999. What would be the DP of the polyester produced?
Degree of polymerisation = $1/(1-p) = 1/(1-0.99999)$

Ex. 7. If 2% molar excess of bisphenol A is used with toluene diisocynate for making a polyurethane, what will be the maximum DP obtained?

(Ans. 9)

Ex. 8. An equimolar mixture of two monomers 1 and 2 is copolymerised. If the reactivity ratios are $r_1 = 4$ and $r_2 = 0.5$, calculate the average sequence lengths for residue 1 and 2:
Here $[m_1]/[m_2] = 1$

$$n_1 = 1 + \{r_1[m_1]/[m_2]\} = 1 + r_1 = 1 + 4 = 5$$

$$n_2 = 1 + \{r_2[m_2]/[m_1]\} = 1 + r_2 = 1 + 0.5 = 1.5$$

Ex. 9. What is the percentage conversion of a monomer $OH(CH_2)_{14}COOH$ to a polymer of average molecular weight 2400? (Ans. 99.0%)

Ex. 10. Determine the degree of polymerisation of a polyester made from equimolar quantities of PTA and ethylene glycol in the presence of 1.5 mol% of acetic acid for $p = 0.999$? (Ans. 133)

2.3 Synthesis and Application of Some Common Industrial Polymers

Polyethylene

The manufacture of polyethylene began around 1930 when ethylene was polymerised at high pressure (1000–300 atm) and temperature of about 250°C using free radicals. The so obtained polymer was a branched one and had less bulk density. It was only in 1953, *Karl Ziegler* polymerised ethylene at pressure just above atmospheric pressure by using organo-metallic catalysts in conjunction with transition metal halides. The polyethylene so obtained was a linear polymer with high molecular weight. The polymer, being linear, is closely packed (crystalline) yielding dense structure (high density) with a greater rigidity and higher softening temperature.

Polyethylene is chemically very inert. Both high density and low density polyethylenes differ in physical properties. The low density polyethylene LDPE (branched polymer), due to its flexibility, high tensile strength and resistance to moisture and chemicals is generally used as film or for foil packaging containers and electrical insulation material as sheathing for wires and cables. The high-density polyethylene HDPE (linear polymer chain synthesised at low pressures) with its stiffness and high tensile strength is used for piping, tubing and a whole range of objects and vessels.

$$CH_2{=}CH_2 \longrightarrow {+}CH_2{-}CH_2{\xleftarrow{}}_n$$
Polyethylene

Polypropylene

High molecular weight, isotactic and highly crystalline polypropylene is generally similar in properties to high density polyethylene. Isotactic polypropylene is harder, stronger and softens at about 160°C. The monomer propylene is inexpensively obtained in large quantities from the cracking of petroleum hydrocarbons and the high molecular weight isotactic polymers are formed in the presence of the stereospecific catalysts.

$$nCH_2{=}\underset{CH_3}{\overset{}{CH}} \xrightarrow[50–120°C]{Ziegler\ catalyst} {+}CH_2{-}\underset{CH_3}{\overset{}{CH}}{\xleftarrow{}}_n$$
Polypropylene

The crystalline polypropylene is about 90–95% isotactic and has high strength, exceptional flexing resistance to stress cracking. Polypropylene is used in preparing moulded and excluded articles and films. Copolymerisation of propylene with ethylene or other monomers is frequently advantageous in order to reduce the tendency towards the cracking at low temperatures, or to enhance the dyeability of fibres.

Polyisobutylene

The monomer isobutene is obtained by cracking petroleum hydrocarbons and can be readily polymerised by cationic process in the presence of acid catalysts, such as boron tetrafluoride or aluminium chloride.

$$CH_2{=}\underset{CH_3}{\overset{CH_3}{C}} \xrightarrow{AlCl_3} {+}CH_2{-}\underset{CH_3}{\overset{CH_3}{C}}{\xleftarrow{}}_n$$
Polyisobutylene

Major use of polyisobutylene is as its copolymer with 2–4% isoprene, which is known as butyl rubber. For its impermeability to gases and resistance to ageing butyl rubber is largely used in automobile tires and tubes.

Polystyrene

Styrene is produced by the dehydrogenation of ethyl benzene, which in turn is obtained by the alkylation of benzene with ethylene as shown in following reaction,

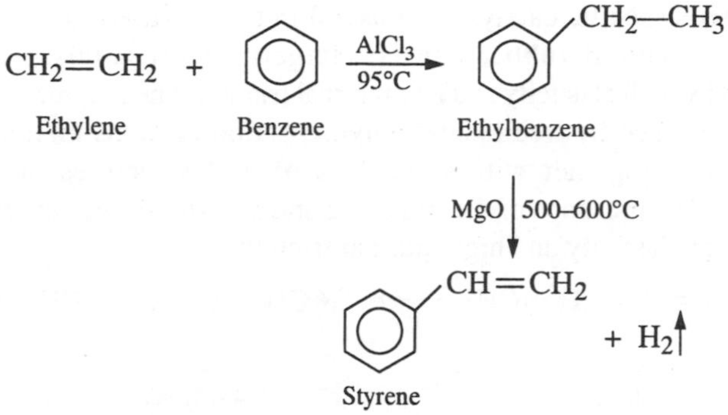

$$CH_2 = CH_2 \ + \ \underset{\text{Benzene}}{\bighexagon} \ \xrightarrow[95°C]{AlCl_3} \ \underset{\text{Ethylbenzene}}{\bighexagon\!\!-\!CH_2\!-\!CH_3}$$

Ethylene   Benzene   Ethylbenzene

$$\Big\downarrow \ MgO \ | \ 500–600°C$$

$$\underset{\text{Styrene}}{\bighexagon\!\!-\!CH\!=\!CH_2} \ + \ H_2\!\uparrow$$

Styrene polymerises to polystyrene,

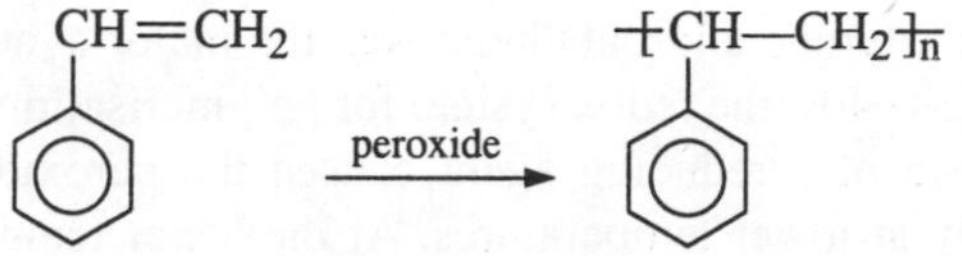

$$\underset{}{\bighexagon\!\!-\!CH\!=\!CH_2} \ \xrightarrow{\text{peroxide}} \ \underset{}{\bighexagon}\ +\!CH\!-\!CH_2\!+_n$$

Polystyrene is a hard, transparent, glass like thermoplastic resin. It is characterised by excellent electrical insulation properties, relatively high resistance to water, high refractive index, clarity, and low softening temperature. Polystyrene may also be fabricated in the form of rigid foam, which is used in packaging, food-service articles and insulating panels.

The high-molecular-weight homopolymers, copolymers and polyblends are used as extrusion and moulding compounds for packaging appliance and furniture components, toys and insulating panels.

Polydienes

Butadiene, isoprene, and 2-chlorobutadiene are widely used dienes. Butadiene can be obtained by the thermal cracking of n-butane and also as a byproduct of other cracking processes.

$$CH_3 - CH_2 - CH_2 - CH_3 \ \xrightarrow[700°C]{Cr_2O_3} \ \underset{\text{1, 3 Butadiene}}{CH_2 = CH - CH = CH_2}$$

Isoprene is readily available by cracking processes. Chloroprene is obtained from acetylene via dichlorobutene as shown below.

$$\underset{\text{Acetylene}}{2HC \equiv CH} \ \xrightarrow{CuCl_2} \ \underset{\text{1, 3 Butadiene}}{CH_2 = CH - CH = CH_2}$$

$$\Big\downarrow Cl_2$$

$$\underset{\text{Chloroprene}}{CH_2 = \overset{\displaystyle Cl}{\overset{|}{C}} - CH = CH_2} \ \xleftarrow{\text{Heat}} \ \underset{\text{Dichlorobutene}}{Cl - CH_2 - \overset{\displaystyle Cl}{\overset{|}{CH}} - CH = CH_2}$$

Butadiene and 2-chlorobutadiene have long been used in the production of synthetic rubbers by free-radical catalysis.

Special stereospecific catalysts are useful in the polymerisation of butadiene and isoprene. Natural rubber consists largely of *cis*-polyisoprene, and is characterised by high elasticity and low internal friction. The polymers of butadiene and isoprene formed by free radical polymerisation contains mixture of the *cis* and *trans* forms together with some 1, 2 or 3, 4 structures in the case of polyisoprene. The presence of the trans 1, 2 and 3, 4 structures causes the rubber to have a lower elasticity and high internal friction

$$CH_2=C-CH=CH_2 \longrightarrow +CH_2-C=CH-CH_2+_n$$

$$\underset{\text{1, 3 diene}}{\overset{|}{R}} \qquad\qquad \underset{\text{1, 4 polydiene}}{\overset{|}{R}}$$

$R=H$, butadiene, $R=Cl$, chloroprene, $R=CH_3$, isoprene

The copolymer of styrene and butadiene was the major synthetic rubber of World War II. During 1940s, the redox system for polymerisation was developed in which the presence of a reducing agent caused the peroxide to yield free radicals more rapidly at lower temperatures. At the lower temperature used in redox polymerisation, a more linear copolymer called cold rubber is obtained in high conversion with improved physical properties. Styrene-butadiene copolymers are still used for automobile tyres and in various rubber articles. High molecular weight styrene-butadiene copolymers (containing more than 50% styrene) are resinous rather than rubbery. The latex as produced by emulsion polymerisation, has achieved wide usage in water based paints.

Copolymers of styrene with 20–30% acrylonitrile may be used when a somewhat higher softening point is desired without sacrifice of transparency. By sulfonation of the copolymer of styrene and divinylbenzene, an insoluble resin is produced. This product in the form of its sodium salt is employed as a cationic exchange resin, which is used for water softening.

Blending of polystyrene with polybutadiene gives high impact polystyrene (HIPS). Other complex polyblends and copolymers of styrene, butadiene, and acrylonitrile in various combinations (ABS resins) are important. Besides many applications of styrene in combination with other materials as in rubber and paints, the homopolymer and the polyblends are employed in the injection moulding of toys, panels, novelty items and also in the extrusion of sheets. The sheets are used for panels or they may be further shaped by vacuum forming for uses such as liners for refrigerator doors.

Polyvinyl Chloride and Polyvinylidene Chloride

The vinyl chloride monomer (VCM) is prepared from chlorine, acetylene and ethylene by any of the following routes.

$$CH_2=CH_2 \xrightarrow{\text{Cl}_2} Cl-CH_2-CH_2-Cl$$

$$\underset{\text{Ethylene}}{} \qquad\qquad \Big\downarrow \text{Dehydrohalogenation}$$

$$CH_2=CH-Cl + HCl$$

$$\underset{\text{Vinyl chloride}}{}$$

$$HC\equiv CH \xrightarrow{\text{HCl}} CH_2=CH-Cl$$

Acetylene   Vinyl chloride

The polymerisation of vinyl chloride gives PVC

$$n\ CH_2=\underset{\underset{Cl}{|}}{CH} \xrightarrow{\text{peroxide}} \underset{\underset{Cl}{|}}{+CH_2-CH+_n}$$

Vinyl chloride   Polyvinyl chloride

Polyvinyl chloride (PVC) is a tough, strong thermoplastic material, which has an excellent combination of physical and electrical properties. The products are usually characterized as plasticized or rigid types. Polyvinyl chloride is one of the most versatile plastics. The plasticized polymer is somewhat elastic material and is used in shower curtains, floor covering, rain coats, dish pans, dolls, wire insulation and films etc.

Rigid PVC is used in the manufacture of photograph records, pipes and chemically resistant liners for chemical reaction vessels etc. The monomer vinylidene chloride is normally prepared by the pyrolysis of 1, 1, 2 trichloroethane. The latter is obtained by the chlorination of 1, 2 dichloroethane which in turn, is formed by the addition of chlorine to ethylene.

$$CH_2=CH_2 \xrightarrow{\text{Cl}_2} \underset{\underset{Cl}{|}}{CH_2}-\underset{\underset{Cl}{|}}{CH_2} \xrightarrow{\text{Cl}_2} \underset{\underset{Cl}{|}}{CH}-\underset{\underset{Cl}{|}}{CH_2} + HCl$$

Ethylene

1, 2 Dichloroethane   1, 1, 2 Trichloroethane

↓ Heat

$$CH_2=\underset{\underset{Cl}{|}}{\overset{\overset{Cl}{|}}{C}} + HCl$$

Vinylidene chloride

Polyvinylidene chloride is a tough, horny thermoplastic with properties generally similar to those of polyvinyl chloride. In comparison with the latter, polyvinylidene chloride is softer and less soluble; it softens and decomposes at lower temperatures, crystallises more readily and is more resistant to burning. Because of its relatively low solubility and decomposition temperature, the material is most widely used in the form of copolymers with other vinyl monomers, such as vinyl chloride. The copolymers are employed as packaging film, rigid pipe, as filaments for upholstery and window screens etc.

Fluoropolymers

Polyvinyl Fluoride

Polyvinyl fluoride is a tough, partially crystalline thermoplastic material, which has a higher softening temperature than polyvinyl chloride. Films and sheets are characterised by high resistance to weathering. The monomer is obtained by catalytic hydrofluorination of acetylene.

$$HC\equiv CH + HF \xrightarrow{HgCl_2-BaCl_2} CH_2=\overset{\displaystyle F}{\underset{\displaystyle }{CH}}$$

Acetylene   Vinyl fluoride

Polymerisation can be effected in the presence of oxygen and peroxidic catalysts. Because of the low boiling point (–88°C) and high critical temperature of the monomer, polymerisation is accomplished by use of pressure techniques similar to those employed in the high-pressure process for polymerising ethylene.

$$n\ CH_2=\underset{\displaystyle F}{\underset{\displaystyle |}{CH}} \xrightarrow{peroxide} -(CH_2-\underset{\displaystyle F}{\underset{\displaystyle |}{CH})_n}$$

Vinyl fluoride   Polyvinyl fluoride

Films of PVF are used in industrial and architectural applications. Coatings made of PVF for example on pipe are resistant to highly corrosive media.

Polyvinylidene Fluoride

The monomer is prepared by the dehydrohalogenation of 1, 1, 1 chlorodifluoroethane or by the dechlorination of 1, 2 dichloro- and 1, 1 difluoroethane. The polymer is inert and possesses low dielectric constant and thermal stability (up to about 150°C). It is used in electrical insulation, piping and process equipment and as a protective coating in the form of a liquid dispersion.

$$CH_3-\overset{\displaystyle F}{\underset{\displaystyle F}{C}}-Cl \xrightarrow{-HCl} CH_2=CF_2$$

$$Cl-CH_2-\overset{\displaystyle F}{\underset{\displaystyle F}{C}}-Cl \xrightarrow{-Cl_2} CH_2=CF_2$$

Polytetrafluoroethylene (PTFE)

The monomer tetrafluoroethylene can be polymerised readily and conveniently by emulsion polymerisation under pressure, using free-radical catalysts such as peroxides or persulfates. The polymer is insoluble, resistant to heat (up to 250°C) and chemical attack, has the lowest coefficient of friction of any solid.

Polytetrafluoroethylene is a useful polymer for applications under extreme conditions of heat and chemical activity. Polytetrafluoroethylene bearings, valve seats, packagings, gaskets, coatings and tubing can withstand relatively severe conditions. Because of its excellent insulating properties, it is useful when a dielectric material is required for service at a high temperature. The non adhesive quality is often turned to advantage in the use of polytetrafluoroethylene to coat articles such as nonsticking frying pans and other kitchenware.

Polyvinyl acetate

The monomer vinyl acetate is conveniently prepared by the reaction of acetylene with acetic acid. Acetaldehyde is produced by the oxidation of olefins or hydrocarbons.

$$\underset{\text{Acetylene}}{HC\equiv CH} + \underset{\substack{\text{Acetic}\\\text{acid}}}{CH_3COOH} \xrightarrow{Hg_2Cl_2} \underset{\text{Vinyl acetate}}{H_2C = \overset{\displaystyle OCOCH_3}{\overset{|}{C}H}}$$

Vinyl acetate is largely polymerised by emulsion (or suspension) polymerisation.

$$\underset{\text{Vinyl acetate}}{\overset{\displaystyle OCOCH_3}{\overset{|}{CH_2}}=CH} \longrightarrow \underset{\text{Polyvinyl acetate}}{-(\overset{\displaystyle OCOCH_3}{\overset{|}{CH_2}}-CH)_{\overline{n}}}$$

Aqueous dispersions (latexes) produced by the emulsion polymerisation are used for treating textiles and paper, as adhesives and as water based paints. The water based paints prepared by pigmenting vinyl acetate polymer and copolymer emulsion achieved wide usage because of low cost, ease of application and resistance to weathering. As water is removed from the latex by evaporation or absorption, the suspended polymer particles coalesce into a tough film.

Polyvinyl acetate is a leathery, colourless thermoplastic material, which softens at relatively low temperatures and which is relatively stable to light and oxygen. The polymers are clear and noncrystalline. The chief applications of polyvinyl acetate are as binders for water based or emulsion paints.

Polyvinyl alcohol

The monomer vinyl alcohol, $H_2C=CHOH$ can not be isolated as it transforms into the tautomeric acetaldehyde. However, the polyvinyl alcohol can be produced commercially from polyvinyl acetate.

$$\underset{\text{Polyvinyl acetate}}{-(\overset{\displaystyle OCOCH_3}{\overset{|}{CH_2}}-CH)_{\overline{n}}} + H\!-\!OH \longrightarrow \underset{\text{Polyvinyl alcohol}}{-(\overset{\displaystyle OH}{\overset{|}{CH_2}}-CH)_{\overline{n}}} + \underset{\text{Acetic acid}}{CH_3\!-\!\overset{\displaystyle O}{\overset{\|}{C}}\!-\!OH}$$

The hydrolysis of polyvinyl acetate can be controlled to get the product with different degree of hydrolysis. Highly hydrolysed polymers may be plasticized with water or glycols and moulded or extruded into films, tubes and filaments which are resistant to hydrocarbons; these can be rendered insoluble in water by cold drawing or heat or by the use of chemical cross-linking agents. It is used in grease resistant coating and paper adhesives, for treating paper and textiles and as emulsifiers and thickeners.

The partially hydrolyzed products crystallise with difficulty. These materials are used as emulsifying solutions in adhesive formulations and textile sizes and in preparation of water soluble films. Polyvinyl alcohol is a tough, whitish polymer which can be formed into strong films, tubes and fibres that are highly resistant to hydrocarbon solvents. Although polyvinyl alcohol is one of the few water soluble polymers, it can be rendered insoluble in water by drawing or by the use of cross-linking agents.

Polyvinyl acetals

These are relatively soft, water insoluble thermoplastic products obtained by the reaction of polyvinyl alcohol with aldehydes. Properties depend on the extent to which alcohol groups are reacted. Polyvinyl butyral is rubbery and tough and is used primarily in plasticized form as the inner layer and binder for safety glass. Polyvinyl formal is the hardest of the group; it is used mainly in adhesive, primer and wire coating formulations especially when blended with a phenolic resin.

Polyvinyl butyral is usually obtained by the reaction of butyraldehyde with polyvinyl alcohol.

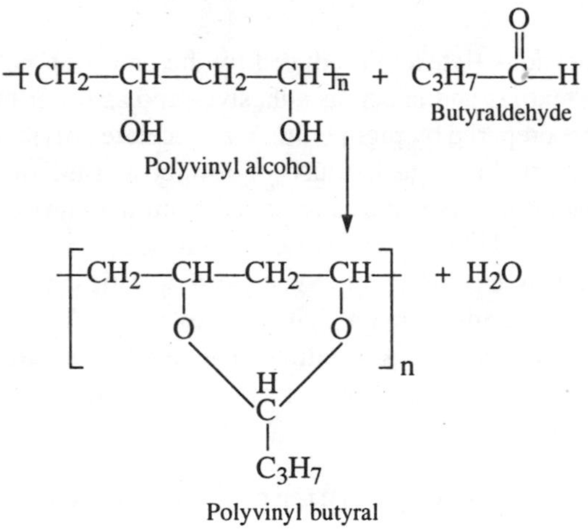

$$-\!\!\left[CH_2\!-\!CH\!-\!CH_2\!-\!CH\right]_{\overline{n}} + \quad C_3H_7\!-\!\overset{\displaystyle O}{\overset{\|}{C}}\!-\!H$$

OH   OH   Butyraldehyde

Polyvinyl alcohol

$$\left[CH_2\!-\!CH\!-\!CH_2\!-\!CH\right]_n + H_2O$$

Polyvinyl butyral

Polyvinyl pyrrolidone

Polyvinyl pyrrolidone is a water soluble polymer of basic nature which has film forming properties, strong absorptive or complexing qualities for various reagents and the ability to form water soluble salts which are polyelectrolytic in nature. The polymer can be prepared by free radical polymerisation in bulk or aqueous solution reaction.

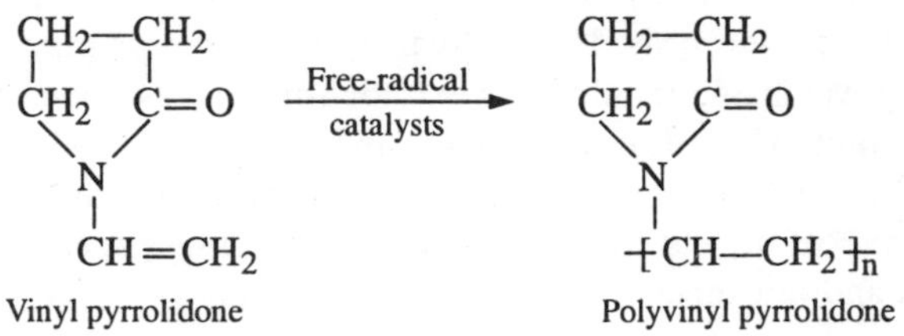

$$\text{Vinyl pyrrolidone} \xrightarrow[\text{catalysts}]{\text{Free-radical}} \text{Polyvinyl pyrrolidone}$$

Isotonic solutions of PVP were used in Germany during World War II as extenders for blood plasma. The main uses of PVP, today however are as solubilizing agents for water based medicinal formulations such as iodine and as semi-permanent setting agents in hair sprays. Certain synthetic textile fibres containing small amounts of vinylpyrrolidone as a copolymer have improved affinity for dyes.

Polyvinyl ethers

The monomers may be prepared by the reaction of alcohols with acetylene in the presence of alkali.

$$HC\equiv CH \xrightarrow[180°C]{C_2H_5OH/C_2H_5OK} CH_2=\overset{\displaystyle OC_2H_5}{\underset{\displaystyle |}{CH}}$$

Acetylene   Vinyl ethyl ether

By careful choice of conditions, it is possible to achieve stereoregular polymerisation, which yields partially crystalline polymers that are tougher than the amorphous products.

$$nCH_2=\underset{\displaystyle OC_2H_5}{\underset{\displaystyle |}{CH}} \xrightarrow[\text{initiators}]{\text{cationic}} \overset{}{\underset{\displaystyle OC_2H_5}{\underset{\displaystyle |}{\text{+}CH_2-CH\text{+}_n}}}$$

Vinyl ethyl ether   Polyvinyl ethyl ether

The alkyl vinyl ether polymers are insoluble in water. Copolymers of vinyl methyl ether with maleic anhydride are useful in textile applications.

Polyvinyl carbazole

Polyvinyl carbazole is a tough, glassy themoplastic with excellent electrical properties and the relatively high softening temperature of 120–150°C. Its uses have been limited to small scale electrical applications requiring resistance to high temperatures.

Vinyl carbazole   Polyvinyl carbazole

Polyvinyl carbonate

Vinyl carbonate polymerizes to polyvinyl carbonate, which is a solid polymer with m. pt. around 200°C. It forms fibers with high tensile strength. Its copolymers with vinyl chloride or vinyl acetate or vinyl thiocyanate are known. Poly (vinyl carbonate-co-vinyl thiocyanate) protects tissues from high energy ionising radiation.

Polyacrylates

Useful polymers are made from a variety of acrylic monomers, such as acrylic and methacrylic acids, their salts, esters, amides and the corresponding nitriles. The most important monomers are shown below.

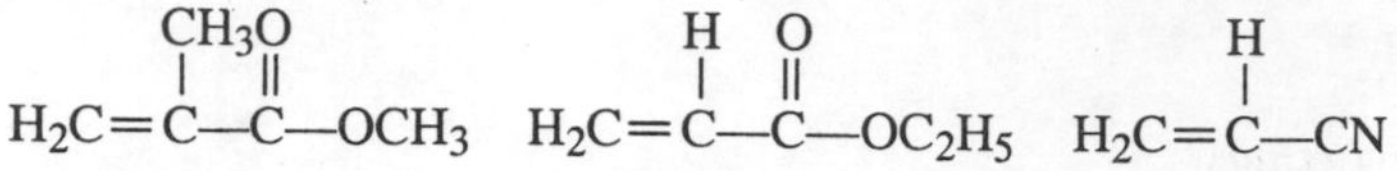

$$
\begin{array}{ccc}
\underset{\text{Methyl methacrylate}}{\text{H}_2\text{C}=\overset{\overset{\displaystyle\text{CH}_3}{|}}{\text{C}}-\overset{\overset{\displaystyle\text{O}}{||}}{\text{C}}-\text{OCH}_3} &
\underset{\text{Ethyl acrylate}}{\text{H}_2\text{C}=\overset{\overset{\displaystyle\text{H}}{|}}{\text{C}}-\overset{\overset{\displaystyle\text{O}}{||}}{\text{C}}-\text{OC}_2\text{H}_5} &
\underset{\text{Acrylonitrile}}{\text{H}_2\text{C}=\overset{\overset{\displaystyle\text{H}}{|}}{\text{C}}-\text{CN}}
\end{array}
$$

Polymethyl methacrylate is a hard and transparent polymer with good resistance to the effects of light and weathering. It and its copolymers are useful for lenses, transparent domes, skylights, dentures and in protective coatings.

The monomer is usually made by the dehydration and methanolysis of acetone cyanohydrin as in reaction,

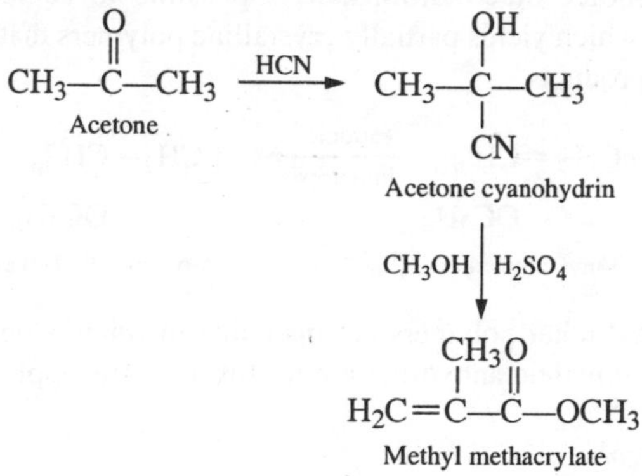

$$
\underset{\text{Acetone}}{\text{CH}_3-\overset{\overset{\displaystyle\text{O}}{||}}{\text{C}}-\text{CH}_3} \xrightarrow{\text{HCN}} \underset{\text{Acetone cyanohydrin}}{\text{CH}_3-\overset{\overset{\displaystyle\text{OH}}{|}}{\underset{\underset{\displaystyle\text{CN}}{|}}{\text{C}}}-\text{CH}_3}
$$

$$
\Big\downarrow \text{CH}_3\text{OH} \mid \text{H}_2\text{SO}_4
$$

$$
\underset{\text{Methyl methacrylate}}{\text{H}_2\text{C}=\overset{\overset{\displaystyle\text{CH}_3}{|}}{\text{C}}-\overset{\overset{\displaystyle\text{O}}{||}}{\text{C}}-\text{OCH}_3}
$$

Polymerisation may be initiated by free radical catalysts such as peroxides, or by organometallic compounds such as butyl lithium. The free radical polymerisation, which is used for commercial production can be carried out in bulk, in solution, and in aqueous emulsions or suspensions. Depending on catalyst, temperature and solvent used in polymerisations, isotactic, syndiotactic, or atactic forms can be obtained. Although solution polymerisation is not commonly used, bulk polymerisation usually with co-monomers is frequently employed in various casting operations such as in the formation of sheets, rods, and tubes in the mounting of biological, textile and metallurgical test specimens and in dental applications. Solutions of polymethyl methacrylate and its copolymers are useful as lacquers. Aqueous latices formed by the emulsion polymerisation of methyl methacrylate with other monomers are useful as water based paints and in the treatment of textiles and leather goods.

Polyethyl acrylate is a tough and somewhat rubbery material. The monomer ethyl acrylate is used mainly as an internal plasticizing or softening component of copolymers. Ethyl acrylate is usually produced by the dehydration and ethanolysis of ethylene cyanohydrin, which can be obtained from ethylene oxide.

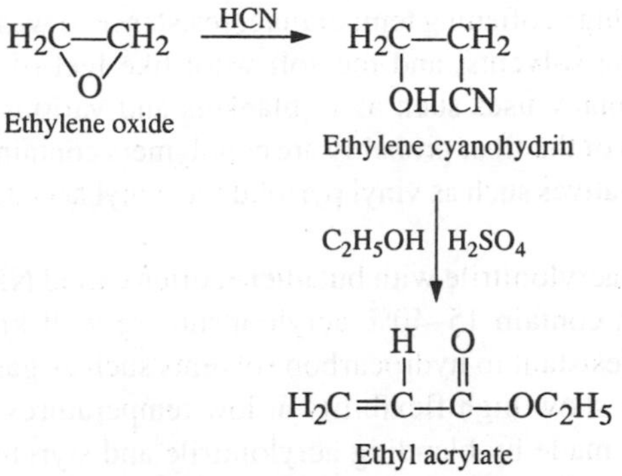

$$H_2C\!-\!CH_2 \xrightarrow{\text{HCN}} H_2C\!-\!CH_2$$

Ethylene oxide

$$\begin{array}{cc} H_2C\!-\!CH_2 \\ |\quad\;| \\ OH\;\,CN \end{array}$$

Ethylene cyanohydrin

$$\xrightarrow{\;C_2H_5OH\;\mid\;H_2SO_4\;}$$

$$\begin{array}{c} H\;\;O \\ |\quad\,\| \\ H_2C\!=\!C\!-\!C\!-\!OC_2H_5 \end{array}$$

Ethyl acrylate

Modified acrylic resins with high impact strengths can be prepared by blending with polyvinyl chloride. Copolymers of ethyl acrylate with 2-chloroethyl vinyl ether are useful rubbers. Butyl and octyl esters of acrylic acid are rubbery materials and when fluorinated have been used as oil-resistance rubbers. Polymers of alkyl, cyanoacrylates make excellent adhesives. Copolymers of methylacrylate or acrylamide are water soluble and useful as sizing and finishing aids. Addition of polylauryl methacrylate to petroleum lubricating oil improves the flow properties of the oil.

Polyacrylonitrile

The monomeric acrylonitrile can be produced using any of the following methods.

(i) Catalyzed addition of hydrogen cyanide to acetylene

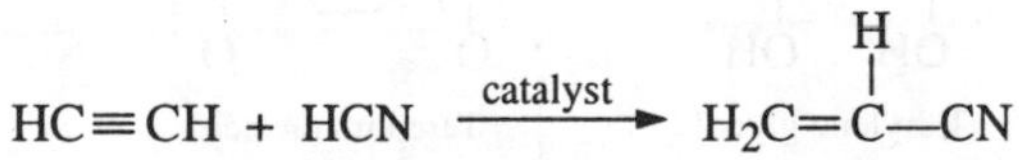

$$HC\!\equiv\!CH + HCN \xrightarrow{\text{catalyst}} \begin{array}{c} H \\ | \\ H_2C\!=\!C\!-\!CN \end{array}$$

(ii) Reaction between hydrogen cyanide and ethylene oxide

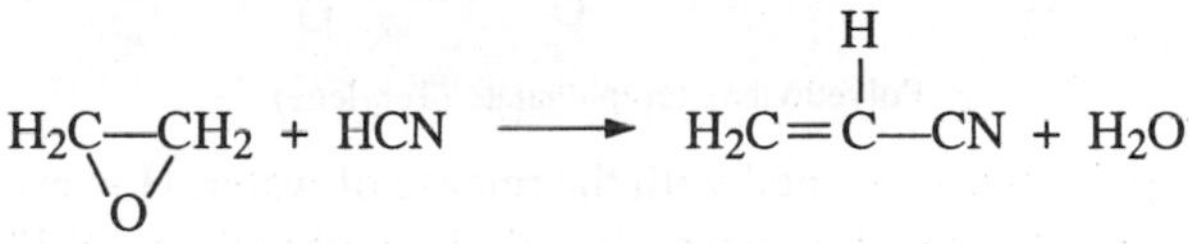

$$H_2C\!-\!CH_2 + HCN \longrightarrow \begin{array}{c} H \\ | \\ H_2C\!=\!C\!-\!CN \end{array} + H_2O$$

(iii) Reaction between ammonia and propylene

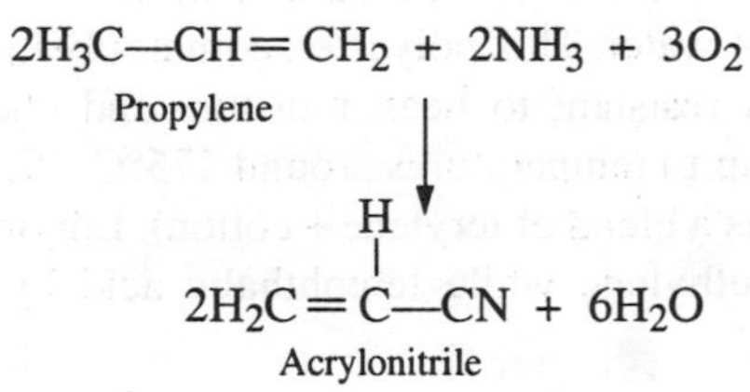

$$2H_3C\!-\!CH\!=\!CH_2 + 2NH_3 + 3O_2$$

Propylene

$$\downarrow$$

$$\begin{array}{c} H \\ | \\ 2H_2C\!=\!C\!-\!CN \end{array} + 6H_2O$$

Acrylonitrile

Polyacrylonitrile (PAN) is largely used in fibers. An acrylic fiber contains at least 85% acrylonitrile; a modacrylic fiber may contain 35 to 85% acrylonitrile.

The high strength, high softening temperature, resistance to weathering, chemicals, water, and cleaning solvents; and the soft wool like feel of fabrics have made PAN popular for many uses such as in blankets and various types of clothing. Commercial forms of the fiber probably are copolymers containing minor amounts of other vinyl derivatives such as vinyl pyrrolidone, vinyl acetate, maleic anhydride and acrylamide.

Copolymers of acrylonitrile with butadiene, often called NBR (formerly Buna N) rubbers, which contain 15–40% acrylonitrile are well known rubbers. The NBR rubbers are resistant to hydrocarbon solvents such as gasoline, to abrasion, and in some cases show high flexibility at low temperatures.

ABS resins are made by blending acrylonitrile and styrene copolymers with a butadiene-acrylonitrile rubber or by interpolymerizing polybutadiene with styrene and acrylonitrile. The ABS resins combine the advantages of hardness and strength of the vinyl resin component on one hand and with toughness and impact resistance of the rubbery component at the other. ABS resins due to their high impact strength are used in pipes and sheets for structural uses such as industrial ducts and components of automobile bodies.

Polyesters

Polyesters are the condensation polymers made by the reaction of an acid and an alcohol each containing at least bi-functionality. The most common example of the polyesters is that obtained from ethylene glycol and terephthalic acid which is known as terylene, dacron and mylar etc.

Ethylene glycol   Terephthalic acid

Polyethylene terephthalate (Terylene)

The linear polyester is formed with the release of water. However, in place of terephthalic acid, its dimethyl ester (dimethyl terephthalate, DMT) is used, as the former is difficult to purify. The polycondensation between ethylene glycol and DMT is a kind of *trans* esterification and in this case methanol is the by product in the place of water. This polyester, commonly known as PET melts at around 265°C and is resistant to heat, moisture and chemicals. It has good mechanical strength up to temperatures around 175°C. PET is extensively used in fibres (e.g. terycot is a blend of terylene + cotton). Ethylene glycol is obtained by the oxidation of ethylene while terephthalic acid by the oxidation of p-xylene.

$$CH_2{=}CH_2 \longrightarrow HO{-}CH_2{-}CH_2{-}OH$$

Ethylene   Ethylene glycol

$$H_3C-\langle O \rangle-CH_3 \longrightarrow HO-\underset{\substack{\| \\ O}}{C}-\langle O \rangle-\underset{\substack{\| \\ O}}{C}-OH$$

p-Xylene

Terephthalic acid

Branched and cross-linked polyesters are obtained if small amount of a polyhydric alcohol is used. For example, alkyd resins called glyptals form when glycerol (trifunctional monomer) is used. Glycerol needed is manufactured by the catalytic hydroxylation or hypochlorination of allyl alcohol. The latter is produced from acryldehyde (acrolein) which is obtained by the oxidation of propylene. Glycerol is also obtained as the byproduct during soap manufacture.

Unsaturated polyesters

Linear unsaturated polyesters often called prepolymers are obtained when the dicarboxylic acid used is unsaturated e.g. maleic acid or its anhydride. These polyesters may be dissolved in styrene and used as cross-linked resins for the production of fibrous glass reinforced plastics. Maleic anhydride is a by product of phthalic acid manufacture by the oxidation of xylene. It can also be obtained by the oxidation of benzene and by the vapour phase oxidation of crotonaldehyde or butene.

Polyamides

Commercially useful polyamides are made by the reaction of dicarboxylic acids with diamines. The most common commercial aliphatic polyamides known by the generic name nylons are nylon 6, 6; nylon 6; nylon 6, 10; nylon 11 and nylon 12.

Nylon 6, 6 is prepared by polycondensation reaction of hexamethylenediamine (6 carbon atoms) with adipic acid (6 carbon atoms). By heating equimolar proportions of the two reactants, a polymeric salt is formed.

$$n\ HO-\underset{\substack{\| \\ O}}{C}+CH_2)_4C-OH + n\ H_2N-(CH_2)_6NH_2$$

Adipic acid   Hexamethylene diamine

$$[N^+H_3-(CH_2)_6 -N^+H_3O^--\underset{\substack{\| \\ O}}{C}-(CH_2)_4\underset{\substack{\| \\ O}}{C}-O]_n^-\ A^+$$

Nylon 6, 6 salt

$$\text{\textasciitilde}\Big[HN-(CH_2)_6NH-\underset{\substack{\| \\ O}}{C}-(CH_2)_4\underset{\substack{\| \\ O}}{C}\text{\textasciitilde}\Big]_n + H_2O$$

Nylon 6, 6

The starting material adipic acid is manufactured from benzene. Oxidation of benzene gives cyclohexane, which oxidizes to cyclohexanone. It on further oxidation by 50% HNO_3 at 55°C undergoes ring scission and produces adipic acid. The other monomer hexamethylenediamine is obtained from the reduction of adiponitrile or adipamide.

Nylon-6 can be obtained by the self condensation of ϵ-aminocaproic acid. Nylon-6 can also be produced by the polymerisation of the lactam of ϵ-aminocaproic acid. The lactam process is generally preferred for commercial operation because it is easier to make and purify the lactam than the ϵ-aminocaproic acid.

$$H_2N-(CH_2)_5-\overset{\displaystyle O}{\underset{\displaystyle \|}{C}}-OH \longrightarrow -(HN-(CH_2)_5-\overset{\displaystyle O}{\underset{\displaystyle \|}{C}})_n-$$

ε-Aminocaproic acid   Nylon-6

The monomer caprolactam is prepared by the following synthetic route,

Nylons are strong and tough materials. Their mechanical properties depend on the degree of crystallinity. Because of good mechanical properties and adaptability to both moulding and extrusion, nylons are often used for gears, bearing, and electrical mountings. Nylon bearings and gears perform quietly and need little or no lubrication. Nylon resins are used extensively as filaments, bristles, wire insulation, appliance parts, and film. Reinforcement of nylons with glass fibers results in increased stiffness, lower creep and improved resistance to elevated temperatures. The aromatic polyamide (aramides) e.g. one prepared from terephthalic acid and 1, 3 phenylene diamine (known as kevlar) have very high melting points and are thus used in places where this property is desired.

Polyimides

Polyimides are polymers that can withstand high temperature (~ 425°C). Pyromellitic anhydride and p, p' diaminodiphenyl ether, for example on polycondensation in dimethylformamide produces polyimide. The process involves two stages. The first stage gives polyamic acid at around 50°C.

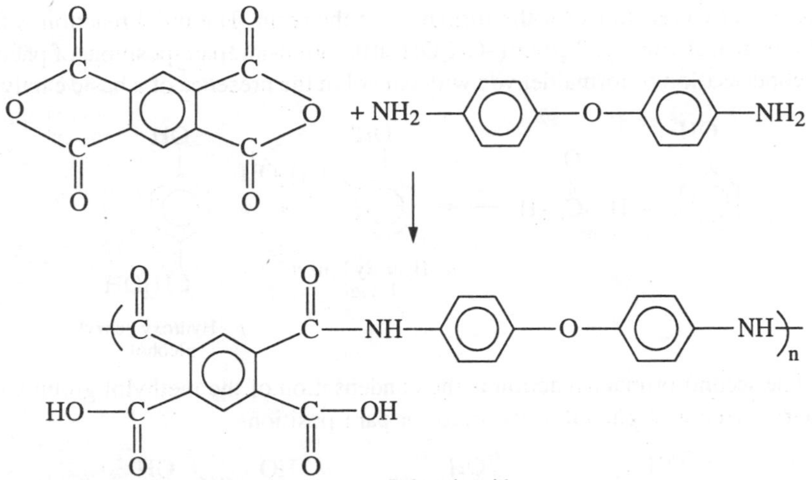

Polyamic acid

Polyamic acid film on heating to 300°C in nitrogen atmosphere results into polyimide by the removal of water molecules.

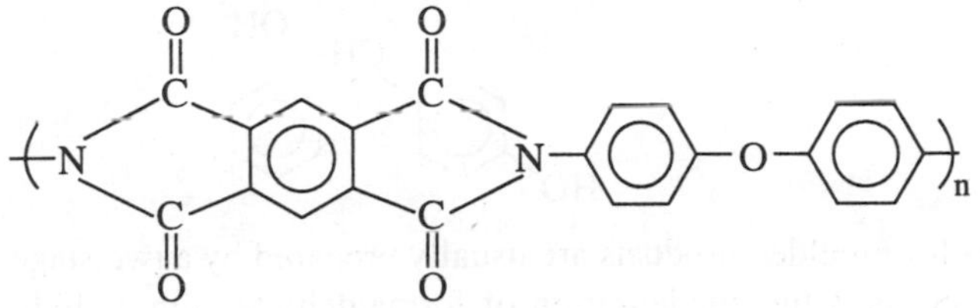

Polyimide

Polyimides are extensively used in electrical industries and in supersonic aircraft coating materials.

Polyformaldehyde

Cationic polymerisation of formaldehyde or trioxane gives polyformaldehyde.

The polymer is highly unstable and depolymerizes to monomer and therefore has to be stabilised. It's high m. pt. (185°C) and fibrous character (50–80% crystallinity) makes it useful as an engineering plastic.

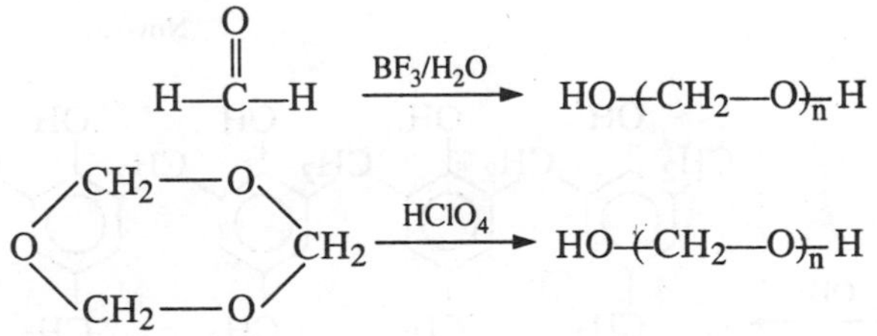

Phenol-Formaldehyde Resin

This important thermosetting resin was first developed by Henry Baekelund and is known as bakelite. It is formed by the condensation of phenol with formaldehyde.

There are two reactions for the formation of the resin. The intial reaction is the introduction of a methylol group (–CH$_2$OH) at the ortho- and para-positions of phenol by condensation of formaldehyde with phenol in the presence of a basic catalyst.

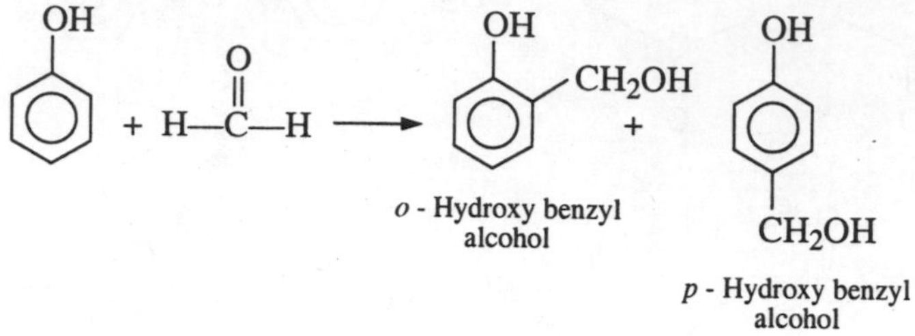

o - Hydroxy benzyl
alcohol

p - Hydroxy benzyl
alcohol

The second primary reaction is the condensation of the methylol group with other molecule of phenol at its ortho- or para-positions.

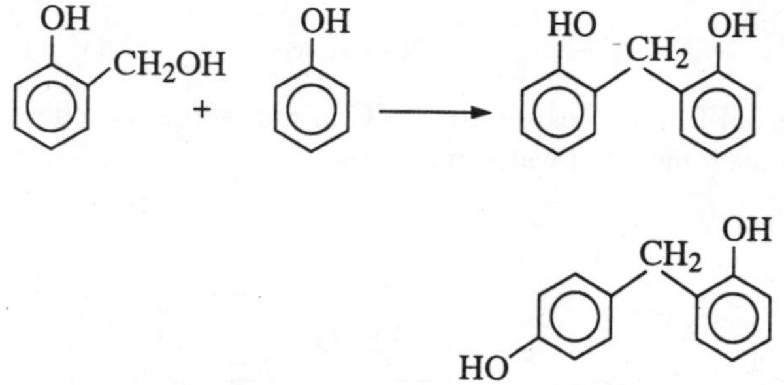

The phenolic resins for moulded products are usually prepared by a two stage process. The initial process is the condensation of formaldehyde with a slight excess of phenol in the presence of an acid catalyst. This reaction produces linear thermoplastic polymer, which is known as novolac. Novolacs contain no methylol groups and probably have the diphenylmethane type of structure.

The monomer phenol is manufactured from benzene (Raschigs process) or from isopropylbenzene (Cumene process).

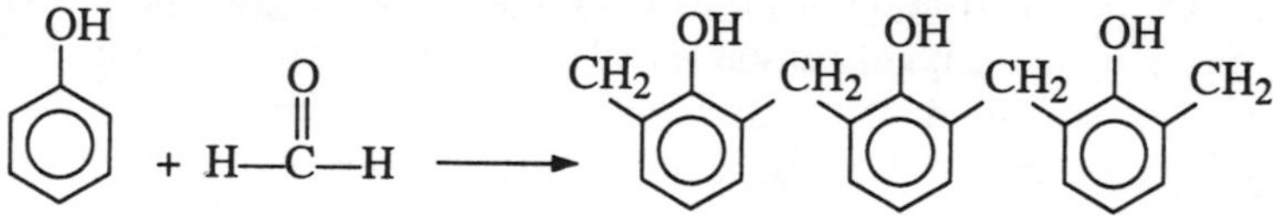

Novolac

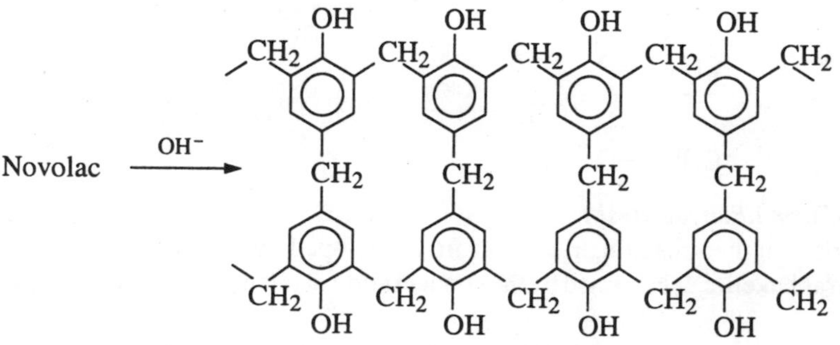

Bakelite

The second stage is the reaction of novolac with more formaldehyde in the presence of a basic catalyst which results in a hard infusible thermoset resin called bakelite.

Urea-Formaldehyde Resin

Urea reacts with formaldehyde under slightly alkaline or neutral conditions to give compounds known by their trivial names such as monomethylolurea and dimethylolurea depending upon the ratio of the two reactants.

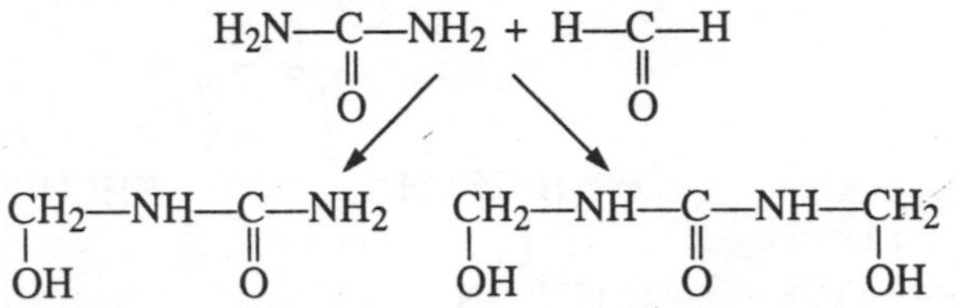

$$H_2N-C-NH_2 + H-C-H$$
(with C=O below each)

$$CH_2-NH-C-NH_2 \qquad CH_2-NH-C-NH-CH_2$$
$$\;\;|\qquad\qquad||\qquad\qquad\quad |\qquad\qquad||\qquad\qquad|$$
$$OH\qquad\quad O\qquad\qquad\quad OH\qquad\quad O\qquad\quad OH$$

Monomethylol urea   Dimethylol urea

These further condense with urea to give final resin in the following manner,

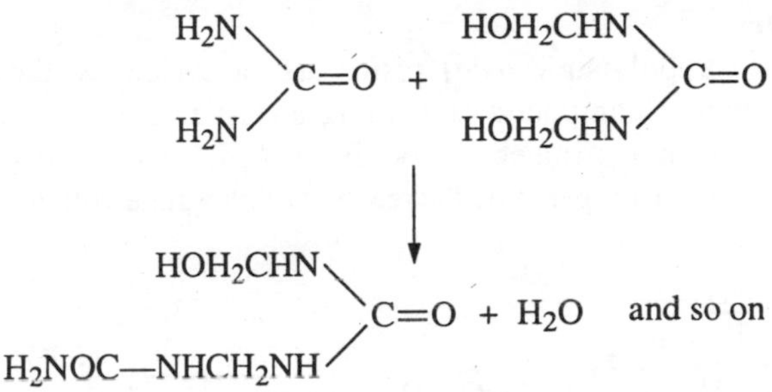

$$H_2N{\textstyle\diagdown}$$
$$\qquad\quad C=O \;+$$
$$H_2N{\textstyle\diagup}$$

$$HOH_2CHN{\textstyle\diagdown}$$
$$\qquad\qquad\quad C=O$$
$$HOH_2CHN{\textstyle\diagup}$$

$$HOH_2CHN{\textstyle\diagdown}$$
$$\qquad\qquad\quad C=O \;+\; H_2O \quad \text{and so on}$$
$$H_2NOC-NHCH_2NH{\textstyle\diagup}$$

or

$$n\; HCHO + n\; NH_2CONH_2 \longrightarrow H-[NH-CO-NH-CH_2]_n-OH$$

The hydrogen atoms in the imide group in this linear polymer can be further replaced with methylol groups by using excess of formaldehyde to finally form a cross-linked resin. Polymeric resins are then formed as a result of polycondensation. The monomer urea is prepared by the reaction of carbon dioxide with ammonia.

Melamine-formaldehyde resin

Melamine and formaldehyde react to give hexamethylol-melamine, which on heating in presence of acids gives cross-linked polymer called melamine resin. The resin is resistant to heat and moisture and is used in making dinnerware and decorative table tops (Formica). The monomer melamine is prepared from calcium cynamide.

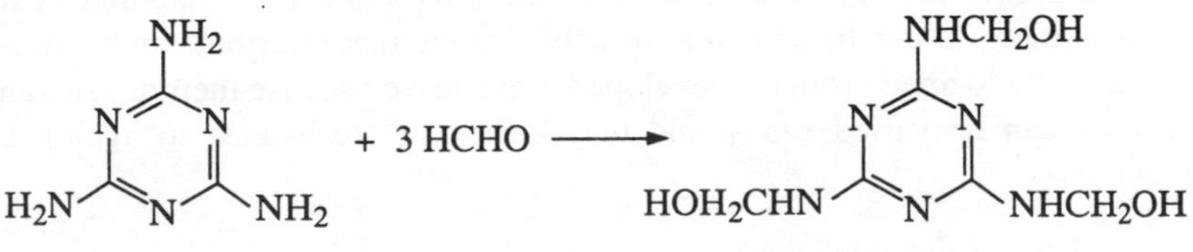

$$+ 3\; HCHO \longrightarrow$$

Melamine   Methylol derivative

The methylol derivative is further condensed with melamine to give a linear polymer. The latter when reacts with excess of formaldehyde produces a cross-linked melamine-formaldehyde resin.

Polyurethanes

Polyurethane (or polyisocyanate) resins are produced by the reaction of a diisocyanate with a compound containing atleast two active hydrogen atoms such as a diol or diamine. Toluene diisocyanate (TDI) is most frequently employed. The diisocyanates are prepared by the reaction of phosgene with the corresponding diamines.

2, 4 Diaminotoluene   Phosgene   2, 4 Toluenediisocyanate

Linear and fiber forming polymers are prepared by the addition of diisocyanates to diols while cross-linking is made possible by the use of polyols or isocyanates having more than two functional groups. Resins can be produced in different forms varying from hard, glossy, solvent resistant coatings, to abrasion and solvent resistant rubbers, fibers, and flexible to rigid foams. The polyurethane foams or U-foams have found the widest use. The more flexible foams are employed as upholstery material for furniture, insulation and crash pads. The more rigid foams are employed as the core material in structural and insulating laminates and as insulation cover in refrigerated appliances and vehicles.

The flexible polyurethanes may be used for coating rubber articles to give them additional resistance to abrasion and solvents. Wire insulated with polyurethane resin can be soldered directly. Among these various applications, the uses of the foamed products developed is extensive because their density and flexibility can be varied easily and they have good resistance to ageing by solvents.

Polyureas

Polyureas are polyamides of carbonic acid and contain —NH CONH—as a part of repeat units. These can be obtained by the reactions of diamines with diisocyanates.

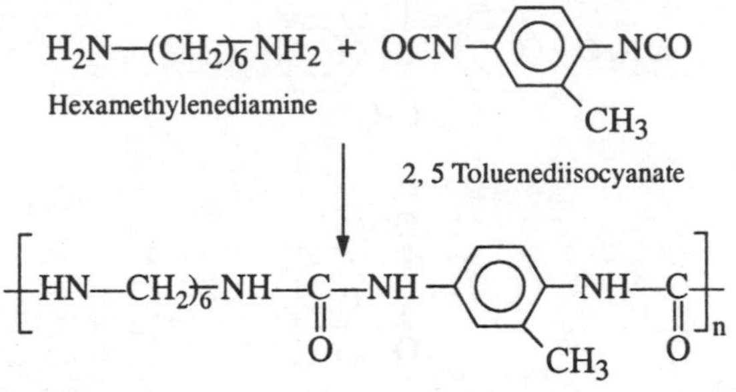

$$H_2N—(CH_2)_6\,NH_2 \;+\; OCN—\!\!\bigcirc\!\!—NCO$$

Hexamethylenediamine

CH_3

2, 5 Toluenediisocyanate

$$\left[\!\!-HN—(CH_2)_6\,NH—\underset{O}{\overset{\displaystyle C}{\|}}—NH—\!\!\bigcirc\!\!—NH—\underset{O}{\overset{\displaystyle C}{\|}}\!\!-\right]_n$$

CH_3

Polyureas have higher melting points than the corresponding polyurethanes and polyamides.

Polysulphides

Alkyl dihalides on reaction with sodium polysulphides produce polysulphide elastomers commonly known a thiokol rubbers,

$$\underset{Cl}{\overset{\displaystyle CH_2}{|}}—\underset{Cl}{\overset{\displaystyle CH_2}{|}} \;+\; Na_2S_x \;\longrightarrow\; +(CH_2—CH_2—S_x)_{\overline{n}} + NaCl$$

Polysulphides have excellent oil resistance and low permeability for the gases. These are mainly used to make sealants, balloons and gaskets etc. Polysulphides on mixing with inorganic oxidisers (e.g. ammonium perchlorate) function as solid propellants for rockets.

Polyethylene Glycols

Polyethylene glycols can be obtained by self-polycondensation of ethylene glycol. The same polymer, usually with high mol. wt. when obtained by ring opening polymerisation of ethylene oxide is called polyethylene oxide. These are water soluble as well as soluble in inorganic solvents. PEGs of different molecular weights are waxy solids commercially known as Carbowax. These are widely used in cosmetics, textile and pharmaceutical industries.

Polyparaphenylene

Polyparaphenylene can be obtained from dibromobenzene, which on reaction with activated copper powder at high temperature yields the polymer.

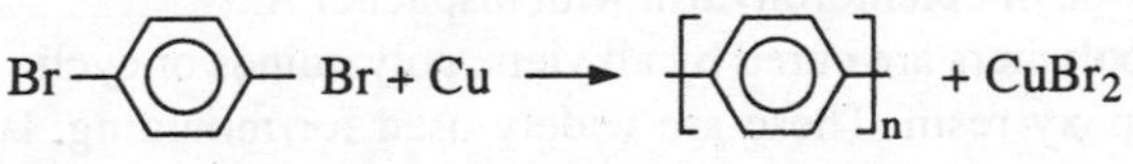

$$Br—\!\!\bigcirc\!\!—Br + Cu \;\longrightarrow\; \left[\!\!\bigcirc\!\!\right]_n + CuBr_2$$

Due to the rigid backbone made of planar aromatic rings, the polymer is very brittle, insoluble and can withstand temperature up to 550°C.

Polysulphones

These are the polymers which contain sulfone groups in the chain, as well as a variety of aromatic or aliphatic constituents such as ether or isopropylidene groups.

$$n(NaO-\underset{}{\bigcirc}-\overset{\overset{CH_3}{|}}{\underset{\underset{CH_3}{|}}{C}}-\bigcirc-ONa)$$

$$+$$

$$Cl-\bigcirc-\overset{\overset{O}{\|}}{\underset{\underset{O}{\|}}{S}}-\bigcirc-Cl$$

$$\left[\bigcirc-\overset{\overset{CH_3}{|}}{\underset{\underset{CH_3}{|}}{C}}-\bigcirc-O-\bigcirc-\overset{\overset{O}{\|}}{\underset{\underset{O}{\|}}{S}}-\bigcirc\right] + NaCl$$

Polysulfones based on aromatic backbone constitute a useful class of engineering plastics owing to their high strength, stiffness, and toughness together with high thermal and oxidative resistance.

Polyethers

Polyethers of polyoxyethylene like polyethylene oxide and polypropylene oxide are well known in detergent industries. The monomers ethylene oxide and propylene oxides are made from ethylene and propylene, respectively.

$$\underset{\underset{CH_3}{}}{\overset{\overset{CH_3}{}}{\bigcirc}}-OH \longrightarrow \left[\underset{\underset{CH_3}{}}{\overset{\overset{CH_3}{}}{\bigcirc}}-O\right]_n$$

Polyphenylene oxide is produced by the oxidative coupling of 2, 6 disubstituted phenols and is used in making airline beverage cases, solar energy collectors and window frames. Polyphenylene oxide is a high temperature resistant polymer.

Epoxy Resins

These resins are commercially known as araldite and epon and can be made by the polycondensation of epichlorohydrin with bisphenol A.

Epoxy prepolymers are cured by alkylene polyamines or cyclic anhydrides to give curved epoxy resin. These are widely used for moulding, laminating and surface coating resins. The monomer epichlorohydrin is prepared by the dehydrochlorination of glycerol.

$$HO-\text{⟨O⟩}-\underset{\underset{CH_3}{|}}{\overset{\overset{CH_3}{|}}{C}}-\text{⟨O⟩}-OH \;+\; Cl-CH_2-CH-CH_2$$

$$\left[O-\text{⟨O⟩}-\underset{\underset{CH_3}{|}}{\overset{\overset{CH_3}{|}}{C}}-\text{⟨O⟩}-OCH_2CHOHCH_2\right]_n$$

Polyetheretherketone (PEEK)

It is a polymer with a T_g of 145°C with the following repeat unit. It finds use as blow moulded containers for nuclear waste, printed circuits, electrical applications and wire coatings.

$$\left[O-\text{⟨O⟩}-O-\text{⟨O⟩}-\overset{\overset{O}{\|}}{C}-\text{⟨O⟩}\right]_n$$

Silicones

These are the polymers of silicon and constitute a very interesting class of commercially available products with wide range of materials as oils, greases, plastics, elastomers and resins. These are discussed in detail in chapter 7.

Polyphosphazenes

Polyphosphazenes are inorganic polymers containing nitrogen and phosphorus and find several biomedical applications. These are described separately under the heading **Inorganic Polymers** (see chapter 7).

Suggested Questions

1. Distinguish clearly addition polymerisation and polycondensation reactions. Give the names of 5 commercial polymers each made by addition polymerisation and polycondensation. Write the structure of monomers and the repeat units in polymers in each case.

2. Write a detailed account on mechanism of polymerisation. Name initiators used in free radical/ionic and co-ordination polymerisation. Explain initiation, propagation, termination and chain transfer steps.

3. Write mechanism of *Ziegler Natta* polymerisation. How is the growing chain terminated? Why are stereo-regular polymers are obtained from it? Write different isomeric forms of polypropylene and polyisoprene.

4. (a) Describe kinetics of chain polymerisation by free radicals. Obtain expression for rate of polymerisation and degree of polymerisation. (b) Describe mechanism of cationic polymerisation. Derive an expression for rate of polymerisation. (c) Name some catalysts for cationic polymerisation. What are its characteristic features?

5. Describe various phase techniques used for polymerisation. Mention merits and demerits of each of the technique. Which type of polymers are obtained by bulk, solution, emulsion and suspension polymerisation? Give two examples for each.

6. What are polycondensation reactions? How are polyurethane, epoxy-polymers, polyester, polyamides, polysulfide, alkyd resins synthesised? Give only the names of monomers used.

7. Write a detailed account on polycondensation reactions describing mechanism. Describe the kinetics of catalysed and non-catalysed polycondensation.

8. (a) What are copolymers? Obtain expression for the kinetics of copolymerisation. How does the reactivity ratio of monomers affect the copolymerisation reactions? Explain what do the following abbreviations stand for, (a) SBR, Buna N, ABS resin and HIPS, (b) What is Copolymerisation? Give some examples of useful copolymers and mention their advantages over the constituting homopolymers. What are different types of copolymers? Explain each one. (c) Discuss various copolymers of styrene. Mention their properties and applications, (d) Describe the methods for the determination of reactivity ratios in general and Q-e scheme of Alfrey and Price in particular.

9. What are block copolymers? How are they synthesised by anionic polymerisation? Write various heterogeneities that exist in block copolymers. Describe their solution behaviour and solid state properties. Give various applications of block copolymers.

10. Explain the mechanism of emulsion polymerisation. Which polymers are usually obtained by this technique?

11. Name two initiators for each; anionic; cationic and free radical polymerisation. Show how do they generate ions/radicals.

12. Identify all the possible structural and stereoisomers that can result from the addition polymerisation of chloroprene.

13. Mention three examples of ring scission polymers. Write their names and structure of monomers and repeating units.

14. Write the synthetic routes for the following condensation polymers; (a) Polyurethanes, (b) Polycarbonates, (c) Epoxide resins, (d) Alkyd resins (e) Melamine-formaldehyde, (f) Polyethylene terephthalate, (g) Nylon 6, 6, (h) Nylon 6, (i) Polyphenylene, (j) Polyphenylene oxide, (k) Thiokol rubber.

15. Show the condensation reaction steps between phenol and formaldehyde resulting to novolac, resole and resitole. How can the bifunctionality in phenol and formaldehyde be ascertained?

16. Answer the following; (a) explain the dependence of mol. mass of the polymer on initiator concentration, (b) explain the dependence of kinetic chain length on initiator concentration, (c) explain the dependence of rate of polymerisation on initiator concentration.

17. Give an account on thermodynamics of polymerisation.

18. Define the terms "Ceiling temperature" and "Floor temperature". Why do these definitions generally apply one to closed systems?

19. Which cyclic monomers are easily polymerised? A three membered or six membered ring compound. Give reasons.

20. Comment on (i) why does anionic polymerisation result to a polymer with low polydispersity? (ii) why is a minimum bifunctionality in the monomers necessary?

21. Show the formation of anion from naphthalene/sodium. How does anionic polymerisation of styrene proceed?

22. Explain the term "Inferfacial polycondensation". Give an example of a polymer made by this process.

23. Write short notes on (a) ring opening polymerisation, (b) coordination

polymerisation, (c) precipitation polymerisation, (d) plasma polymerisation, (e) photopolymerisation, (f) living polymerisation, (g) Trommsdorf effect, (h) telomerisation, (i) redox polymerisation, (j) chain transfer, (k) inverse emulsion polymerisation.

24. Construct tables to show clearly the comparison in (a) bulk, solution, suspension and emulsion polymerisation, (b) addition and step polymerisation, (c) free radical, anionic and cationic polymerisation (d) melt, solution and interfacial polycondensation

25. Comment on: (a) Living anionic polymerisation yields polymers with low polydispersity and tailored block copolymers, (b) Micelles but not dispersed monomer droplets are the centres where polymerisation takes place in emulsion polymerisation, (c) Polyvinyl alcohol is not made by the polymerisation of vinyl alcohol but is made by the hydrolysis (or alcoholysis) of polyvinyl acetate, (d) Free radical polymerisation of ethylene is more easier than that of polyisobutylene.

26. In the monomer $R-CH = CH_2$, if R is an electron donating group, cationic polymerisation is favoured whereas for R is an electron withdrawing group anionic polymerisation is favoured. Why?

27. Discuss the manufacturing process of: (a) HDPE, (b) stereoregular polypropylene, (c) nylon 6, (d) polyimides, (e) carbon fibres

28. Show using chemical equations how the following monomers are synthesised industrially: (a) hexamethylene diamine, (b) toluene diisocynate, (c) vinyl chloride, (d) ω-caprolactam, (e) adipic acid, (f) bisphenol-A, (g) phenol

29. Show the mechanism of generation of free radicals by the decomposition of following initiators: (a) benzoyl peroxide, (b) azobisisobutyronitrile, (c) persulphate and (d) Fe^{+2}/H_2O_2.

30. Define the following terms: (a) functionality of monomer, (b) degree of conversion, (c) kinetic chain length, (d) initiator efficiency, (e) branching coefficient.

31. How are the following polymers made? Write the chemical equations: (a) carbon fibres, (b) polyimides, (c) polyvinylbutyral and (d) cross-linked polystyrene sulphonate.

Suggested Further Readings

General Fundamentals:
Allen, P.E.M. and C.R. Patrick, *Kinetics and Mechanism of Polymerisation Reactions*, New York: Halsted-Wiley, 1974.
Brunnelle, D.J. (Ed.), *Ring-Opening Polymerisation*, Munich: Hanser, 1993.
Elias, H.-G., *Macromolecules, Synthesis, Materials and Technology*, 2nd ed., New York: Plenum, 1984.
Erunsalimskii, B.L., *Mechanisms of Ionic Polymerisation*, New York: Plenum, 1986.
Kucern, M., *Mechanism and Kinetics of Addition Polymerisation*, Amsterdam: Elsevier, 1982.
Leng, R.W., *Organic Chemistry of Synthetic High Polymers*, New York: Interscience, 1967.
Odian, G., *Principles of Polymerization*, 3d ed., New York: Wiley, 1992.
Remp, P., E.W. Merrill, *Polymer Synthesis*, 2d ed., Basel: Hultig and Wept, 1991.

Free Radical Polymerisation:
Eeisenberger, J.A., D.A. Sebastian, *Principles of Polymer Engineering*, New York: Wiley, 1983.

Blackley, D.C., *Emulsion Polymerisation: Theory and Practice*, New York: Halsted, 1975.

Cationic Polymerisation:
Matyjaszewski, K., (ed.), *Cationic Polymerisation*, New York: Dekker, 1996.

Coordination Polymerisation:
Frink, G., R. Mulhaupt, H.H. Britzinger, (eds.), *Zeigler Catalysts*, Berlin: Springer, 1995.
Kissin, Y.V., *Isospecific Polymerisation of Olefins*, Berlin: Springer, 1986.

Copolymerisation:
Ham, G.E. (ed.), *Copolymerisation*, New York: Interscience, 1964.

3

Polymer Analysis and Characterization

3.1 Identification

Synthetic polymers are widely used today as plastics, rubbers, fibers, adhesives and surface coating materials with or without trade names. It is important to identify the polymer in a commercial product. After the polymer is identified and analyzed, its physical characteristics in general and molecular weight (MW) and molecular weight distribution (MWD) in particular are to be determined. The knowledge of the last two parameters is highly essential because they greatly influence the physical properties of the polymers. Table 3(a) describes important parameters that need to be determined for the total analysis and characterization of polymers either in their virgin form or in an end product form.

Table 3(a)   Parameters for the analysis and characterization of polymers

Structural parameter	Information obtained
Chemical composition	Homopolymer or copolymer, percentage of the monomers and their arrangement in the copolymer
MW and MWD	Number and weight average mol. wt. and thus the polydispersity index
Stereochemistry	Microstructure, *cis/trans*, isotactic/syndiotactic/atactic
Topology	Linear/branched/cross-linked
Morphology	Structure in solid state, crystallinity and crystallizability
Miscellaneous	Identification of additives present

The following discussion concerns with the analysis and characterization of the polymers. The analysis of the polymers mainly involves the physical and spectral identification of either virgin polymers or a polymer in end-use or finished product. Besides this examination and isolation of polymers, they are to be characterized for the type of molecular weight distribution present and evaluation of their total molecular weight. One is always confused with words analysis, identification, testing and characterization as applied to polymers, because all these attempts are based on measuring either some property or monitoring the behavior of polymers. This discussion on individual properties such as solid state or solution state is closely linked to the analysis of polymers. One more difficulty associated with the analysis of polymers especially in finished products is the presence of additives, which can interfere with the methods used and hence the results of analysis. Similarly, other factors such as solvent quality, temperature, concentration and other conditions such as moisture and humidity have large effects on a given measurement. Thus one should be cautious in drawing conclusions.

3.1.1   Physical Testing

It involves the examination of polymer material for knowing about its appearance, feel, hardness and density and monitoring the effects of heat, combustion and solubility etc. A very quick guess can be made using the foregoing statements. Pure polystyrene, polymethyl methacrylate, polycarbonate are transparent polymers. Teflon is milky opaque. A light colored material can not be a phenol-formaldehyde polymer. Polyetheylene and polytetrafluoroethylene have soft waxy feel. Rubbers and thermoplastics like polystyrene, PMMA, polyethylene, PVA, PVAc are soft enough to be scratched by a nail. A sharpened knife gives a soft cut to thermoplasts but a hard cut to thermosetts. Elastomers like natural rubber, SBR, butyl rubber, nitrile rubber are compressible and can be stretched.

Polymers such as polyethylene (both LDPE and HDPE), natural rubber have density less than 1 g/cm^3 and float over water surface. The density of phenolformaldehyde, PMMA, PVC, polyvinyl acetate, polystyrene ranges between 1–1.34 g/cm^3 and these polymers sink in water but float over saturated magnesium chloride solution. PVC, UF resin, cellulose nitrate/acetate possess density around 1.5–2 g/cm^3 and sink even in saturated magnesium chloride but float over saturated ZnCl$_2$ solution. PTFE and urethane polymers have density more than 2 g/cm^3.

Linear polymers melt on heating. Thermosetts are infusible plastics. Polyethylene and polypropylene burn like a wax candle. PVC and PVDC are self-extinguishing. Polystyrene burns with a sooty flame. Acrylic polymers are hard to decompose and burn slowly. Polyesters burn with smoky flame, melt and darken. Nylons also melt and darken.

Chemical tests for the detection of elements like nitrogen, sulphur and halogen (as done in organic qualitative analysis) using sodium fusion extract can be conducted to ascertain the presence of these elements and their proportion.

Solubility of a polymer in different solvents can provide some definite clue to identify it. The general behavior 'like dissolves like' is followed by polymers. Usually, a polymer whose solubility parameter value is close to that of a liquid, will dissolve in that liquid. The solubility parameter concept will be discussed in detail while discussing the polymer solutions. Table 3(b) lists the solubility parameter values for some common polymers and solvents.

Cross-linked polymers do not dissolve in any solvent and may swell only to a limited extent. Crystalline polymers are often soluble at temperature above their melting temperature, T_m, or in strongly interacting solvents i.e. good solvents. Thus, polyvinyl alcohol, polyethylene oxide, polyvinyl pyrrolidone, polyacrylic acid with polar pendant groups are soluble in water. Polystyrene, polyolefins etc. dissolve in hydrocarbon solvents. Nylons, which have strong intermolecular

Table 3(b)   Solubility parameters of some common
polymers and solvents

Polymers/Solvents	Solubility parameter, H
Polystyrene	9.1
Polypropylene	9.2
PVC	9.7
PMMA	9.5
PET	10.7
Nylon 6, 6	13.6
Acetone	10.0
CCl_4	8.6
Cyclohexane	8.2
Dioxane	9.3
Chlorobenzene	9.5
Diethyl ether	7.4
n-Hexane	7.3
Acetonitrile	11.9
DMF	12.1
Xylene	8.3
Pyridine	10.7

hydrogen bonding (which imparts crystallinity) dissolve in strongly interacting solvents like phenols, conc. H_2SO_4.

Table 3(c) and 3(d) provide some important information about the physical properties of polymers useful in their identification. The test methods that can be used for identifying and characterizing the polymers or polymeric materials are mostly based on the measurements of either typical responses shown by them to a given condition or by measuring literally all the physical properties which may include thermal, electrical, crystallinity and mechanical properties. In fact the characterization of polymers is mainly based on the structure-property relationship. Following tests based on mechanical property measurements are routinely done to predict the performance of polymers and these can also be used to identify and characterize them. These tests are briefly described below:

Tensile Strength
It measures the ability of a polymer to withstand pulling stresses. It is usually determined by pulling a dumb-bell shaped specimen of the plastic. The ultimate tensile strength is equal to the load that caused failure divided by minimum cross-sectional area (ASTM D638).

Flexural Strength
It is a measure of the bonding strength or stiffness of a test bar specimen used as a sample beam. The flexural strength is based on the load required to rupture a sample beam before its deflection is 5% (ASTM D790).

Impact strength
It is a measure of toughness or ability of a specimen to withstand a sharp blow

Table 3(c)   Physical properties of polymers and their possible identity

Type of measurement	Observation	Possible identification of polymer
1. Appearance	transparent	PS, PMMA, PC
	milky opaque	PTFE
	light	PF absent
·2. Feel	soft and waxy	PE and PTFE
3. Hardness	scratching by nail cutting by sharp	Rubbers and soft thermoplasts
	knife	PS, PMMA, PE, PVC and PVAc
	soft cut	Thermoplast
	hard cut	Thermosett
	hot needle melt	Thermoplast
	compress to stretch	NR, SBR, Butyl R, NBR
4. Specific gravity	<1	PE, NR
	1–1.34	PF, PMMA, PVC, PVAc, PS, (float over sat. $MgCl_2$)
	1.34–2.01	PVC, UF, cellulose acetate and cellulose nitrate (sinks in sat. $MgCl_2$ but float in a sat. $ZnCl_2$)
	>2.01	PTFE
5. Heating and combustion		Linear polymers melt on heating Thermosets are infusible plastics PE, PP burn like a wax candle PVC, PVDC are self extinguishing, PS burns with sooty flame. PET burns with smoky flame, darkens upon heating. Nylons melt and darken

such as the ability to withstand a given object being dropped from a specific height (ASTM D256).

Shear strength
It is a measure of the load required to cause the failure in the area of the sheared specimen (ASTM D732)

3.1.2   Spectral Methods
Spectroscopic methods that have been successfully applied to polymers are ultraviolet (UV), infrared (IR), nuclear magnetic resonance (NMR), electron spin resonance (ESR), x-ray, raman, mass and microwave spectroscopy. Commonly employed methods for the analysis and identification of polymers, however, are based on IR and NMR spectroscopy. UV-visible (UV-VIS) and Raman spectroscopy are also used sometimes to provide useful information. ESR is used to a limited extent; it is principally used to detect the free radicals generated during degradation or pyrolysis.

Table 3(d)   Densities of some polymers

Polymer	Density g/cm^3
Unfoamed PS	1.04–1.06
Teflon	2.2
Polyacrylates	~ 1.2
Nylons	0.97–1.15
Polyesters	1.35
Alkyds	1–1.4
Polycarbonate	1.25
Cellulose	1.5
Cellulose acetate	1.3
Cellulose nitrate	1.4
Natural rubber	0.92
Ebonite (30% S vulcanized rubber)	1.1–1.3
Polybutadiene	0.9
Butyl rubber	0.93
UF, MF resin	1.5
PF resin	1.2–1.35
Epoxy resin	1.2
Polyacrylonitrile	1.2
Silicones	0.96–1.5

Infrared Spectroscopy (IR)

Most polymers absorb electromagnetic radiation in the IR region (1–500 μm) because their molecules undergo transitions between vibrational states of different energies causing absorption and emission. The infrared spectra are generally used as qualitative finger prints for the identification of polymers, since they rarely identify specific bonds. The infrared spectrum of polymers is often surprisingly simple. The simplicity has its origin both in the unique nature of molecular arrangement in the polymer chain and also on the utility of selection rule in identifying the bands in the IR spectrum. Since polymer chains are made of same repeating units spread monotonously along the chain, the many expected IR vibrations occur almost at the same frequency. The use of IR technique is relative in nature because, the spectrum of a given polymer with unknown structure is often compared with the reference collection of spectra of several other polymers of known structures. Some of the wave numbers that can be assigned with reasonable certainty are given in Table 3(e).

The bands in the range of 8.7 to 9.7 μm are characteristic for hydrocarbon polymers and are called finger print region. Typical C—H stretching vibrations are observed at 3.8, 3.4 and 3.5 μm, and out of plane bending of aromatic C—H bonds are at 11.0 and 14.3 μm. Characteristic C—C stretching vibrations are noted at 6.2 and 6.7 μm. The C—H bond stretching vibrations of the methylene (—CH$_2$—) group appear close to 3000 cm^{-1} which may be symmetric or asymmetric as illustrated in Fig. 3.1.

These vibrational frequencies appear in the spectra of almost all hydrocarbon polymers. Thus, they are of no use for diagnostic purpose. The deformation

Table 3(e)   Useful IR bands for the identification of polymers

Wave number, cm^{-1}	Chemical bond
3550–3070	O---H or N---H
3100–3000	C—H (aromatic)
3000–2800	C—H (aliphatic) – absent in PTEE
1820–1680	C—O
1680–1620	C—C
1650–1500	N—H
1470–1370	C—H, absent in PTEE
1420–1330	S—O
1440–1250	O—H
1300–1000	C—O
1250–1100	C—F
700–800	C—D

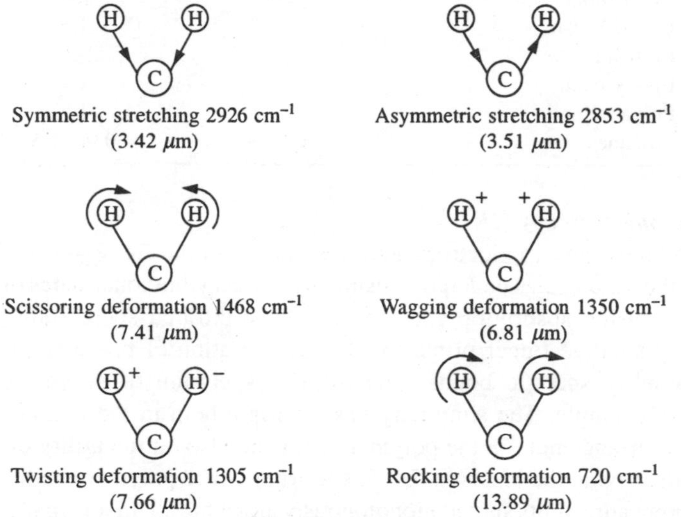

Symmetric stretching 2926 cm^{-1}
(3.42 μm)

Asymmetric stretching 2853 cm^{-1}
(3.51 μm)

Scissoring deformation 1468 cm^{-1}
(7.41 μm)

Wagging deformation 1350 cm^{-1}
(6.81 μm)

Twisting deformation 1305 cm^{-1}
(7.66 μm)

Rocking deformation 720 cm^{-1}
(13.89 μm)

Fig. 3.1   Stretching and deformation vibrational modes of methylene group

vibrations due to the bending of bond angles or scissoring motion appear at lower frequencies, giving a large band near 1500 cm^{-1}. Wagging and twisting modes of CH$_2$ occur near 1300 cm^{-1} and rocking deformations occur at the low energy end of the spectrum. The carbonyl (C=O) stretching band near 1700 cm^{-1}, the C=C stretching band near 1600 cm^{-1} and the olefinic C—H bending bands between 900 and 1000 cm^{-1} are also particularly important.

The polymer sample for IR analysis can be prepared in various forms i.e. in the form of a thin film, a pellet of the powdered compound with KBr, a nujol mull or as a solution in chloroform or carbon tetrachloride held in a NaCl cell. In a double beam infrared spectrophotometer having matched cells, one can compensate the absorption for the solvent also, if a solution has been used for recording the spectrum.

Most monomers and several polymers may be identified by IR spectrocopy in which the energy is associated with molecular vibrations and vibration-rotation spectra of polymer molecules. IR spectra for polymers may sometimes be complex to interpret due to presence of repeat units. IR spectra of some polymers are shown in Fig. 3.2.

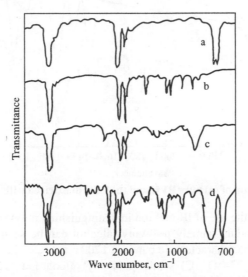

Fig. 3.2 Transmission IR spectra of (a) polyethylene, (b) polypropylene, (c) polyisoprene and (d) polystyrene (the ordinate in all the spectra is to be taken as 0–100%)

It can be seen that the —CH—stretching vibrations of methylene (—CH_2—) group are found in the narrow frequency range from 2880 to 2900 cm^{-1} in all the four polymers (polyethylene, polypropylene, polyisoprene and polystyrene). The local chemical environment effects the position of a principal absorption band in the spectra. The simple wagging and wagging deformation of (—CH_2—) group appear in the frequency range 1345–1470 cm^{-1}. The band due to the rocking deformation of (—CH_2—) group at a frequency of 720 cm^{-1} is apparent in polyethylene, but interestingly shifted to left hand side in polypropylene, polyisoprene and from (—CH_2—) to —$\overset{|}{\underset{CH_3}{CH}}$— and to $\overset{H_3C}{\diagdown}C=$ and to

$\diagup CHC_6H_5$ from PE to PP, to PI and to PS respectively. The IR spectrum of polystyrene possesses some characteristic features and is used as standard for checking instrumental operation. The repeat unit in polystyrene is C_8H_8 and thus contains 16 atoms. Since it has no symmetry, all vibrations are active i.e. 3° each of rotational and translational freedom and 42° of vibrational freedom (3n–6). Fig. 3.3 shows the IR spectra of pure polyvinyl chloride and dioctylphthalate plasticized PVC sample. The spectrum of pure PVC shows the characteristic C—Cl band which is apparent near to the right hand edge and the C—H band is also evident near to the left hand limit. The spectrum of plasticized polyvinyl chloride using dioctylphthalate shows the characteristic C=O band at 1740 cm^{-1}.

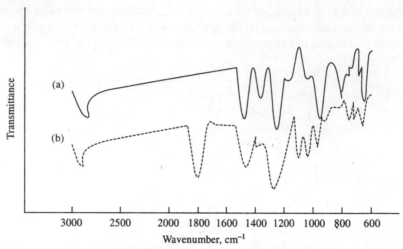

Fig. 3.3   **IR spectrum of (a) pure PVC and (b) PVC plasticized with dioctylphthalate**

An example of the use of IR method in distinguishing polyvinyl acetate from its hydrolysed product namely polyvinyl alcohol can be seen from Fig. 3.4. Polyvinyl acetate has characteristic bands at 1700 cm^{-1} (C=O) and at 1240 cm^{-1}, $1010–1040$ cm^{-1} (C—O—C) while polyvinyl alcohol at $3200–3500$ cm^{-1} (O—H). IR absorption of a polymer is significantly altered when it crystallizes. Thus, information on crystallinity can be obtained from IR. Usually, intensity of some IR bands increases and some others decreases as the crystallinity increases.

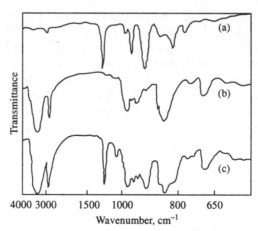

Fig. 3.4   **IR spectra of (a) PVA, (b) PVAc and (c) PVA + PVAc**

The ratio of intensity of a band to that of some reference band thus provides a measure of crystallinity. For example, the ratio of intensity of a band at 846 cm^{-1} to that of a band at 1171 cm^{-1} for polypropylene as a function of temperature slowly decreases and drops to zero at the melting point of the polymer (around 170°C) when the crystallinity in the polymer vanishes. Similarly, the ratio of absorption bands at 635 cm^{-1} to 615 cm^{-1} in PVC has been found to increase with the increase in crystallinity (as found from density measurements) (see Fig. 3.5).

Nuclear Magnetic Resonance (NMR)

It is an excellent method for resolving the environment of protons in molecules and can also be used to detect ^{13}C, ^{15}N, ^{17}O and ^{19}F.

Studies based on nuclear magnetic spectroscopy are made on polymers with the objective of ascertaining and establishing chain configuration,

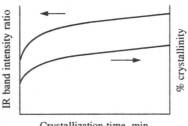

Crystallization time, min.

Fig. 3.5 Plot showing ratio of intensities of IR bands at 635 cm^{-1} to 615 cm^{-1} and the percentage of crystallinity as a function of crystallisation time

sequence distribution and microstructure. ^1H NMR technique is particularly advantageous in that it allows study of the motion and location of protons, which is not that readily accomplished by other approaches of study and analysis. As a distinctive and powerful analytical tool, this technique allows independent confirmation of isomeric structure (isotactic and syndioctatic sequence and their random combination giving atactic structure) and evaluation of copolymer composition including the composition of selected copolymers.

In NMR spectroscopy, a strong magnetic field is applied to a material containing above said nuclei which splits the energy level into two states with the spin parallel and antiparallel to the field. The chemical shift (in the scale of 0 to δ in the form of unit ppm), is helpful in determining the amount of the respective nuclei in the component. A good illustration of the use of NMR measurements to determine polymer microstructure is polymethyl methacrylate (PMMA). The NMR spectra of β-methylene protons for two samples of PMMA as obtained in CDCl$_3$ at 220 MHz are shown in Fig. 3.6.

If the polymeric chain is syndiotactic, the two protons on the β-methylene groups are equivalent and therefore there should be a single resonance. However, if the chain is isotactic, the local environment of the two protons varies due to electron cloud shielding and one expects four principal resonances due to spin–spin interaction. On similar lines the NMR spectra of amorphous and crystalline polypropylenes in which the polymer chains have predominantly atactic or predominantly isotactic configurations, are expected to be different (as shown in Fig. 3.7).

Polymers particularly containing conjugated multiple bonds also absorb in UV-VIS region and thus UV-VIS spectroscopy can be used to identify them. It is also used to identify the additives like antioxidants present in the polymeric material.

Raman spectroscopy is frequently used as a complementary technique to IR because certain vibrational modes which are inactive in IR are found in Raman spectra.

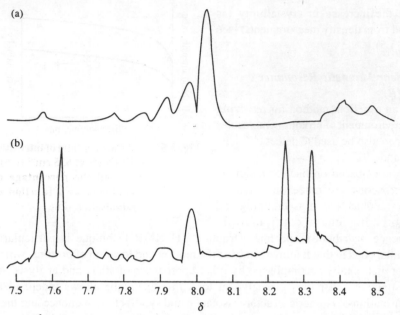

Fig. 3.6 ^{1}H NMR spectra of β-methylene protons for (a) predominantly syndiotactic and (b) predominantly isotactic PMMA samples

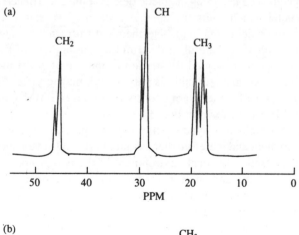

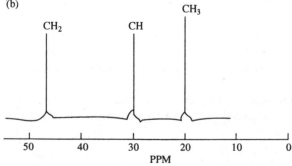

Fig. 3.7 ^{13}C NMR spectra of (a) amorphous and (b) crystalline polypropylene

3.1.3 Chromatographic Methods

Pyrolysis Gas Chromatography

It is another important technique used in the identification of polymers and is often coupled with mass spectrometry. The polymeric material on strongly heating degrades into smaller gaseous components, which are detected by the gas chromatograph. This technique may be used for qualitative and quantitative analysis. Later requires calibration with known amount of standard polymer, pyrolyzed under the same conditions as used for the unknown. Further, this procedure enables to identify the presence of volatile additives in polymers.

In gas chromatography, the mobile phase is a gas and solutes are separated as vapours. The separation is feasible by partitioning of sample between mobile gas phase and thin layer of nonvolatile high boiling liquid held onto solid support. In contrast gas–solid chromatography employs a solid sorbent. In GC, the volume of carrier gas necessary to convey a solute band through full length of a column is retention volume and is the fundamental quantity measured in gas chromatography.

3.1.4 Identification of Typical Plastic Materials

Thermoplastics

Acrylonitrile–butadiene–styrene (ABS)

It is an amorphous terpolymer with a specific gravity of 1.04. It shows very good solubility in toluene and ethylene chloride. IR spectroscopy can be used to identify the ABS because one can detect the vibrational bands due to characteristic groups such as C = N, —CH = CH— and ofcourse C—H stretching vibrations of methylene (—CH$_2$—) group. The percentage of each monomer in the terpolymer can be estimated from IR spectroscopy method. ABS polymer burns with a yellow flame and it continues to burn after the removal of heat source with black smoke. Soot and drippings are formed in the air.

Acrylics

Acrylic is a common generic name used to represent whole lot of polymers formed by methyl methacrylate and its analogues. Most of the acrylic polymers are amorphous with a specific gravity ~ 1.18. These polymers burn with a blue flame with a yellow tip. They produce no smoke while they burn. Upon burning, acrylics give off fruity odour. Acrylic polymers are soluble in acetone and aromatic hydrocarbons. IR spectroscopy is the best means to identify the acrylic polymers.

Polyfluorocarbons

This class of polymers include FEP (elastomeric copolymer of tetrafluoro-ethylene and hexafluoro propylene), CTFE (polychlorotrifluoro-ethylene), PTFE (polytetrafluoroethylene) and PVF (polyvinyl fluoride) and PVDF (polyvinylidene difluoride). The unique characteristic features of all these polymers are that (i) they do not burn when exposed to flame, (ii) they are practically insoluble in any organic or inorganic solvents, (iii) they have high melting point, (iv) they have specific gravities > 2.0 and (v) they can be easily identified by a copper wire test–in this test, the tip of the copper wire should be heated to red-hot temperature in a flame and with the hot wire small amount of polymer is picked up and then heated in the flame. A green coloured flame is observed when polymers contain chlorine or fluorine atoms.

Polycarbonate

It is one of the most transparent and toughest polymers. It has got a specific gravity around 1.2. It is soluble in methylene dichloride and ethylene dichloride. The polymer burns with an yellow or orange flame and gives off a black smoke along with soot formation in the air. The polymer has an odour of phenol and can be easily identified by IR method.

PVC

Polyvinyl chloride is a self extinguishing amorphous polymer with specific gravity of 1.2–1.7. When burnt, PVC gives a yellow flame and white smoke. It gives off an odor of hydrochloric acid when burnt. Since PVC contains Cl atom, copper wire test gives a bright green flame. PVC is highly soluble in solvents such as tetrahydrofuran and methyl ethyl ketone.

PE and PP

Polyethylene (PE) is one of the most common crystalline plastic in use. The specific gravity of PE ranges from 0.91–0.96 and it floats on water. PE is insoluble in most of common organic solvents but is soluble in hot aromatic hydrocarbons such as toluene and benzene. PE burns like a candle and small drips are formed with a paraffinic odor when burnt. IR method is most useful in identifying the polymer. Polypropylene (PP) has characteristics similar to PE, but burns with an odor similar to diesel fumes.

PS

Polystyrene (PS) is also another widely used amorphous plastic with a specific gravity of 1.09. It burns like a candle with drips formation and thick smoke and soot in the air. The drops also burn continuously. Polystyrene is soluble in wide variety of solvents such as benzene, toluene, methyl ethyl ketone, acetone and carbon tetrachloride.

PPO

Polyphenylene oxide (PPO) is another phenyl group containing polymer and has self extinguishing property. It gives off phenolic odor when burnt and does not form drippings when flamed. A yellow-orange flame is observed upon exposing the polymer to flame. IR spectroscopy method can be easily used for identifying the polymer. The polymer dissolves in toluene and dichloroethylene.

PU

Polyurethanes (PU) ignite quickly upon burning and a yellow flame with a faint odor of an apple is given out. PU burns with a black smoke. PUs are soluble in tetrahydrofuran and dimethylformamide.

Nylons

Different types of nylons (nylon 6; nylon 6,6; nylon 6,10; nylon 11 and nylon 12 etc.) have varying specific gravities in the range of 1.04–1.17. They have different melting points and solubility characteristics. All nylons burn with blue flame with a burnt wool or hair odor. Interestingly nylons burn only when they are flamed and self extinguish when the flame is removed away. Phenol, m-cresol and formic acids are solvents for nylons. The solubility and specific gravity data combined with IR method are used for identifying different nylons.

Thermosetts

Epoxy resins

The various epoxy resins have specific gravity ranging from 1.10–2.10. These resins produce yellow flame with a pungent amine odour upon burning. A black smoke is given off when they are burnt. Both thermal and IR spectral methods are useful in identifying the epoxies.

PF resins

Phenol-formaldehyde (PF) resins are soluble in acetone and acetic acid. They burn with yellow flame and pungent odour of phenol. Different PF resins have specific gravity ranging from 1.30–1.90.

UF resins

Urea-formaldehyde (UF) resins burn with yellow flame releasing typical strong odor of formaldehyde. The brittleness of UF resins causes swelling and cracking as they burn. They are not self extinguishing.

MF resins

The melamine-formaldehyde (MF) resins are self-extinguishing. These burn with yellow flame and characteristic fish like odor. The material also swells, cracks and turns white while burning.

3.2 Testing Methods

3.2.1 Thermal

The thermal behavior of polymers, especially plastics is rather complex. Unlike metals, plastics are extremely sensitive to temperature changes. The ultimate performance of a plastic for a particular use depends upon its response to the changes in temperature. Several factors, structural as well as molecular, contribute to the final thermal response of a given plastic material. These factors can be listed as crystallinity, molecular orientation, the spatial arrangement, molecular weight, intermolecular forces, cross-linking and polymer compositional variations (viz. copolymers, graft polymers and branched polymers etc.).

The design and selection of a plastic material for specialty application purpose is always based on the available thermal property data. Mainly, the thermal evaluation of plastics is done by two types of test methods. These methods are based on long term and short-term effects of temperature on thermal response properties of plastic materials. The results of these tests help polymer technologist to select a suitable plastic material for elevated temperature performance.

Short-term Tests
In these tests plastic specimens are exposed to elevated temperatures for shorter duration of time and the changes in the thermal responses are noted. Three main tests of this category are widely employed by plastic technologists. They are:

(a) Heat deflection temperature
(b) Vicat softening temperature
(c) Torsion pendulum test

Heat Deflection Temperature (HDT)
Heat deflection temperature is defined as the temperature at which a standard test specimen deflects 0.010 inches under a stated load of either 66 or 264 psi. The basic requirements for this test are standard test bar, oil bath with heating and cooling devices such as dial recorder etc. The test bars are either molded or cut from extruded sheets and should have 5 in. (length), 0.125–0.5 in. (width) and 0.5 in. (depth). Care should be taken to keep the surface of the cut specimen flat and smooth and free from excessive ink marks. The desired pressure on specimen during the testing is usually applied by means of a dead weight. Common laboratory manual heat deflection measuring devices and automatic heat deflection temperature testers are available.

The actual test procedure is described and recorded in ASTM D 648. The specimen mounted on standard support, temperature and deflection measuring devices are lowered into the oil bath, kept initially at room temperature. Then the desired load value of 66 or 264 psi is applied by a fiber stress. After 5 min. the dial pointer is adjusted to zero and the temperature of the bath is raised at the rate of $2 \pm 0.1°C/min$. The temperature in the oil bath at which the test bar has deflected 0.010 in. on the dial indicator is noted as heat deflection temperature. There are several precautions, one needs to take care before accurately recording a HDT for a plastic material. Also, the residual stress (normally expected in high degree molded-in articles), thickness of the specimen, and the type of molding etc. have large influence on test results. Usually accurate test results are recorded by taking pre-annealed, thin and compression molded specimen.

Heat deflection temperature is a single point measurement and does distinguish between those materials, which become flexible in narrower temperatue range. The behavior of plastic material at elevated temperatures and for longer period of time cannot be predicted from HDT data.

Vicat Softening Temperature
The Vicat softening temperature is defined as the temperature at which a flat ended needle of 1 mm^2 of circular cross section will penetrate a thermoplastic

specimen to a depth of 1 mm under a specified load using a selected uniform rate of temperature rise. The test apparatus designed for deflection temperature under a specified load can be used for Vicat softening temperature with minor modification in the assembly.

A needle rod or indenting tip is allowed to rest on the specimen, which is molded or cut from a sheet with minimum thickness and width of 0.12 and 0.50 inch, respectively. The test procedure is recorded in ASTM D1525. The specimen mounted on the steel support is exposed to a needle rod, which rests initially on the former before test procedure begins. The temperature of the bath is increased at the rate of 50°C or 120°C/hr uniformly and the temperature at which the needle rod penetrates into the specimen by 1 mm is noted and recorded as Vicat softening temperature. All the precautions in the preparation of test specimen for HDT measurements are needed to be taken. The Vicat softening temperature data are highly useful for comparing the heat softening qualities of thermoplastic materials. However, this test is not recommended for flexible plastic materials such as PVC or ethylcellulose etc. which have a wider softening range.

Torsion Pendulum Test

The behavior of plastic materials at elevated temperatures is too complex and too varied. They undergo deformation at elevated temperatures and their properties profoundly vary at the temperature range of glass transition in case of amorphous materials and by the melting temperature in case of semicrystalline plastics. Mechanical and thermal properties are very sensitive to deformation and hence to temperature. The performance of plastic material at elevated temperatures can not be judged by simple tests such as heat deflection and Vicat softening temperatures carried under load. In order to understand the elevated temperature capability of plastic materials, the knowledge of full temperature-modulus plot especially, dynamic modulus is highly essential and useful. Dynamic mechanical analysis which determines the temperature dependence of shear modulus by forced oscillations or by damping free oscillations, offer best means of testing the plastic specimens at elevated temperatures. Of all the experimental techniques available to study the dynamic mechanical properties of plastics, torsion pendulum technique is the simplest and widely used in plastic industries. This technique is based on obtaining a plot of short time modulus as function of temperature (from room temperature to the temperature of melting and degradation).

The apparatus for the torsion pendulum test is designed as per the ASTM procedure D 2236. It consists of an insulated chamber with heating and cooling facilities and a rigidly fixed clamp at one end. The movable clamp is attached to an inertial disk or a rod with a known moment of inertia. The angular displacement of the oscillating specimen versus time is recorded by using differential transformer and stip chart recorder. Specimens for the test are prepared either in a rectangular or cylindrical shape and different length. The thickness of the specimen needs to be specified because it influences the dynamic results. To carryout the test, the clamped specimen is put into oscillation, after placing the assembly into the chamber, which was preconditioned to a desired temperature of test. The period and rate of decay of amplitude of oscillations are measured. The elastic shear modulus is calculated from specimen dimensions, moment of inertia of the movable member, I and the frequency of oscillation. The shear modulus is given by;

$$G = \left(\frac{12\, \mu^2 \times I \times f^2 \times l}{b \times h^3 \times C} \right) - \left(\frac{m \times g \times b}{4 \times h^3 \times C} \right) \qquad (3.1)$$

where l = free length of test species, b = test piece width, h = test piece thickness, m = tensile force on test piece, g = acceleration due to gravity and C = correction factor (or shape factor) defined by the ratio of both. A quantity log decrement, Λ is defined as the natural logarithm of the ratio of the amplitude of successive oscillations in the same direction i.e. $\Lambda = \ln (A_n/A_{n+1})$. The log decrement-temperature profile is shown in Fig. 3.8.

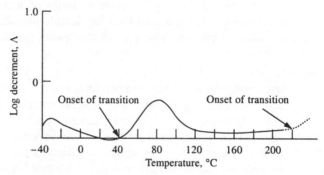

Fig. 3.8   The log decrement-temperature profile of a typical semicrystalline thermoplastic (nylon-6)

It is evident that a dramatic drop in shear modulus occurs at a particular temperature (corresponding to glass transition region). This temperature is regarded as the maximum usable load bearing temperature as applicable to test plastic material. In addition to these two other long term tests at elevated temperatures are also recommended; (i) long term heat resistance test (ASTM D794) and (ii) underwriters laboratories index.

Thermal Conductivity

The replacement of metals and ceramics with synthetic plastics has been occurring at a faster rate. The tremendous success of plastics as a material for cooking utensils, thermal automobile parts and cellulose plastic insulation devices etc. is mainly due to their low thermal conductivity. This property measures the quantity of heat flow in unit time across unit area of a sheet of substance of unit thickness with unit temperature gradient between its faces. In the thermal insulation devices, the observed thermal resistance is not due to plastics but for the entrapped gases. There are several methods for measuring thermal conductivity of plastic materials. These are divided into two main groups: steady state and transient methods.

Steady state methods employ steady state conditions, i.e. the temperature at any point does not change with time. Based on sample geometry, four steady state methods are used for determining the thermal conductivity of plastic materials.

These are unguarded hot plate, guarded hot plate, coaxial cylindrical and quasi-stationary methods. In transient methods, the time dependent temperature distribution is measured. It encompasses line source and plane source methods. Among all these methods, the guarded hot plate method is widely used. The equipment used in this method is quite complex and expensive. ASTM C 177 lists the procedure for guard hot plate method. A schematic representation of the general arrangement in guarded hot plate method is shown in Fig. 3.9.

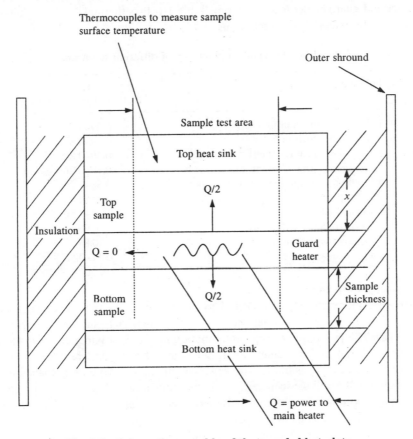

Fig. 3.9 Schematic assembly of the guarded hot plate

The assembly shows that the heat source is surrounded by a guard heater, which is connected to an independent power supply. The test sample is then sandwiched between the heat source and heat sinks from the above and below. The temperature in the heat sink is controlled by circulating a liquid from a constant temperature thermostatic bath.

A known power is supplied to heat source and at the same time the power of the guard is adjusted such that the heat source and the guard are maintained at the same temperature. This ensures that the heat from the heat source flows normally through test samples. Two thermocouples are used to record the temperature drop across the samples. Then thermal conductivity, K of test specimen is calculated using equation $K = qx/A\ \Delta T$, where K is the thermal conductivity (BTU/(h) (ft^2) (°F/in.); q the time rate of heat flow (BTU/h); x the thickness of test specimen (in.); A the area under the test (sq.in.) and ΔT the temperature difference between hot T_1/°F) and cold surfaces (T_2/°F) of the specimen.

ASTM procedure recommends specifications for accurate thermal conductivity measurements. Accordingly, the apparatus dimensions are usually 30 mm square, the sample thickness range from 6 mm to 50 mm, and the gap between the heat source and guard heater is about 2 mm. Table 3(f) lists thermal conductivities of some plastic materials of common use.

Table 3(f)   Thermal conductivity of different materials

Polymer	K BTU/(h) (ft^2) (°F/in.)
Polystyrene (medium impact)	0.29–0.87
Polystyrene (high impact)	0.29–0.87
Polystyrene (heat resistant)	0.29–0.87
Polystyrene (general purpose)	0.70–0.96
Polyethylene-polypropylene copolymer	0.58–1.20
Polypropylene	0.81
Polycarbonate	1.40–1.50
Polyvinyl chloride (flexible)	0.99
Polyvinylidene fluoride	0.87
Polyurethane	0.11–0.52

Thermal Expansion

Polymeric materials when subjected to thermal changes tend to expand and contract from five to ten times more than other materials such as metals and ceramics. The possibility of segmental motion in a long polymer chain and also temperature induced movement along the chain as a whole, are the two main factors that are responsible for the sensitive changes in volume (expansion or contraction) of a polymeric material with the application of heat. The temperature dependent volume changes in a polymeric material are expressed in terms of (i) coefficient of linear thermal expansion and (ii) coefficient of volume expansion. The coefficient of linear thermal expansion, α, is defined as the fractional change in length or volume of a material after application of an unit change of temperature. Following expression gives the values of α.

$$\alpha = 1/l\ (\partial l/\partial T)_p \qquad (3.2)$$

where l is the length of the specimen, T the temperature and p is the pressure.

The range of values of coefficient of thermal expansion observed for polymer materials means the expansion develops internal stresses in the polymer. The coefficients of volume expansion or coefficient of thermal expanision β is given by the expression

$$\beta = 1/V\ (\partial V/\partial T)_p \qquad (3.3)$$

where V is the volume of specimen

ASTM D 696 provides a method for determining the coefficient of linear thermal expansion. This method uses a fused quartz tube dilatometer, temperature controlled bath, linear volume displacement transducer (LVDT). Wide range of temperatures can be covered. Fig. 3.10 illustrates a schematic assembly of quartz tube dilatometer.

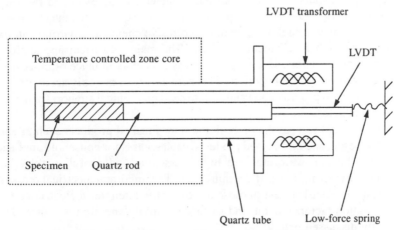

Fig. 3.10 Schematic assembly of quartz tube dilatometer

The test measurement involves the placement of a preconditioned polymeric specimen of length between 2 and 5 inch in the dilatometer. The dilatometer containing the specimen and the measuring devices are lowered below the liquid level in the bath. The temperature of the liquid bath is varied as desired and the change in the length is recorded. The coefficient of linear thermal expansion is calculated as follows;

$$\alpha = \Delta l / l_o \, \Delta T \qquad (3.4)$$

where Δl = change in length of the specimen due to temperature increase or decrease, l_o = length of the specimen at room temperature and ΔT = temperature difference in °C over which the change in length of the specimen is measured.

A liquid-in-glass dilatometer (ASTM D 864–52) is used for measuring coefficient of cubical thermal expansion of polymeric materials. The sample is placed in the dilatometer and then a known amount of liquid (mercury) is filled. The dilatometer is then suspended in a temperature controlled bath. After reaching the initial thermal equilibrium, the height of the liquid meniscus in the capillary of the dilatometer is noted by using cathetometer. Then the temperature of the bath is changed and the changes in the height of the mercury meniscus as a result of expansion of polymeric material is recorded at predicted temperatures. After making necessary corrections to the expansion of dilatometer and mercury liquid, the volume expansion of the polymeric specimen is measured.

3.2.2 Electrical
The electrical testing of polymeric materials is based on some measurable dielectric and electric properties. Polymeric materials especially plastics have been used in

many electrical applications which are related to their insulation capacity. Although plastics can be made to conduct electricity by adding dopants such as carbon black and iodine etc, the extensive use of polymers as materials in electrical application is due to the following factors; ease of fabrication, low cost, light weight and of course excellent insulation properties. The emergence of high and super performance engineering plastic materials has extended the use of polymeric materials to specialty applications, which requires the material to show extreme resistance to high temperatures, corrosive chemical environment, high humidity and high mechanical stress etc. The major electrical application of polymers is as insulators. Use of polymers as insulators in place of other media viz. air or vacuum, is highly preferred because polymers not only act as effective insulators but also provide good and well needed mechanical support for the field carrying conductors.

Thus electrical test results on both the semi finished products such as sheet, film and tape etc. and on finished products, cables, cases for household appliances and coated kitchen ware etc., are highly essential and useful in selecting a particular polymer for a specific requirement. The important electrical properties of interest for electrical test purpose are dielectric strength, dielectric constant and dissipation factor, volume and surface resistivity and arc resistance. These are briefly discussed below.

Dielectric Strength

It is a very important characteristic property of an insulating material, because it measures the electrical strength of a material. Higher the electrical strength or resistance of a material for electrical flow, the larger the insulating capacity. The dielectric strength of the material is defined in terms of maximum voltage the material withstands and beyond which the dielectric nature of the material is lost. It is thus expressed in voltage per unit thickness i.e. V/mm. ASTM D 149 procedure is routinely used for testing the dielectric strength of an insulator. This method recommends three basic procedures; (i) the short time method, (ii) slow rate of rise method and (iii) step by step method. In all these three procedures, the basic set up remains the same. A test specimen is sandwiched between the electrodes, connected to a variable voltmeter. The thickness of the specimen is not fixed but commonly 0.05–0.06 inch thickness is preferred. A schematic representation of set up used for dielectric strength is shown in Fig. 3.11.

Short Time Method

In this method the voltage is raised uniformly from zero to maximum value (till the breakdown i.e. actual rupture or decomposition of the material occurs). The rate of rise in the voltage generally employed is 100, 500, 1000 or 3000 V/sec until the failure occurs.

Slow Rate of Rise Method

In this method an initial voltage equal to half of the breakdown voltage (as determined from short time method) is applied and is then followed by further uniform increase in voltage until the breakdown occurs.

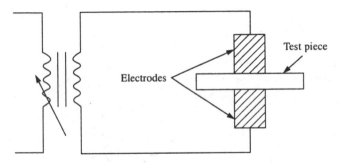

Fig. 3.11   Block diagram for dielectric strength test

Step by Step Method

In this method, initially the voltage equals to fifty per cent of breakdown voltage as established by short time test results is applied. Then, the increase in voltage is affected not only in uniform increment values but also for specific period of time. The dielectric strength (*V*/mm) is calculated as the ratio of breakdown voltage (*V*) to specimen thickness (mm).

Several factors affect the test results. These can be listed as specimen thickness, temperature, humidity, composition and geometry of electrodes, time of application of voltage and other factors such as residual mechanical stress and defects (if any) during the processing of material.

Dielectric Constant and Dissipation Factor

Dielectric constant of an insulating material is defined as the ratio of the capacitance induced by two metallic plates with an insulator sandwiched between them and the capacitance of the same plates without the material i.e. with a vacuum between them. The knowledge of dielectric constant of an insulating polymer material is highly desirable because, it measures the storage capacity of electrical energy. Thus the suitability of a given insulator for a given electrical application can be judged priorily, if its dielectric constant is known.

The test procedure for dielectric constant measurement is quite simple. It consists of placing a thin film or a pelette of a polymeric dielectric material between two metal electrodes and the capacitance is measured by an LCR bridge or a Dielectrometer which can cover wide range of capacitances. Similarly, the capacitance of metallic plates without sample i.e. with air as medium is measured. Then the dielectric constant or relative permittivity of the material is calculated from the ratios of capacitance observed when specimen and air are placed between electrodes. A schematic block diagram or dielectric test is shown in Fig. 3.12.

Dissipation factor

The dissipation factor of an insulating material measures the amount of electric loss and in other words the dissipation factor is defined as the ratio of the conductance of a dielectric material to its susceptance. The product of dissipation factor and dielectric constant of the same insulating material is known as loss factor.

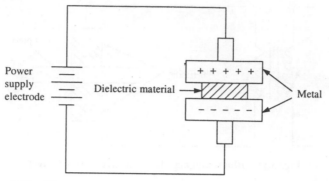

Fig. 3.12   Block diagram for dielectric constant test (ASTM D 150)

Arc resistance test

Most of the insulating polymeric materials are exposed to high voltages carried by the conducting carriers around which the insulator material is strapped. The action of high voltage on an insulator is measured in terms of time till the surface of the insulator material gets carbonized or burnt and rendering it to loose its insulating property. The applied high voltage causes arcing on the surface of the insulator material, leaving a carbonized path in the form of a wiry line along the surface and between the electrodes. The more time a material withstands the arcing, the better arc resistance it has. The arc resistance is thus capacity of the insulator material to withstand high voltages for a specified time without getting damaged. Resistance to arcing by the polymer materials depends upon the type of plastics used. For example, PF-resins tend to get carbonized easily and on the other hand alkyd resins, melamine resins and fluorocarbon polymers have excellent arc resistance.

Acrylic plastics though do not leave any tracks on arcing but get decomposed to ignitable gases and thus show a failure at high voltage in a short time. Since the arcing is the phenomenon initiated at the surface of the material and the plastic materials have fillers as additives, the determination of arc resistance has always been a problem.

ASTM D495 procedure recommends the measurement of arc resistance of an insulating material in clean and dry laboratory conditions, by applying the voltage intermittently and in a stepwise increase till the failure occurs. However, the result of this test may help in initial screening of material, its reliability is not high since the actual use conditions are affected by so many factors such as moisture, fog, dust and other wet conditions. That is why tracking on the surface of an insulator material has also been monitored (i) in conducting the test in a fog chamber and after coating the surface of the material with standardised dust, (ii) in partially immersing the specimen in an aqueous solution of ammonium chloride and a wetting agent, and (iii) in developing tracks or erosion on the surface by discharging electrolyte on the surface of the material.

3.2.3   Chemical

These test methods are basically meant for estimating the resistance of the synthetic polymers against attack by chemicals. The resistance of polymers especially plastics to chemical (the typical chemical substances of interest and importance are staining agents, coloring agents, surface active agents, hydrogen sulfide and other organic solvents etc.) mainly depends upon the (i) structural characteristics of its repeat units (type of bond linkage, bond length and bond energy), (ii) nature of polymer chain i.e. linear or branched and (iii) the packing of polymer chains i.e. degree of crystallinity.

Chemical resistance tests are conducted mainly in four different ways; (i) immersion test, (ii) strain-resistance test, (iii) solvent stress cracking resistance and (iv) environmental stress cracking resistance. Besides the results from these tests, one should consider other factors such as test temperature, solvent media, exposure time, nature and composition of additives in polymers etc. before reporting the final assessment.

Immersion Test

This is one of the simplest methods used in chemical industry to screen the unsuitable material for a given end use application. The test consists of preparing a specimen, of any size and shape depending upon the material at hand, immersing it in a bath containing a particular solvent and maintained at controlled atmosphere and temperature. ASTM D 543 specifies that the specimen should be kept totally immersed for seven days. Then the specimen is taken out and changes ranging from its physical appearance to mechanical properties are recorded. The physical appearance includes loss of weight, change in shape and other surface characteristics such as transparency etc. The results of a test specimen are compared with another reference standard of the same material but not immersed in the bath. The contrasts obtained for immersed and non-immersed specimen would help observe the following: loss of gloss, swelling, clouding, tackiness, crazing and bubbling etc. Also, the results reported are highly dependent upon duration of immersion, temperature/atmosphere of the container bath and concentration of reagent/solvent media.

Staining Resistance of Plastics

Plastic materials are not only composites but also heterogeneous in nature. The end used form of thermoplastics/thermosetts consists of the polymer, filler, coloring agents, stabilizers and other additives. The wide usage of plastic materials in domestic applications exposes them to different staining reagents found in the domestic consumables. These agents are chemical substances that can react with some of the additives in the plastic material and cause staining or development of a color-coat on its surface different than the original. The staining of a plastic material may result in loss of its aesthetic appearance. The staining agents present in common consumable items such as food items, beverages, cosmetics, solvents, detergents, jelly, tea, coffee, bleach, shoe-polish, lipstick and nail polish remover and other solvents cause the staining resulting in the decoloration of plastic container in which they are used. Hence the staining resistance tests for the plastic has been usually carried out before the plastic containers are sent to the market.

The test specimen can have any size but it should have a flat, smooth and large surface, so that visual examination of a stain caused can be made. The specimen should be thoroughly cleaned and dried before it is kept in an applicator container in which the staining agent is applied so that a thin coat is formed on the specimen. Usually, the specimen is dipped in the staining agent taken in a glass container. The container in closed condition and along with the specimen is kept in an oven, at $50 \pm 2°C$ for 16 hours. After the exposure period is over the excess solvent is removed and the specimen is visually examined for appearance of any stain against an unexposed standard specimen of the identical dimension and from the same specimen. This test is more effective, when conducted on the finished products, which directly enter the market.

One of the very reactive staining agents that is found in industrial fumes and rubber items is the sulfide. When sulfide in the form of hydrogen sulfide comes in contact with the plastic material, which usually contains the salts of lead, copper and antimony as stabilizers, fillers and pigments, the former stains the plastic materials very easily. ASTM D 1712 recommends a procedure for testing resistance of plastics to sulfide staining. In this procedure, the plastic test specimen is immersed in a freshly prepared hydrogen sulfide solution for 15 min. along with control specimen with a known susceptibility to sulfide stain. After 15 min., both the specimens are removed and a comparison is made between control and test specimen to judge the stain resistance of a given plastic material against sulfide agents.

Solvent Stress Cracking Resistance

The selection of a proper plastic material to a given application depends on its resistance to chemicals besides other considerations. It has been observed that plastic materials are attacked by chemicals when they are under stress. The chemical agents may not have any effect on unstressed parts. The effect of chemicals on plastic materials under stress is called stress cracking or solvent stress cracking. There is an inherent stress left in the finished plastic product during its processing. Thus results from simple immersion tests are not always sufficient to measure the solvent resistance of a plastic material. Solvent stress cracking has been found to occur in almost all plastic materials. The stress in a plastic material can be due to external or internal conditions. Though the complete elimination of internal stress is impossible, the stress effects can be minimised by using proper design of the mold and processing method.

Under stress, the cohesiveness of polymer-polymer bond is weakened and due to solvent attack new polymer-solvent bonds are developed. This causes disturbance in the packing and arrangement of polymer chains. If the disturbance caused exceeds the cohesive strength within the polymer chains, the rupture of polymer-polymer bonds can take place. It has been observed that plastic materials undergo solvent stress cracking only when they are subjected to either minimum internal and external stress, below which the solvent has no effect or does not penetrate into the bulk of the polymer. This minimum stress is known as 'critical stress'. Critical stress is defined as the amount of stress at which cracking of polymer occurs due to solvent attack in a defined environment. There are two methods through which critical stress for a given polymer under a specified chemical attack can be measured. These are known as calibrated solvent stress and solvent stress cracking tests.

Calibrated solvent test uses a specimen cut from a plastic specimen, that is used in real conditions, preferably in bar form (dimensions such as thickness and the area depends upon the type of plastic to be tested) and subjected to a known tensile stress by stretching. Just after stretching, the specimen is exposed to chemical either by a spraying or dipping process. The exposure time is fixed for one minute and thereafter the specimen is examined for any crazing that might have been developed due to solvent/chemical penetration. If no crazing is noted, a fresh specimen of the plastic material is cut into identical dimensions and is once again subjected to increased tensile stress and exposed to solvent for observing any craze. The test is repeated every time with a fresh specimen and increasing tensile stress till the crazing on the surface is recorded. The material is considered to show high solvent stress cracking resistance, if no crazing is observed till its yield point i.e. the maximum stress value at which the material is broken.

One of the major disadvantages of this test is that it requires large number of pieces of the specimen and the test results are highly dependent upon the time of exposure of the material to a given chemical. Long exposure time is not suitable for visible examination of cracking. Thus another alternate method is suggested. In this method, known as standard solvent stress cracking test, a specimen of $4 \times 1 \times 0.03$ inch dimension is strapped in an ellipsoidal jig and the specimen along with the jig is immersed in the solvent or a chemical reagent. Because of the geometrical variations in the jig, the specimen experiences different stress levels. Thus a contrast in the crazing from different parts of the specimen can be developed. The test is repeated at different time intervals. The time at which the crazing stops in the specimen is noted and it is considered as critical Stress Point.

Environmental stress cracking

This test is specially designed for polyethylene samples, which are highly susceptible to agents such as detergents, water, sunlight and oil etc. It has been found that polyethylene specimens, which perform well in the ordinary laboratory conditions, very rapidly develop severe cracks, when they are exposed to lubricants. It has also been established that the chemical agents have severe effects on polyethylene, when the specimens are under high stress. So following test method is developed for judging the failure limits of a polyethylene sample against chemical agents. A specimen of $1.5 \times {}^1/_2$ inches dimensions is cut precisely into rectangular shape. A nick of a fixed length and depth is inflicted onto the specimen using a nicking jig. Then the specimen is bent exactly by 180° so that the nick made is at right angles to the line of bend. Then the bent specimen is mounted onto a sample holder. The holder with the specimen fixed is lowered into a test tube filled with fresh solution of a non ionic detergent or any other lubricant. The test tube is placed in a thermostat bath maintained at a constant temperature of 50°C or 100°C. Then the test specimen is removed after a specified time and is examined for crazing. The time at which crazing is observed is noted. The stress crack resistant polyethylene samples can be prepared by increasing molecular weight, decreasing the molecular weight distribution and increasing the crystallinity in the virgin polymer.

3.3   Characterisation

3.3.1   Molecular Weight Distribution

Many properties of a polymeric material e.g. solubility, melt and solution viscosity, moldability etc. depend on its molecular weight. The knowledge of the mol. wt. of polymers is thus very important. Several experimental methods are available to determine the molecular weight of polymers and these will be described here. However, unlike low molecular weight substances, a particular polymer can have different molecular weights. For example, polyethylene of varying mol. wt. can be made. Furthermore, it is also important to know that a polymer sample always contains molecules of different sizes and thus never has a fixed value of mo. wt. The different sized chains in a polymer sample result as the polymerization is a statistical process and the growing chain can terminate at any instant. If termination happens soon after the chain starts to grow, then obviously, the chain will be short. If the growing chain evades termination for a while, it will be longer. Because of this, polymers are usually characterized by a distribution of molecular weights. This distinguishes them from their low molecular weight analogues. Thus for these polydispersed polymers, it is necessary to describe their molecular weights as relative mol. wt. or an average molecular weight. There can be in principle different ways to average the relative mol. wt. for describing the mol. wt. in a justified manner, the two most common ones are number average and weight average mol. wt. The various average molecular weight expressions are shown in Table 3(g).

Table 3(g)   The different molecular weight averages in polymers

Mol. Wt. average	Expression
$\overline{M}_n$	$\dfrac{\Sigma N_i M_i}{\Sigma N_i} = \dfrac{\Sigma W_i}{\Sigma W_i / M_i}$
$\overline{M}_w$	$\dfrac{\Sigma N_i M_i^2}{\Sigma N_i M_i} = \dfrac{\Sigma W_i M_i}{\Sigma W_i}$
$\overline{M}_z$	$\dfrac{\Sigma N_i M_i^3}{\Sigma N_i M_i^2} = \dfrac{\Sigma W_i M_i^2}{\Sigma W_i M_i}$
$\overline{M}_v$	$\left[\dfrac{\Sigma N_i M_i^{1+\alpha}}{\Sigma N_i M_i} \right]^{1/\alpha}$

In the expressions given in Table 3(g), the terms N_i and W_i represent the number of moles of molecules with a molecular weight of M_i and weight of molecules with molecular weight M_i, respectively. The weight, W_i of molecules with molecular weight M_i is then equal to $W_i = N_i M_i$. Before we understand the concept of $\overline{M}_n$ and $\overline{M}_w$, let us consider the foregoing definitions. It has already been mentioned that the polymerisation process yields several chains each having a definite number of molecules and thus a definite molecular weight. Such non-uniformity in the chains prevents us from describing the polymers in terms of a single degree of polymerisation. So one resorts to average degree of polymerisation. The averaging of degree of polymerisation is done over the molecules with various degree of polymerisation Dp_i according to their number based statistical entities. These statistical entities can be the number n_i, the amount of substance $w_i = n_i / N_A$, where N_A is the Avogadro number. Or the number fraction

$$x_i = n_i / \Sigma_i \, n_i = w_i / \Sigma_i \, w_i$$

Thus, one can write number average degree of polymerisation as

$$Dp_n = \frac{\Sigma_i n_i Dp_i}{\Sigma_i n_i} = \frac{\Sigma_i w_i Dp_i}{\Sigma_i w_i} \tag{3.5}$$

$$= \Sigma_i x_i Dp_i$$

where x_i is now the mole fraction of i molecules.

From the above definition for expressing an average quantity, let us now consider that a polymer sample has total, N number of molecules out of which a fraction N_1 has molecules with molecular weight M_1, another fraction N_2 has molecules with molecular weight M_2 and similarly ith fraction has molecules N_i with a molecular weight M_i. Then by simple mathematical consideration we can write the total number of molecules $N = \Sigma_i N_i$ and the number of molecules in fraction $1 = N_1$ and in fraction $2 = N_2$ and in fraction $3 = N_3$ and in fraction $i = N_i$

The molecules in a given fraction relative to the total number of molecules can be expressed in terms of molecular number fraction.

Then for the fraction 1, the molecular fraction would be equal to $N_1 / \Sigma_i N_i$ and for fraction 2, the same is written as $N_2 / \Sigma_i N_i$, and so finally for fraction i, the molecular fraction is given as $N_i / \Sigma_i N_i$. When a molecular weight of a polymer sample is determined by any method, the value obtained considers all the fractions present in the polymer. Thus one needs to calculate the molecular weight contribution of every fraction present to the total molecular weight of polymer.

The molecular weight contribution of each individual fraction will be the product of molecular weight M_i and molecular number fraction $N_i / \Sigma_i N_i$, i.e. $N_1 M_1 / \Sigma_i N_i$, $N_2 M_2 / \Sigma_i N_i$, $N_3 M_3 / \Sigma_i N_i$, and so on till $N_i M_i / \Sigma_i N_i$. Thus the total molecular weight of polymer sample containing several fractions would be simply an additive quantity, i.e.

$$\overline{M}_n = \frac{\Sigma_i N_i M_i}{\Sigma_i N_i} \tag{3.6}$$

Similarly, the molecular fractions can be expressed in terms of weight fraction i.e. weight of each ith fraction having a molecular weight M_i. The weight of fraction1 will be W_i, fraction 2, W_2 and for ith fraction, W_i. The weight fractions W_1 to W_i are simply defined as N_1M_1 to $N_i M_i$ because, by definition, weight is equal to no. of moles multiplied by molecular weight. Then we can write

Weight of fraction $W_1 = N_1M_1$

Weight of fraction $W_2 = N_2M_2$

Weight of fraction $W_i = N_iM_i$

The weight fraction for the individual polymer fractions, 1 to i relative to total weight of polymer $W = \Sigma_i N_i M_i$ would be

$$\frac{N_1M_1}{W}, \quad \frac{N_2M_2}{W} \quad \text{and} \quad \frac{N_iM_i}{W}$$

$$\frac{N_1M_1}{\Sigma_i N_i M_i}, \quad \frac{N_2M_2}{\Sigma_i N_i M_i} \quad \text{and} \quad \frac{N_iM_i}{\Sigma_i N_i M_i}$$

Before one expresses the total molecular weight of polymer containing several fractions in terms of weight average molecular weight, $\overline{M}_w$, what needs to be done is simply calculate the molecular weight contribution of each fraction and summate all the contributions. Thus the molecular weight contribution of weight fractions, W_1, W_2 and W_i etc. are

$$\frac{N_1M_1}{\Sigma N_i M_i} M_1, \quad \frac{N_2M_2}{\Sigma N_i M_i} M_2 \quad \text{and} \quad \frac{N_iM_i}{\Sigma N_i M_i} M_i$$

or

$$\frac{N_1M_1^2}{\Sigma N_i M_i}, \quad \frac{N_2M_2^2}{\Sigma N_i M_i} \quad \text{and} \quad \frac{N_iM_i^2}{\Sigma N_i M_i}$$

whose overall summation defines weight average molecular weight of polymer $\overline{M}_w$ as

$$\overline{M}_w = \frac{\Sigma N_i M_i^2}{\Sigma N_i M_i} \tag{3.7}$$

As shown in Table 3 (g), the average molecular weights of polymers are also expressed by another two ways i.e. $\overline{M}_z$ and $\overline{M}_v$, in addition to above discussed $\overline{M}_n$ and $\overline{M}_w$. The terms $\overline{M}_z$ and $\overline{M}_v$ are measured based on sedimentation equilibrium experiments and flow behavior (or viscosity) of polymer solutions. The symbol z in $\overline{M}_z$ comes from German word Zentrifuge = centrifuge. Both $\overline{M}_z$ and $\overline{M}_v$ are used for expressing the molecular weights of larger molecules. Since $M_i > 1$, the expression given in Table 3 (g) implies $\overline{M}_z > \overline{M}_w \geq \overline{M}_v > \overline{M}_n$.

The superscript, α, in $\overline{M}_v$ expression is a variable with its value ranging from 0.5 to 1 depending upon the solvent quality. The α value is defined by Mark-Houwink equation, which will be discussed in the later part of this chapter.

Thus it is clear from foregoing discussion that the molecular weight of polymer is not an absolute quantity and the value of a given polymer may vary with the method of its determination. The different experimental methods are described later in the chapter. Each method or group of methods defines the molecular weight depending upon the statistical weight that is measured. Suppose that the molecular weight of two polymers of same chemical constitution is determined to be same, but both of them may possess different properties. This is because the two samples might consist fractions that are not identical in terms of number of molecules/moles, weight of molecules and of course in molecular weight. But still, the averaging of all the fractions yield same final value. That is why not only the averaged molecular weight but the dispersity of molecular weight from smaller to larger fraction is to be known. Such an analysis is normally done by expressing or mapping the molecular weight distribution curve. Such a curve is developed easily from the experimentally derived data and simply is a plot showing the variation of number (or mole fraction) N_1 of molecule (having a molecular weight M_1) against the corresponding molecular weight M_1. Such a curve for polymer having a hypothetical distribution of fractions with different molecular weight is shown in Fig. 3.13.

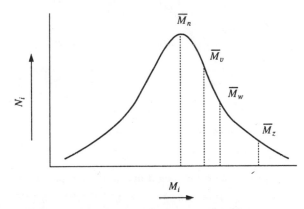

Fig. 3.13 The different molecular weight averages

A perusal of the shape of the curve in the figure reveals not only interesting but also important information. It can be seen that the number of molecules (or moles) having particular molecular weight passes through a maximum i.e. it increases initially with respect to molecular weights. This means the statistical weighing factor predominantly shifts to weight and size of fraction at high molecular weights. So this explains that the experimental methods need to be changed for determination of molecular weight of a same type of polymer but differing in molecular weight. The pattern shown in the above figure also reveals that $\overline{M}_n < \overline{M}_v \leq \overline{M}_w < \overline{M}_z$.

Molecular Weight Distribution

The subject of mol. wt. distribution is of great practical as well as fundamental importance. For example, a small amount of a fraction of either very high or very low mol. wt. can greatly change the solid and solution state properties, and thus affects the processing characteristics of a polymer. It is therefore necessary to develop quantitative ways to obtain mol. wt. distribution. Figure 3.14 shows the mol. wt. distribution of three polymer samples with almost same number average mol. wt. It can be seen from the curves that polymer sample 1 is highly polydispersed than polymer 2, while polymer sample 3 is less polydispersed than the other two. The polydispersity index, P.I. for polymers is defined as the ratio $= \overline{M}_w / \overline{M}_n$. As mentioned earlier, the polydispersity of the molecular weight in the polymer fractions is mainly due to variation of degree of polymerisation in each fraction formed during polymerisation process. Hence the values of P.I. are highly dependent on the type of polymer process. For a monodispersed polymer, P.I. value is unity.

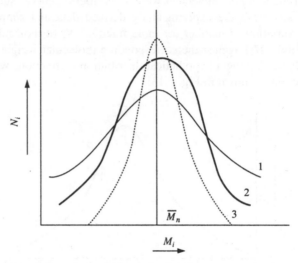

Fig. 3.14   Molecular weight distribution in polymers

The polydispersity index in polymers prepared using different techniques are shown in Table 3(h).

Table 3(h)   The polydispersity index (P.I.) using different polymerization methods

Technique/conditions	$\overline{M}_w / \overline{M}_n$ = P.I.
Free radical polymerization	
with precise temperature control	1.5–2
moderate temp. control (e.g. in bulk polymerization)	2–5
without temp. control (e.g. with autoacceleration)	8–10
Ionic polymerization	
with homogeneous catalyst	< 1.5
with heterogeneous catalyst	> 10
Polycondensation	2–3
Coordinate polymers and polymers with many branches	>20

3.3.2 Fractionation
The fractionation of polymer means separating a polydispersed polymer sample into several fractions each with a molecular weight distribution as narrower as possible. Most methods used for this purpose are based on the fact that polymer solubility decreases with increase in mol. wt. Some common fractionation methods are described below.

Fractional Precipitation
In this method, the polymer sample is dissolved in some suitable solvent. The concentration of polymer solution is fixed around 0.1% (g dl^{-1}). To this dilute solution of the polymer, a non-solvent is added drop wise with vigorous stirring. As a result of the addition of a non-solvent, the portion of the dissolved polymer that contains molecules with largest size becomes insoluble first. The insoluble material is removed as the fraction of highest molecular weight. To the clear solution, more non-solvent is added when the portion with next highest mol. wt. precipitates out and is separated as the second fraction. This procedure is repeated several times to obtain various fraction with decreasing molecular weight.

Fractional Elution
In this method, polymer is extracted from solid into solution. The method used is to pack a column with, for example, glass beads which are coated with polymer. The column is eluted with solvent/non-solvent mixtures of gradually increasing solvent power. Here, the lowest mol. wt. portion will elute first from the column, followed by fractions containing gradually increasing mol. wt. as the proportion of the solvent in the solvent/nonsolvent mixture is increased.

Gel Permeation Chromatography (GPC)
An elegant method used for the fractionation of polymers is gel permeation chromatography. The technique when applied by biologists for the separation of proteins from a mixture is called gel filtration chromatography and by colloid chemists to separate colloidal dispersions according to size is called size exclusion chromatography.

It is a liquid-solid elution chromatography that separates polydispersed polymer into several fractions by means of the sieving action of a cross-linked polystyrene gel. The gel material acts as stationary phase and is commercially available with a wide distribution of pore size (1–10^6 nm). This is an expensive technique but provides a very quick information on fractionation and MWD. A single run can yield the value of all the averages together with the distribution of relative molecular weights. The GPC column is packed with above mentioned insoluble polystyrene or polystyrene–polydivinylbezene gels. The solvent is forced continously through the sets of columns at a controlled rate of about 1 ml/min by a high pressure pump. The polymer solution is injected at a position upstream of the column. The detector is located downstream of the column and responds sensitively to the presence of the polymer in a small volume of eluent. A variety of detector techniques are used. The common ones measure differences in refractive index, UV absorption or light scattering.

The basic principle underlying the separation of different fractions of a polydispersed sample is based on the size of individual polymer molecules that explore the pore system of the column material. Large molecules are excluded from small pores and can only diffuse into a restricted part of the pore system within the beads while smaller ones would enter into the pores of the bead. Thus large molecules would have less residence time and would emerge first. Fig. 3.15 schematically illustrates the principle of GPC.

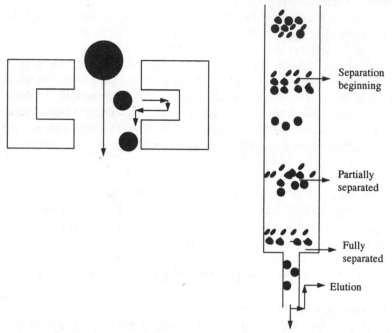

Fig. 3.15   Schematic representation of the basic principle involved in GPC

A calibration curve is necessary to obtain information from GPC. Several samples of narrow dispersed polymers like polystyrene, polyethylene oxide etc. of varying mol. wt. (these are available commercially) are first eluted separately from the column and the retention time (or volume) for each sample is determined. This procedure provides a calibration plot, which can be used to obtain the mol. wt. of an unknown polymer fraction. A GPC run of the sample provides a plot between record output versus elution volume. Another plot between log mol. wt. versus elution volume is then constructed which is linear as shown in Fig. 3.16.

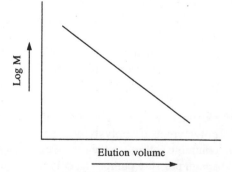

Fig. 3.16   Calibration plot of log molecular weight versus elution volume for a narrow dispersed polymer of known molecular weight

Now the unknown polymer is run in the column and with the help of different fractions eluting and the calibration plot MWD curves can be constructed.

3.3.3 Determination of Molecular Weight of Polymers

Different methods that enable the mol. wt. determination of polymers are based either on the colligative properties or estimation of size of polymer chains. As expected, the former methods will yield number average mol. wt. and while the later give weight average molecular weight. In addition to these two methods another method based on estimation of chemical functional groups present as end groups in polymer chains is also used to get $\overline{M}_n$. The details of the above mentioned methods are discussed as follows.

End Group Analysis

This method concerns with the determination of the number of the end groups present in a given mass of the polymer (provided it has a given number of functional groups) using any standard analytical method. For example, if a polyester contains one end COOH group per chain which can be estimated by the titration using standard solution of NaOH solution. The procedure is for example as follows. Suppose 1 g of the solution of the polyester needs 'a' ml of 0.01 M NaOH, then moles of NaOH needed to neutralise the polyester solution will be, ['a'/1000] × 0.10 = 10^{-3} × 'a' mole. Therefore the number of moles of COOH groups will be (10^{-3}) × 'a'. Since there is only one functional group per polyester chain, the number of mole of the polymer in 1 g of it would be 10^{-3} × 'a'. So the number average molecular weight of the polymer will be 10^3/'a'.

Colligative Properties and Molecular Weight

The two common methods for the determination of the mol. wt. of low molecular weight substances are based on properties viz. depression in freezing point (cryoscopy) and elevation in boiling point (ebulliometry). As these two properties are colligative (i.e. depend only on the number of moles of a solute present in a liquid and not on their nature). the measurement in the decrease in freezing point or increase in boiling point of the liquid on addition of a given mass of a solute can simply be used for the determination of the mol. wt. of the solute material. However, the magnitude of this change in b.pt. or f.pt. is very small and can be measured precisely using Beckmann thermometers, which measure small changes in temperature. These two techniques are highly sensitive to the mol. wt. Substances with low mol. wt. would give the measurable changes in b.pt. or f.pt. For polymeric substances, particularly when the mol. wt. is more than few thousands, these techniques are seldom used.

The other two colligative properties viz. osmotic pressure and lowering in vapour pressure can however be conveniently used for the mol. wt. determination in polymers. The later property is again limited to mol. wt. up to few thousands only. Since all the methods mentioned above are based on the number of molecules, the experimentally determined mol. wt. they would offer will be the number average mol. wt. Table 3 (i) shows the comparison between changes in various colligative properties upon dissolution of a low mol. wt. (1000) and high mol.

wt. (10,00,000) substances in a suitable solvent. It is clear that for high mol. wt. sample, the only useful method that can provide number average mol. wt. is membrane osmometry which is based on measurement of osmotic pressure.

Table 3(i)   The subitability of methods based on colligative properties versus mol. wt

Property	$\overline{M}_n = 10,000$	$\overline{M}_n = 10,00,000$
Depression in f.pt.	0.012 K	0.00012 K
Elevation in b.pt.	0.006 K	0.000006 K
Osmotic pressure	600 mm solvent	6 mm
Vapour pressure lowering	0.018 mm Hg	0.00018 mm

The details of osmometric methods (vapour pressure osmomety, VPO and membrane osmometry) are described below.

Vapour Pressure Osmometry

This technique is governed by the principle that the vapour pressure of a liquid is lowered on addition of a solute and this lowering is dependant on the number of the solute molecules present and is based on the idea of isothermal distillation. The construction principle of a vapour pressure osmometer is schematically shown in Fig. 3.17.

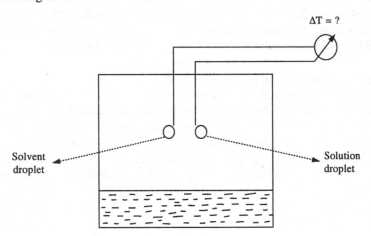

Fig. 3.17   Block diagram of a vapour pressure osmometer

In this method, drops of solvent are placed using syringes in an insulated chamber as in proximity to thermistors T_1 and T_2. As both the drops are of the solvent, equilibrium is achieved. The two thermistors form a part of a Wheatstone bridge and the balance point is established. Now on one of the thermistors, a drop of solution is introduced with the help of syringe. The thermistor probe would register the temperature difference, ΔT arising due to the faster evaporation of the molecules of liquid from solvent drop than from solution drop.

This temperature difference can be accurately measured as a function of the bridge imbalance output voltage, ΔV and is related to the molecular weight of the polymer by,

$$\Delta V / C = K / \overline{M}_n + K B C \tag{3.8}$$

where K is the calibration constant that can be determined by doing the similar experiment with any polymer sample of known molecular weight. Thus a plot between $\Delta V/C$ versus C gives a linear behavior, the intercept of which would provide $\overline{M}_n$. This method needs calibration, which involves the determination of calibration constant K. It is obtained by measuring ΔT for a relatively low mol. wt. polymer of known mol. wt. and low polydispersity. The measurements at different concentrations when plotted as $\Delta V/C$ versus C would give the intercept as $K/\overline{M}_n$. Substituting the value of $\overline{M}_n$, K is estimated. The advantage of the method is its quick determination of molecular weight using only a small amount of the sample. The disadvantage is that it is limited to determination of low molecular weights (less than 30,000).

Membrane Osmometry
The membrane osmometer apparatus basically measures osmotic pressure of polymer solutions of known concentration, say 1 g dl^{-1}, which can be correlated to the number average mol. wt. of the polymer. Since polymer solutions even in dilute regime usually behave nonideally, the osmotic pressure is measured at several concentration, C, and an extrapolation to zero concentration is made from the linear plot of reduced osmotic pressure (π/C) versus C. Using the van't Hoff equation, the intercept of this linear plot (see Fig. 3.18) can be considered equal to $RT/\overline{M}_n$.

A static membrane osmometer works on the principle that if the polymer solution is separated from the pure solvent through a semipermeable membrane (that allows passage of only small solvent molecules but not large polymer molecules), then due to the difference in the chemical potetial in the two compartments, the solvent molecules transport from the solvent compartment to the solution compartment. If the two compartments have columns containing the capillaries in each compartment and if initially the two compartments have liquids at the same heights in the capillaries, then due to the passage of solvent molecules across the membrane (osmosis) the level in the capillary in the solvent compartment will show a gradual decrease while that in the solution compartment will show an increase. It would continue for some time till the equilibrium is reached due to the osmotic head working in the solution compartment does not allow any more solvent molecules to migrate from solvent compartment to solution compartment. In principle, the difference in the level of capillaries of the two compartments, after the equilibrium is achieved can be taken as the osmotic pressure.

The membrane osmometry (static) suffers with a disadvantage that often long time (several hours for a single measurement of mol. wt.) is required as the achievement of equilibrium is a slow process. Also, since no membrane can be perfectly ideal, small sized molecules of a polydispersed polymer sample can pass through the membrane causing error in the mol. wt. determination. A modified version, the dynamic osmometer reduces the time problem. In this method, the counter pressure required to prevent the transport of solvent molecules through the membrane is measured. In high speed osmometers, an optical system in the solvent chamber detects flow through the membrane and automatically adjusts pressure by an electro-mechanical device to prevent any flow. In the absence of any flow the whole process completes in few minutes. Also since the time is reduced, there is virtually no possibility of the migration of even smallest polymer molecules that causes errors in static method.

As already mentioned the van't Hoff law forms the basis for the determination of number average molecular weight, $\overline{M}_n$. Following equation gives the relation ship between osmotic pressure, π of polymer solution to $\overline{M}_n$ of the polymer,

$$\pi/C = RT \, [A_1 + A_2C + A_3C^2 + A_4C^3 + \ldots] \tag{3.9}$$

where A_1, A_2 and A_3 are the first, second and third virial coefficients, C_2 is the concentration of polymer solution and R and T are the gas constant and temperature respectively. The first virial coefficient, A_1 is equal to $1/\overline{M}_n$. Thus the above equation can be written as,

$$\pi/C = RT \, [1/\overline{M}_n + A_2C + A_3C^2 + A_4C^3 + \ldots] \tag{3.10}$$

For very dilute solutions (typically less than 1 g dl^{-1}), the concentration terms containing higher order powers can be neglected and hence it can be written that

$$\pi/C = RT[1/\overline{M}_n + A_2C] \tag{3.11}$$

Thus a plot of $(\pi/RTC]$ versus C should yield a straight line with an intercept of $1/\overline{M}_n$ and slope equal to A_2 (see Fig. 3.18b). The value of A_2, the second virial coefficient is often used as a measure of polymer-solvent interaction. As the quality of the solvent worsens the slope of the line decreases. The slope is practically zero i.e. the line is parallel to the concentration axis in solvents called theta solvents. In order to have better results on mol. wt. by osmometry, the solvent used should not be very good for the polymer as a large increase in the osmotic pressure would be observed with concentration of the polymer and this

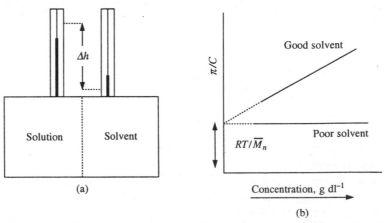

(a)

(b)

Fig. 3.18 (a) **The working principle of a membrane osmometer and (b) a plot of reduced osmotic pressure versus concentration**

would cause difficulty in extrapolation of the line to obtain precise intercept. In Table 3(j) are given some solvents, which behave as theta solvents for some polymers at the specified temperature. In membrane osmometry, determination of mol. wts above a million is essentially impossible because there are too few particles in a given weight of polymer. On the other hand, polymers with mol. wt. less than 20,000 can cause problems by their diffusion through the membrane. Thus membrane osmometry is ideally suitable for determination of molecular weights in the range from 20,000 to 1,00,000.

Table 3(j)   Theta solvents for some polymers

Polymer	Solvent	Temperature, °C
Polystyrene	Decalin	31
Polypropylene	Isoamyl acetate	34
PVC	Benzyl alcohol	155
PVAc	3-Heptanone	29
Polyacrylic acid	1,4-Dioxane	30
Polyisoprene	Propyl ketone	~20
PMMA (atactic)	n-Butyl chloride	35.4
Polyisobutylene	Benzene	25
Polydimethyl siloxane	Butanone	20

Dilute solution viscosity method for molecular weight determination
This method involves the measurements of viscosity of dilute polymer solutions over a concentration range (usually up to 2 g dl^{-1}). The measurements do not require any sophisticated instrumentation and are thus used routinely in molecular weight determinations. The presence of small amount of a polymeric material dissolved in a solvent significantly increases viscosity of the solvent. Such an increase in viscosity is dependent on the concentration and molecular weight of the polymer and its interaction with solvent. In order to determine the mol. wt. of a polymer, viscosity of polymer solutions at different concentrations relative to that of the solvent is measured. Since only relative measurements are required, capillary viscometers are well suited. The most convenient viscometers are Ostwald viscometer and Ubbelohde viscometer. In the latter the additional advantage is that dilutions of the solution can be done in the viscometer itself. The design of an Ubbelohde viscometer is shown in Fig. 3.19.

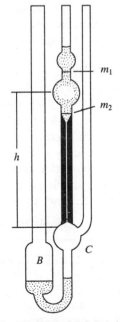

Fig. 3.19   The Ubbelohde viscometer

A polymer solution of about 2 wt % (2 g dl^{-1}) concentration is transferred into bulb B after meticulously cleaning the viscometer. The viscometer is then vertically immersed in a constant temperature bath. The end of the tube attached to bulb C is closed by a finger and suction is applied at A end so that the solution comes above level m_1, Tube C is then opened till the solution in the bulb A is drained and the pressure at this end of the capillary remains at atmospheric levels. Thereafter, the suction at A end is released and the time (t) taken for the solution to fall from level m_1 to m_2 is measured by a stopwatch. The flow occurs only along the walls in bulb C and flooding is totally avoided by the bell-shaped design of the capillary end. After reproducible results are obtained, a known amount of solvent is added to bulb B to dilute the solution and the viscometer is gently shaken by hand. The flow times for polymer solutions at several concentrations are thus obtained with a single filling only. Ubbelohde viscometers are available in several sizes and the one which gives flow times of about 200 to 400 s for the solution to be studied should be chosen to reduce shear errors. The viscosity is then given by the equation $\eta = \rho\,[At - B/t]$, where the first term on the right represents the Hagen-Poiseuille law and the second term gives the correction for the kinetic energy of the fluid leaving the capillary. Usually, the constants A and B are either supplied by the manufacturers or can be obtained by calibration using standard liquids like water and cyclohexane at two different temperatures. Normally, B is a small number and in case A and B are unknown, one takes η as proportional to t. The relative viscosity can be obtained by dividing the flow time of the solution by the flow time of the solvent (since the solutions of polymers are very dilute, the density of the solution and that of solvent can be considered same). Also, since $t > t_0$, the relative viscosity would always be greater than 1. Thus a new term specific viscosity is defined as a positive quantity by subtracting 1 from relative viscosity. Specific viscosity is then divided by the corresponding concentration to give reduced viscosity. When the reduced viscosity is plotted on ordinate (y-axis) against polymer concentration on abscissa (x-axis), a straight line behavior is observed. The extrapolation of reduced viscosity to zero concentration gives intrinsic viscosity. Another way is to plot inherent viscosity versus concentration, which also shows linear behavior with extrapolated value to zero concentration as intrinsic viscosity. The different viscosity terms are shown in Table 3(k). For calculating different terms given in the above table, the concentration of the polymer is usually in g per 100 ml i.e. g dl^{-1} or wt. %, t and t_o are the times of efflux for the solution and the solvent through the viscometer. Intrinsic viscosity of a polymer solution at a particular temperature is obtained by the use of either Huggins or Kraemer equation. The forms of the equations are as follows:

Table 3(k) Viscosity terms and their definitions

Viscosity term		Definition		Unit
Relative viscosity	η_{red}	η/η_o	t/t_o	
Specific viscosity	$\eta_{sp} = \eta_r - 1$	$(\eta - \eta_o)/\eta_o$	$(t - t_o)/t_o$	
Reduced viscosity	η_{red}	(η_{sp}/C)		dl g^{-1}
Inherent viscosity	η_{inh}	$(\ln \eta_{rel})/C$		dl g^{-1}
Intrinsic viscosity	$[\eta]$	$(\eta_{sp}/C)_{C \to o}$ or $(\ln \eta_{rel})/(C)_{C \to o}$		dl g^{-1}

$$\eta_{sp}/C = [\eta] + k' \, [\eta]^2 C \quad \text{Huggins equation} \qquad (3.12)$$

$$(\ln \eta_{rel})/C = [\eta] + k'' \, [\eta]^2 C \quad \text{Kraemer equation} \qquad (3.13)$$

The representative plots using both the equations are shown in Fig. 3.20.

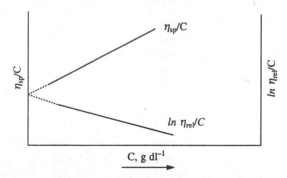

Fig. 3.20   The typical Huggins and Kraemer viscosity plots

If the higher terms of C are neglected, both the lines should intersect at the ordinate to correspond to zero concentration and the constants should satisfy the condition, $k' + k'' = 0.5$. The intrinsic viscosity is related to the mol. wt. by the Mark-Houwink-Sakurada equation,

$$[\eta] = kM^{\alpha} \qquad (3.14)$$

where k and α are empirical constants and are characteristic for a polymer-solvent pair at a given temperature. Typical values of k are of the order 10^{-4} and α about 0.7. The values of k and α are obtained from the known values of molecular weights for well defined fractions of given polymer dissolved in a solvent and measured intrinsic viscosities for the same pair. When log $[\eta]$ is plotted against log molecular weight, a linear relation results (based on Mark-Houwink and Sakurada equation). The slope of the plot gives α and the intercept yields the value of log k. Table 3(1) provides the values of k and α for several polymer-solvent pairs at 30°C.

Although, viscometry measurement provides an indirect method (as it needs the values of constants k and α for the determination of molecular weight of a polymer), it is still the most convenient way for routine determination of molecular weights of polymers, due to the simplicity of the set up. Wide range of molecular weights for all the polymers having solubility in any of the solvents are obtained by this method. However, there are some limitations of the method. It gives errors in molecular weight in branched polymers and copolymers. Because, the values of k and α are highly dependent upon the degree of branching in branched copolymer and compositional variations in the statistically regular or irregular copolymers such as block and graft copolymers or random copolymers. One more limitation of the method is that the polymer solutions used have to be very dilute so that the chain entanglements, drag effect and adsorption onto capillary inner surfaces are eliminated maximally.

Table 3(l) The constants k and α of the Mark-Houwink-Sakurada equation at 30°C

Polymer	Solvent	$k \times 10^3$ ml/g	α
Polypropylene (atactic)	benzene	27.0	0.71
Polyvinyl alcohol	water	45.3	0.64
Polyvinyl acetate	acetone	10.2	0.72
	benzene	56.2	0.62
	methanol	31.4	0.60
Polystyrene	benzene	11.5	0.73
	toluene	11.0	0.72
Polyvinyl chloride	THF	83.3	0.83
	cyclohexanone	16.3	0.77
Natural rubber	benzene	18.5	0.74
SBR	toluene	37.9	0.71
	cyclohexane	31.6	0.70
Polyethylene oxide	water	16.6	0.82
	benzene	40.0	0.70
	acetone	32.0	0.67
Polyacrylamide	water	68.0	0.66
Polyvinylpyrrolidone	water	39.3	0.59
	methanol	23.0	0.65
Polymethyl methacrylate	acetone	6.2	0.72
	benzene	4.0	0.77
Polyacrylic acid	1,4 dioxane	76.0	0.50
Polyacrylonitrile	DMF	20.9	0.75
Nylon 6	m-cresol	320.0	0.62
Nylon 6,6	m-cresol	240.0	0.61

Light Scattering

When a beam of light is passed through a colloidal solution, it is scattered. This is well-known Tyndall effect, which results from the scattering of a part of the beam of light by the colloidal particles in all directions (see Fig. 3.21). Since polymer solutions can be considered as colloidal (lyophilic) solutions and as the intensity of light scattered depends on the size of colloidal particles (or polymer molecules), the scattering phenomenon can be used for the determination of molecular weight of polymers. Light scattering measurements thus provide the weight average molecular weights of polymers.

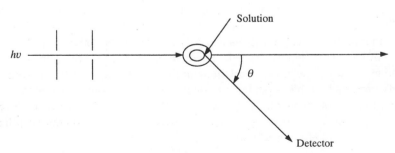

Fig. 3.21   Light scattering from a polymer solution

Debye, in 1944, correlated molecular weight of polymer with the intensity of light scattered by its solution by the equation (now known as Debye equation) which holds good only for particles which are smaller than the wavelength of light used for the scattering experiment as

$$KC/R_{90} = HC/\tau = 1/\overline{M}_w + 2\,BC \tag{3.15}$$

where B is the second virial coefficient, C is the concentration of the solution, and R_{90} is the Rayleigh ratio at $90°$ observation angle. This ratio in a generalized case is represented as R_θ i.e. the Rayleigh ratio is determined at an observation angle of $90°$, $R_\theta = R_{90}$.

$$R_\theta = i_\theta\, r^2/I_0\, V \tag{3.16}$$

where i_θ is the intensity of the scattered light per unit volume, V, of scattering material which is observed at a distance r and at an angle of θ with reference to the incident beam, and I_0 is the intensity of the incident beam. The turbidity τ is defined as, $e^{-\tau x} = I/I_0$. The constants K and H are light scattering calibration constants and are defined by the following equations,

$$K = \frac{2\pi^2 n^2\,(dn/dC)^2}{\lambda^4 N_A} \tag{3.17}$$

where n is the refractive index of the solution, (dn/dC) is the specific refractive index increment, i.e. the change of refractive index with concentration, λ is the wavelength of the incident light and N_A is the Avogadro number.

$$H = \frac{32\pi^3 n^2\,(dn/dC)^2}{3\lambda^4 N_A} \tag{3.18}$$

For polymer solutions where the molecules are larger in size compared to the wavelength of light, the equation needs to be modified to,

$$\frac{KC}{R_\theta} = \frac{HC}{\tau} = \frac{1}{\overline{M}_w \cdot P(\theta)} + 2\,BC \tag{3.19}$$

where $P(\theta)$ is the particle-scattering factor. It represents the angular dependence of the scattered light by following equation,

$$\frac{1}{P(\theta)} = 1 + \left[\frac{16\pi^2}{3\lambda^2}\right] \langle \overline{S}^2 \rangle \sin^2\left[\frac{\theta}{2}\right] \tag{3.20}$$

where $\langle \overline{S}^2 \rangle$ is the mean square radius of gyration of the polymer random coil. Substituting this value of $P(\theta)$ in eq. 3.19 we get,

$$\frac{KC}{R_\theta} = \frac{HC}{\tau} = \frac{1}{\overline{M}_w} + \left[\frac{16\pi^2}{3\lambda^2 \overline{M}_w}\right] \langle \overline{S}^2 \rangle \sin^2\left[\frac{\theta}{2}\right] + 2\,BC \tag{3.21}$$

$P(\theta)$ is considered as a correction factor to the scattered intensity at different values of θ, such that $P(\theta) = 1$ at $\theta = 0$. Similarly, $BC = 0$ at $C = 0$. Now, above equation can be rewritten in the following forms:

$$\left[\frac{KC}{R_\theta}\right]_{C\to 0} = \frac{1}{\overline{M}_w} + \left[\frac{16\pi^2}{3\lambda^2 \overline{M}_w}\right]\langle \overline{S}^2\rangle \sin^2(\theta/2) \qquad (3.22a)$$

$$\left[\frac{KC}{R_\theta}\right]_{\theta\to 0} = \frac{1}{\overline{M}_w} + 2BC \qquad (3.22b)$$

$$\left[\frac{KC}{R_\theta}\right]_{C,\theta\to 0} = \frac{1}{\overline{M}_w} \qquad (3.22c)$$

Light scattering technique has been developed as fastest tool for molecular weight determination of polymer due to the availability of laser radiation sources. Light scattering measurements traditionally require extremely clean, dust-free samples in order that erroneous scattering may not be generated. The presence of any gel like, semi-dissolved or associated polymers cause large errors. The light scattering measurements are usually valid between molecular weights of 10,000 and 10,00,000. Below 10,000, there are interference effects from solvent molecules and dust particles. Above 10,000,000, there are interference effects from other polymer molecules.

Use of equations 3.22(a) and 3.22(b) is made either by Debye method or Zimm method to determine the molecular weights of polymers. Debye method is also known as disymmetry method. It requires measurements of the scattered intensities at three angles, which typically are 45°, 90° and 135° and at several different concentrations. A disymmetry ratio, z can be defined as $z = I$ (45°)/I (135°). Since z value is concentration dependant, in usual practice the value of z at different concentrations is plotted in the form of $(z-1)^{-1}$ versus concentration plot and the intercept is obtained as value of z at zero concentration. Then the particle scattering factor $P(\theta)$ i.e. $P(90°)$ and $\langle \overline{S}^2\rangle$ are calculated from the known values using models developed for describing the scattering systems. These models are derived for a particular conformational state i.e. sphere, random coil, rod and disk etc., Once $P(90°)$ is known, $\overline{M}_w$ can be calculated from the intercept of plot of $KC/R_{(90°)}$ versus concentration (eq. 3.22a) by simple extrapolation to zero concentration.

The main disadvantage of this method however is that a model for the polymer chain in dissolved state is to be pre assumed. That is why another method is developed by Zimm, which does not need any advance information on the conformational state of polymer chains in solution.

Zimm method is based on a double extrapolation procedure (see Fig. 3.22). Both the extrapolations of plots of KC/R_θ versus $\sin^2(\theta/2) + kC$ (where k is a constant chosen arbitrarily) to zero angle and to zero concentration, are made. Then from the intercept and slope, the molecular weight of polymer and second virial coefficients can be determined by use of equations 3.22(a)–3.22(c).

Low Angle Laser Light Scattering (LALLS)

The conventional light scattering instruments use light from high intensity mercury lamp. In recent years, the mercury lamp source is replaced by a laser source.

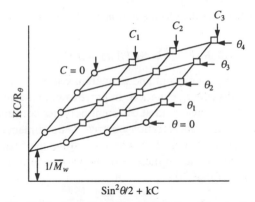

Fig. 3.22   The double extrapolation Zimm plot

Mostly, helium – neon (He – Ne) lasers (with $\lambda = 6328$ Å) are used. The advantage of the laser source is that, the high intensities of the light emerging from them permit scattering experiments at much smaller or low angles (which typically may vary from 2° to 10°) and also solutions with much lower polymer concentrations can be studied. At very low angles and low concentrations, eq. 3.22 effectively reduces to the classical Debye equation derived for small spherical particles,

$$\frac{KC}{R_\theta} = \frac{1}{\overline{M}_w} + 2BC \qquad (3.23)$$

Thus, one can construct linear plots between KC/R_θ versus C at a single θ value and different concentrations to get intercept which yields the reciprocal of molecular weight, $\overline{M}_w$. The slope of the line gives the value of second virial coefficient, B. Such a general plot of KC/R_θ versus C at low angles is shown in Fig. 3.23.

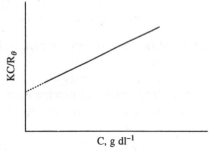

Fig. 3.23   Plot of Rayleigh ratio versus concentration from low angle laser light scattering

Ultracentrifugation
Polymer solutions are lyophilic colloids and these can be made to settle down on the application of centrifugal force. The rate of the settling (sedimentation rate) of polymer molecules depends on the size of the molecules. Then, this property can be applied to experimentally determine the molecular weight of polymers. Two methods viz, sedimentation velocity method and sedimentation equilibrium method are used for this purpose. The sedimentation velocity method involves the determination of sedimentation coefficient, S on applying a very high gravitational field using an ultracentrifuge which can provide spins as high as 65000 rpm. With the help of the predetermined paramters like diffusion coefficient, D, specific volume of the polymer in solution, V_{sp} and the density of solvent, d, the mol. wt of the polymer is obtained using

Stokes law which correlates the sedimentation coefficient and diffusion coefficient by

$$\text{Mol. wt.} = SRT/D(1 - d\,V_{sp}) \qquad (3.24)$$

Where R and T are the gas constant and temperature in Kelvin scale. The above equation is known as Svedberg equation. In practice both the sedimentation and diffusion coefficients are measured at different polymer concentrations and the values extrapolated to zero concentration are used. The sedimentation equilibrium method as the name suggests concerns with the setting up of equilibrium during sedimentation. The merits and limitations of the various mol. wt. determination methods are shown in Table 3(m).

Table 3(m)   Different molecular weight determination methods: A comparison

Method	Type of average	Mol. wt. range
Ebbuliometry or Cryoscopy	$\overline{M}_n$	< 10,000
Vapour pressure osmometry*	$\overline{M}_n$	< 25,000
Memberane osmometry	$\overline{M}_n$	1500–100,000
Dilute solution viscosity**	$\overline{M}_v$	15,000–1000.000
Light scattering	$\overline{M}_w$	2000–10,000,00
Ultracentrifuge	$\overline{M}_w, \overline{M}_z$, MWD	2000–10,000,00
Gel permeation chromatography*	$\overline{M}_w, \overline{M}_z$, MWD	up to 50,00,000

*Relative methods and need calibration from standard polymers samples
**Indirect method and needs values of k and α for particular polymer – solvent system

3.3.4   Molecular Weight Distribution (MWD) Curves

The mapping of a MWD curve for a given polymer system can be done by following steps: (i) synthesis of polymers, (ii) fractionation of polymers (by any methods from fractional precipitation, partial extraction or size exclusion chromatography), (iii) determination of molecular weight of each of the fraction and its weight and finally (iv) by plotting either simple or integral or differential distribution curves. In the fourth step, a simple distribution curve can be obtained by using information obtained in step (iii). i.e. by plotting the weight of each fraction against its molecular weight. If the fraction form and separated form differ in molecular weight by same value successively, one can easily obtain a simple and smoothed distribution curve as shown in Fig. 3.24.

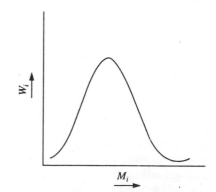

Fig. 3.24   Plot of weight fraction W_i versus its molecular weight M_i for simple representation of MWD

But in real experiment and practice, getting such uniform fractions is extremely difficult. When one plots a distribution curve between the weight of each fraction and its molecular weight (as obtained in actual experiments), the curve looks like as shown in Fig. 3.25.

Figure 3.25 shows that the spread of fractions with a given weight and molecular weight is uneven. The MWD shown in Fig. 3.25 has least utility and convey very little information on the dispersity. Tackling of the data as shown in Fig. 3.25 is done either by (i) plotting the cumulative weight fractions (i.e. each successive fractions are added

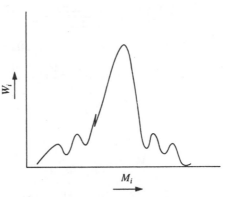

Fig. 3.25   Actual MWD curve

continuously) against a molecular weight of preceding fraction. The curve thus resulted is known as integral distribution curve and looks as shown in Fig. 3.26, or (ii) by taking the slopes at as many points as possible from the integral curve and the slopes at each point corresponding to a particular W_i are plotted against M_i (see Fig. 327). This method yields the differential weight percentage of molecules of particular molecular weight in the whole polymer sample.

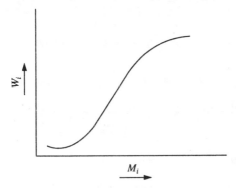

Fig. 3.26   Integral distribution curve

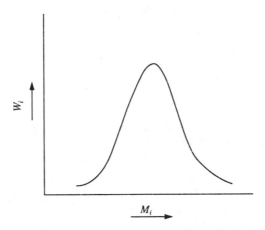

Fig. 3.27   Differential distribution curve of a real polymer sample

Numerical exercises

1. Equal number of molecules with $M_1 = 1,0,000$ and $M_2 = 1,00,000$ are mixed. Calculate $\overline{M}_n$ and $\overline{M}_w$.

    Solution Let $n_1 = n_2 = 10$ (say), then,

    $$\overline{M}_n = \frac{n_1 M_1 + n_2 M_2}{n_1 + n_2} = \frac{(10 \times 10,000) + (10 \times 100,000)}{(10 + 10)}$$

    $$= \frac{10^5 + 10^6}{20} = \frac{10^5 (1 + 10)}{20} = \frac{11 \times 10^4}{2} = 55,000 \text{ g mol}^{-1}$$

    $$\overline{M}_w = \frac{n_1 M_1^2 + n_1 M_2^2}{n_1 M_1 + n_2 M_2} = \frac{[10 \times (10,000)^2] + [10 \times (1,00,000)^2]}{[10 \times 10,000] + [10 \times 1,00,000]}$$

    $$= \frac{10^9 + 10^{11}}{10^5 + 10^6} = \frac{10^9 (1 + 100)}{10^5 (1 + 10)} = \frac{101 \times 10^4}{11}$$

    $$= 91,818 \approx 92,000 \text{ g mol}^{-1}$$

2. Equal masses of polymer molecules with $M_1 = 10,000$ and $M_2 = 1,00,000$ are mixed. Calculate $\overline{M}_n$ and $\overline{M}_w$.

    Solution Let $m_1 = m_2 = 2,00,000$ (say), then number of moles n_1 and n_2 are

    $$n_1 = \frac{\text{mass } m_1}{\text{molar mass } M_1} = \frac{2,00,000}{10,000} = 20$$

    $$n_2 = \frac{\text{mass } m_2}{\text{molar mass } M_2} = \frac{2,00,000}{1,00,000} = 2$$

    $$\overline{M}_n = \frac{n_1 M_1 + n_2 M_2}{n_1 + n_2} = \frac{[20 \times 10,000] + [2 \times 1,00,000]}{20 + 2} = \frac{10^5 + 10^5}{11}$$

    $$= \frac{2 \times 10^5}{11} = 18,182 \approx 18,000 \text{ g mol}^{-1}$$

    $$\overline{M}_w = \frac{n_1 M_1^2 + n_1 M_2^2}{n_1 M_1 + n_1 M_2} = \frac{[20 \times (10,000)^2] + [2 \times (1,00,000)^2]}{[20 \times 10,000] + [2 \times 100,000]}$$

    $$= \frac{10^9 (1 + 10)}{2 \times 10^5} = \frac{11}{2} \times 10^4 = 55,000 \text{ g mol}^{-1}$$

3. Calculate $\overline{M}_n, \overline{M}_w$ and $\overline{M}_z$ for a polymer consisting of three fractions with molecular weights, 1×10^5, 2×10^5 and 3×10^5. The mole fractions of each of these fractions are found to be 0.1, 0.5 and 0.4, respectively.

    Solution $\overline{M}_n = \Sigma n_i M_i / \Sigma n_i$, $\overline{M}_w = \Sigma n_i M_i^2 / \Sigma n_i M_i$ and $\overline{M}_z = \Sigma n_i M_i^3 / \Sigma n_i M_i^2$. Introducing the values of n_i and M_i namely 0.1, 0.5 and 0.4 and 1×10^5, 2×10^5 and 3×10^5 into the relation of $\overline{M}_n$, we get

    $$\overline{M}_n = \frac{[(0.1 \times 1 \times 10^5) + (0.5 \times 2 \times 10^5) + (0.4 \times 3 \times 10^5)]}{[0.1 + 0.5 + 0.4]}$$

$$= \frac{(1,00,000) + (10,00,000) + (12,00,000)}{1}$$

$$= 2.30 \times 10^6 \text{ g mol}^{-1}$$

4. Calculate $\overline{M}_n$ and $\overline{M}_w$ and $\overline{M}_z$ for a polydispersed polymer composed of the following mixture of fractions (mass % and molecular weight of each of the fractions are given):

Mass %	20	30	50
Mol. weight	50,000	1,00,000	2,00,000

Solution The mass fraction W_i for each of the fractions would be 0.2, 0.3, 0.5 respectively. In terms of mass fractions,

$$\overline{M}_n = \Sigma \, W_i / \Sigma \, (W_i/M_i), \ \overline{M}_w = \Sigma \, W_i M_i / \Sigma \, W_i \text{ and } \overline{M}_z = \Sigma \, W_i M_i^2 / \Sigma \, W_i M_i \text{ and so}$$

$$\overline{M}_w = \frac{[(0.2 \times 50,000) + (0.3 \times 1,00,000) + (0.5 \times 2,00,000)]}{0.2 + 0.3 + 0.5}$$

$$= 1,40,000 \text{ g mol}^{-1}$$

$$\overline{M}_n = \frac{[0.2 + 0.3 + 0.5]}{[(0.2/50,000) + (0.3/1,00,000) + (0.5/2,00,000)]}$$

$$= 1,05,263 \approx 1,05,260 \text{ g mol}^{-1}$$

and $\overline{M}_z$ can also be calculated in the similar way.

5. Calculate $\overline{M}_n$ and $\overline{M}_w$ for a polymer which contains a mixture of following amount of fractions (with respective mol. wts. given in parenthesis). 1 g (20,000), 2 g (50,000) and 0.5 g (1,00,000).
Hint Calculate the mass fractions of each of them as $1/(1 + 2 + 0.5)$; $2/(1 + 2 + 0.5)$; $0.5/(1 + 2 + 0.5)$, i.e. 0.285 (≈ 0.30), 0.571 (≈ 0.6) and 0.142 (≈ 0.1). Then estimate $\overline{M}_n$ and $\overline{M}_w$ in the usual manner.

6. Two fractions of polymers with mol. wts. of 10,000 and 1,00,000 are mixed in the ratio 0.145: 0.855 by mass. Calculate the ratio $\overline{M}_w / \overline{M}_n$. Comment on the nature of distribution.

7. The intrinsic viscosity of myosin is $217 \text{ cm}^3 \text{ g}^{-1}$. Calculate the approximate concentration of myosin in water, which would have a relative viscosity of 1.5.
Solution $[\eta] = (\eta_{sp}/C)_{C \to 0} = 217 \text{ cm}^3 \text{ g}^{-1}$ and $\eta_{rel} = 1.5$, Since $\eta_{sp} = \eta_{rel} - 1 = 1.5 - 1 = 0.5$, so introducing η_{sp} as 0.5 in (η_{sp}/C) and equating the whole term to $217 \text{ cm}^3 \text{ g}^{-1}$, we get $0.5.C = 217 \text{ cm}^3 \text{ g}^{-1}$; $C = 0.5/217 = 2.30 \times 10^{-3} \text{ g cm}^{-3}$.

8. The intrinsic viscosity of a solution of polyisobutylene at 20°C is $180° \text{ cm}^3 \text{ g}^{-1}$. If $[\eta]$ is related to the viscosity average molecular weight $\overline{M}_v$ by the expression, $[\eta] = 3.60 \times 10^{-4} (\overline{M}_v^{0.64})$, calculate the molecular weight $\overline{M}_v$ of the polymer.
Solution $[\eta] = 3.60 \times 10^{-4} (\overline{M}_v)^{0.64}$ or $180 \text{ cm}^3. \text{ g}^{-1} = 3.60 \times 10^{-4} (\overline{M}_v)^{0.64}$, $(\overline{M}_v)^{0.64} = 180 \text{ cm}^3 \text{ g}^{-1}/3.60 \times 10^{-4} = 5.0 \times 10^5$ taking the log form, $0.64 \log \overline{M}_v = \log (5.0 \times 10^5)$ and hence $\overline{M}_v = 8.03 \times 10^6 \text{ g mol}^{-1}$.

9. The following data were obtained for intrinsic viscosity of some polyisobutylene samples in CCl_4 solutions at 30°C. Verify by a suitable plot that the data fit to the equation $[\eta] = k (\overline{M}_v)^{\alpha}$. Determine the constants k and α

$[\eta]$, $\text{cm}^3 \text{ g}^{-1}$	$\overline{M}_v$, g mol^{-1}
430	12,60,000
206	4,63,000

78	1,10,000
73	92,700
43	48,000
15.1	10,000
13.8	9,550
11.5	7,080

Solution Since $[\eta] = k(\overline{M}_v)^\alpha$, $\ln [\eta] = \ln k + \alpha \ln \overline{M}_v$, one can plot $\ln [\eta]$ versus $\ln \overline{M}_v$. It will be found that slope $= \alpha = 0.70$ and intercept $= \ln k = -8.245$ and hence, $k = 2.63 \times 10^{-4}$.

10. Calculate the intrinsic viscosity of a polystyrene sample in toluene from the following relative viscosity data obtained at 25°C:

$C \times 10^3$, kg m^{-3}	0.002	0.004	0.006	0.008	0.010
η_r	1.102	1.208	1.317	1.430	1.548

Calculate $\overline{M}_v$ if the Mark-Houwink constants are $\alpha = 0.69$ and $k = 1.7 \times 10^{-3}$ m^3 kg^{-1}.

Solution Mark-Houwink equation is given as $[\eta] = k (\overline{M}_v)^\alpha$ where, $\overline{M}_v$ is the viscosity-average molecular weight of the polymer, $[\eta]$ = the intercept for the plot of $(\eta_r-1)/C$ versus C. From the above data of η_{rel}, calculate $\eta_{rel}-1 = \eta_{sp}$ and η_{sp}/C versus C. It would be equal to 50.1×10^{-3} m^3 g^{-1}. Then $\overline{M}_v$ can be calculated from the equation $50.1 \times 10^{-3} = 1.7 \times 10^{-3} \overline{M}_v^{0.69} \cdot \overline{M}_v$ would be 135 kg mol^{-1} or 1,35,000 g mol^{-1}.

11. For a 2% aqueous solution of a polymer with molecular weight 50,000, calculate at 27°C (a) the depression in freezing point (ΔT_f), and (b) the elevation in boiling point. Given that molal depression in freezing point of water $= K_f = 1.85$ and molal elevation in boiling point of water $= K_b = 0.52$.

Solution molality of 2% solution i.e. 2 g in 100 ml or 20 g in 1 lit. $= m = 20/50,000 = 4.0 \times 10^{-4}$ m. Thus calculate $\Delta T_f = K_f \cdot m = 1.85 \times 4.0 \times 10^{-4} = 0.00074$ and $\Delta T_b = K_b \cdot m = 0.52 \times 4.0 \times 10^{-4} = 0.00021$.

12. The following data were obtained on the osmotic pressure of solutions of β-globulin in 0.15 M NaCl at 37°C:

C, g/100 ml	19.27	12.53	5.81
π, mm H$_2$O	453	253	112

Calculate the molecular weight of the polymer.

Solution 100 ml = 0.1 litre = 0.1 dm^3. Obtain π/C values (e.g. 453 mm H$_2$O/192.7 g dm^{-3} = 2.35 mm H$_2$O dm^3 g^{-1}) and tabulate the data for all the three concentrations as below:

C(g dm^{-3})	192.7	125.3	58.1
π/C (mm H$_2$O dm^3 g^{-1})	2.35	2.05	1.93

Plot π/C versus C to obtain a straight line. It would be seen from the plot that the intercept $(\pi/C)_{C\to0} = 18.6$ mm H$_2$O dm^3 g^{-1}. Given that $(\pi/C)_{C\to0} = RT/\overline{M}_n$; $\overline{M}_n = RT/(\pi/C)_{C\to0}$ where R = gas constant = 0.08206 lit atm K^{-1} mol^{-1} = 0.08206 × 760 × 13.56 mm H$_2$O dm^3 K^{-1} mol^{-1} and hence $R = 845.67$ mm H$_2$O dm^3 K^{-1} mol^{-1}, then

$$\overline{M}_n = \frac{845.67 \times 310}{1.86} = 1.409 \times 10^5 \text{ g mol}^{-1}$$

13. Use the following data to calculate the number average molecular weight and the second virial coefficient of the solutions of collagen at 25°C.

$C(g\ m^{-3})$	2.4	4.1	5.0	5.5	6.4
$\pi/CRT\ (10^{-3}\ mol.\ kg^{-1})$	3.7	4.2	4.2	4.2	4.9

Solution According to the van't Hoff osmotic virial equation

$$\pi/CRT = \left[\frac{1}{\overline{M}_n} + B_2C + B_3C^2 + \ldots \right]$$

Where, $\overline{M}_n$ is the number-average molecular weight. Thus, a plot of π/CRT versus C will be a straight line with intercept = $1/\overline{M}_n$ and slope = B_2. If we plot the data, we find that $\overline{M}_n = 1/(3.1 \times 10^{-3}\ mol\ kg^{-1}) = 320\ kg\ mol^{-1}$ $B_2 = 2.5 \times 10^{-4}\ mol\ m^3\ kg^{-2}$

14. The excess Rayleigh ratio, $R_{(\theta)}$ of a polymer dissolved in dioxane was determined as a function of concentration by low angle laser light-scattering measurements per the data given below;

$C \times 10^3\ g\ ml^{-1}$	$R_\theta \times 10^5\ cm^{-1}$
0.503	0.239
1.007	0.440
1.510	0.606
2.014	0.790
2.517	0.902

If the refractive index, (n_0) of dioxane is 1.4199, the refractive index increment (dn/dC) for the polymer in dioxane is $6.297 \times 10^{-2}\ cm^3\ g^{-1}$, the wavelength (λ) of the laser light is 6328 Å, calculate the weight average molecular weight of the polymer and second virial coefficient (A_2). The following equation is employed for determining the molecular weight from light scattering data;

$$\left[\frac{KC}{R_\theta} \right]_{C \to 0} = \frac{1}{M_w} + 2BC$$

and the constant K is calculated from the relation, $K = \dfrac{2\pi^2 n^2\ (dn/dC)^2}{\lambda^4 N_A}$

By introducing the values of $n_0 = 1.4199$, $dn/dC = 6.297 \times 10^{-2}$ and $\lambda = 6328$ Å (6328×10^{-8} cm we get $K = 1.66969 \times 10^{-9}$. Then calculate KC/R_θ for different concentrations and obtain a plot between KC/R_θ versus C. It would be seen that the intercept of the above plot is 0.33×10^{-6} and hence $\overline{M}_w$ will be equal to 1/intercept and that is equal to $3.030 \times 10^6\ g\ mol^{-1}$. The slope of the line would be 0.05333 and the second virial coefficient A_2 would than be one half the slope that is $2.66 \times 10^{-2}\ cm^3\ mol\ g^{-2}$.

15. (a) What is the value of the exponent in Mark-Houwink equation in theta condition? (b) The solubility parameter of water is 23.4 H. What is the value of CED? (c) What would be the value of ΔG at theta condition? (d) What is the value of the exponent in Mark-Houwink equation for a rigid rod? (e) What is the value of the ratio of weight average mol. wt. to the number average mol. wt. for proteins? (f) If we use a 2% molar excess of bisphenol with TDI for making polyurethane, what would be the maximum DP of the polymer? (g) How many amino groups

are present in each molecule of nylon 6,6 made with an excess of hexamethylenediamine? (h) What would be the second virial coefficient at theta temperature? (i) What would be the value of interaction parameter in poor and good solvents?

Ans. (a) 0.5 (b) $(23.2)^2 = 543.6$ cal cm^{-3} (c) 0 (d) 2 (e) 1 (f) 9
(g) 2 (h) 0 (i) 0.5, less than 0.5.

Suggested Further Readings

Identification and Testing of Polymers

Bark, L.S., and N.S. Allen (eds.) *Analysis of Polymer Systems*, New York: Elsevier Applied Science, 1982.

Braun, D., *Simple Methods for Identification of Plastics*, 2d ed., Cincinnati, Ohio: Hanser-Gardner, 1986.

Brown, R.P. (ed.), *Handbook of Plastic Test Methods,* 2d ed., London: George Goodwin Ltd., 1981.

Craver, C.D., and T. Provider (eds.), *Polymer Characterisation: Spectroscopic, Chromatographic and Physical Instrumental Methods,* Washington D.C.: ACS, 1990.

Shah, V., *Handbook of Plastics and Testing Technology,* New York: Wiley-Interscience, 1984.

Characterization Methods

Barth, H.G., and J.W. Mays, *Modern Methods of Polymer Characterisation,* New York: Wiley, 1991.

Bern, B.J. and R. Pecora, (eds.), *Dynamic Light Scattering,* New York: Plenum, 1976.

Chu, B., *Laser Light Scattering,* New York: Academic Press, 1974.

Dillinghan, N.C., *Molar Mass Measurements in Polymer Science,* New York: Wiley, 1977.

Flory, P.J., *Principles of Polymer Chemistry,* Chap 8, Ithaca, New York: Cornell Univ. Press, 1953.

Frankuskiewicz, E., *Polymer Fractionation,* Berlin: Springer, 1994.

Gloskner, G., *Polymer Characterization by Liquid Chromatography,* New York: Elsevier Applied Science, 1987.

Higlin, M.B. (ed.), *Light Scattering from Polymer Solutions*, New York: Academic Press, 1972.

Kratochvil, P., *Classical Light Scattering from Polymer Solutions,* New York: Elsevier Applied Science, 1987.

Peebles, L.H., Jr., *Molecular Weight Distributions in Polymers,* New York: Wiley-Interscience, 1971.

Smith, C.G., W.C. Buzanowski, J.D. Graham, and Z. Iskandarani, (eds.), *Handbook of Chromatography, Polymers,* Boca Raton, Fla: CRC Press, 1982.

Slade, P.E., Jr., *Polymer Molecular Weights,* (2 vols.), New York: Dekker, 1975.

Wu, C.S., (ed.), *Handbook of Size Exclusion Chromatography*, New York: Dekker, 1995.

Yau, W.J., J.J. Kirland, and B. Bly, *Modern Size Exclusion Liquid Chromatography,* New York: Wiley, 1979.

4

·Behaviour of Polymers

4.1 Crystalline Behaviour

High molecular weight substances due to the entanglements of the long chains and polydispersity are not perfectly crystalline unlike low molecular weight crystalline solids. Generally, no polymer can be regarded to have 100% crystallinity. The degree of crystallinity varies from polymer to polymer and many polymers are non-crystalline (i.e. amorphous). The presence of crystalline and amorphous regions together in polymers can be accounted by "Fringe micelle model" as depicted in Fig. 4.1. The model presumes that the ordered and disordered domains are intermingled with no clear cut demarcation between the two regions. The ordered and disordered arrangements shown in the figure can be equated to the crystalline and amorphous regions as observed in packing of small molecules. The ordered crystalline region in a polymer physical structure are called as crystallites.

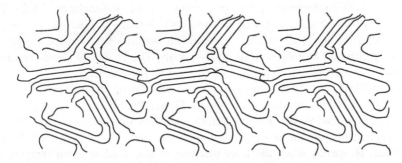

Fig. 4.1 The fringe micelle model showing crystalline and amorphous regions

The crystallites along with the inseparable amorphous regions in polymer chain packing, form shallow pyramid like structures called "spherulites". Besides these morphologies, it is also observed that ordered polymers may form single (or lamellar) crystals with a thickness of 10–20 nm in which the polymer chains are folded back upon themselves. Single crystals are perfect and possess internal order that can be revealed from x-rays or electron diffraction studies.

The crystallizability i.e. ability to undergo crystallisation in polymers depends on several factors such as regularity in the chains, thermodynamic chain flexibility and possibility of close packing. The pre-thermal history of the amorphous polymers plays not only an important but also decisive role in deciding the ordered arrangement in polymer solids. Higher degree of crystallinity in polymers can be developed by using the proper method of polymerisation. Polymerisation of ethylene at ordinary temperature and pressure can be carried out using special catalysts (Ziegler – Natta catalysts). As discussed in the second chapter, the Ziegler Natta catalysts consist of a combination of metal halides (e.g. titanium tetrachloride and an organometallic compound e.g. trialkylaluminium) dissolved in inert solvent such as hexane. Polyethylene obtained using Ziegler's catalyst at low pressures is a linear polymer (with few branches) and has higher degree of regularity or ordered arrangement in the chain. The polyethylene thus obtained has high density and melting point and upon crystallisation forms very good single crystals. The crystalline polyethylene (HDPE) is similar to crystalline hydrocarbon wax (petroleum product). It is interesting to note that polyethylene obtained by free radical polymerisation and at high pressure, is strongly branched which results into lower crystallinity and density. This form of polyethylene is known as low density polyethylene (LDPE). The copolymers of ethylene and other alkenes such as butene, hexene etc. hardly have linear structures and can not be crystallised. When the proportion of ethylene in these copolymers is major, they are called as LLDPE (linear low-density polyethylenes).

The replacement of H atom of ethylene by other hydrocarbon groups yields polymers e.g. polypropylene, polyisobutylene or polystyrene, in which the close packing of the chain becomes difficult due to the presence of the bulky groups and thus these polymers are amorphous with very low crystallinity, if any. The presence of a polar group along the back bone of the hydrocarbon chain increases inter-chain attraction and makes the polymers hard and brittle. These polar polymers have high melting points (e.g. polyvinyl chloride melts at 273°C). The introduction of amide linkages in the hydrocarbon chain (e.g. nylons) increases inter-molecular attraction due to the hydrogen bonding between the oxygen of –CO– group and hydrogen from –NH– group. The chains of these polymers crystallise with ease and higher degree of ordering is obtained.

Crystallinity of polymers has a direct bearing on their physical properties and application as fibers, plastics or elastomers. Generally, fibers possess high degree of crystallinity and tensile strength. Plastics are often amorphous and un-oriented. The crystallites present in a crystalline polymer such as nylon are usually un-oriented. However, when the fiber material is stretched at room temperature

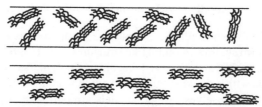

Fig. 4.2   Crystallites oriented upon cold drawing

(cold drawn), the molecules slip across one another and become oriented. Stretching of amorphous polymers may also induce some crystalline regions in them.

Elastomers are amorphous polymers with very low glass transition temperature. However some crystallinity can be induced in elastomers under the influence of stress. For example the crystallinity in a stretched rubber band can be observed from its opaqueness and warmth it produces when touched by lips. Similarly, the fiber forming polymers which are predominantly in amorphous form can be stretched along the axis of their orientation under stress. Then the fibrous chains are drawn in a desired direction of their orientation. Fig. 4.3 depicts the effects in an elastomer upon its stretching.

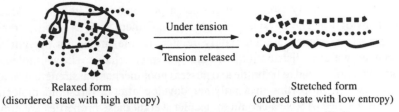

Relaxed form	Stretched form
(disordered state with high entropy)	(ordered state with low entropy)

Fig. 4.3   An elastomer on stretching develops crystallinity

The crystallinity in elastomers has pronounced effect on its mechanical properties. In Fig. 4.4, the general variation of mechanical properties in an elastomer is shown as function of crystallinity. The mechanical strength, for example, increases sharply after the development of some ordered arrangement and shows a continuous increase at higher crystallinity.

Isotactic polymers are more crystalline and thus have high melting point and density as compared to the corresponding atactic forms. Syndiotactic polymers have a tough cohesive structure because of the interlocking of the side chains which project above and below the carbon backbone.

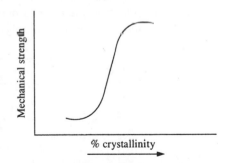

Fig. 4.4.   Mechanical properties of natural rubber with different degree of crystallinity

X-ray diffraction method is commonly used to study crystallinity in polymers. Concentric rings are observed in the x-ray diffraction powder photograph of crystalline polymers similar to low molecular weight solids. Since most polymers have some amorphous regions, these rings are not usually so sharp. For amorphous polymers, the x-ray diffraction pattern is quite diffused and an amorphous halo (characterised by dark shaded portion in the centre of the photograph) has been invariably observed. It is possible to obtain information about the conformation adopted by the polymer chains in the crystallites. Many polymers crystallise in helical conformations. The percentage of crystallinity of a partially crystalline polymer can be determined by x-ray crystallography, calorimetry, dilatometry and infrared spectroscopy.

Factors affecting crystallinity and thus physical properties of polymers
There are several factors which affect the crystallinity in polymers and thus their physical properties. These are the molecular length of the chain, the secondary valence forces, polymer configuration and conformation, and the nature of chain packing in polymer molecules. Some of these factors have already been considered earlier. These factors influence important physical properties such as impact and tensile strength, melting point, melt viscosity and even the solution properties of polymers.

Molecular weight

One of the most important properties of a polymer is its molecular weight. This characteristic property alone distinguishes a polymer from its low molecular weight analogues. The physical properties of polymers such as tensile strength and impact resistance are intimately related to the molecular weight and molecular weight distribution of the polymer chains. Low molecular weight substances tend to be brittle and possess poor mechanical strength. Usually, the mechanical properties in a polymer develop after a threshold molecular weight is reached. Above a certain molecular weight these properties are almost unaltered. However, the melt viscosity at first increases slowly and then rapidly after the polymer has attained a certain molecular weight. The value of threshold molecular weight varies from polymer to polymer. For example, polyolefins exhibit such behaviour at a DP value around 5000 whereas for polyamides the value is only 200 on account of hydrogen bonding between the different chains. The effect of molecular weight of polymer on its mechanical strength is shown in Fig. 4.5.

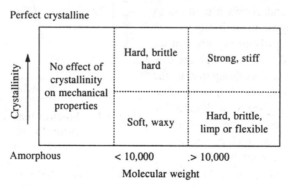

Fig. 4.5   Effect of mol. wt. and crystallinity on the mechanical strength of polymers

It may be pointed out that not all the physical properties of polymers are dependent on the molecular weight or for that matter on the magnitude of the intermolecular interactions. For example, the refractive index, colour, density, hardness and electrical properties are independent of the molecular weight of the polymer. Due to very high melt viscosity, extremely high molecular weight polymers become tough and intractable and can not be easily handled. Thus, such extremely high molecular weight polymers are not desired for end use purpose.

Secondary valence forces

The secondary valence forces e.g. van der Waals and dipole-dipole forces in polymers are similar to those present in small molecules. In polymers, however, many types of electrostatic forces may be present that act among different parts of the same chain. The strength of these forces increases with the increasing polarity and decreases sharply with increasing distance between such polar groups along the chain. The cumulative effect of thousands of the bonds spread along the polymer chain results in large attraction. The inter-molecular bonds formed in polymers are described in the introduction part of this book. The physical properties of polymers thus drastically change with inter-molecular forces. This has been illustrated using some examples. Polyethylene, due to the presence of only very weak van der Waals forces between the chain segments (as it is non-polar) is a soft waxy solid. Nylons and cellulose chains have extensive inter-molecular hydrogen bonding and are strong fibers. Dipole-dipole interactions in polyester and PVC make them tough plastics. Markedly different properties can be seen in polyethylene and nylon 11, which may be considered some what structurally similar as shown in the Fig. 4.6.

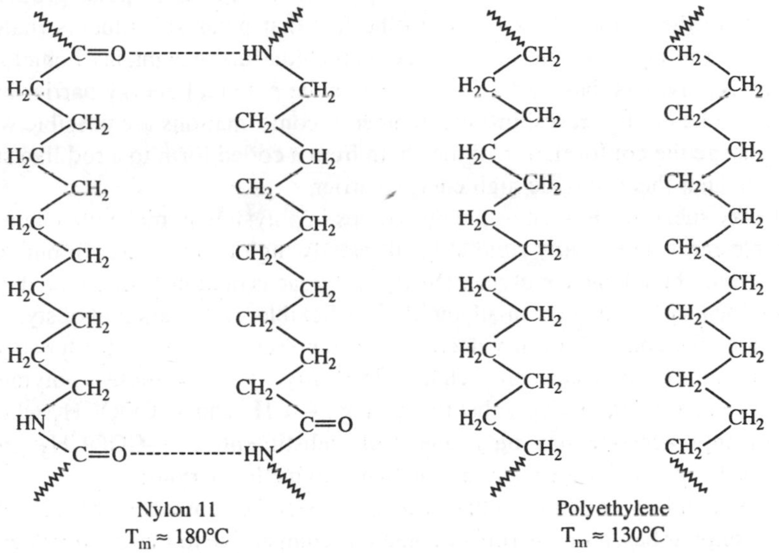

Nylon 11
$T_m \approx 180°C$

Polyethylene
$T_m \approx 130°C$

Fig. 4.6 Structural similarities in nylon 11 and polyethylene

Chain flexibility

A polymer chain is considered to be flexible if the chain segments can rotate with respect to each other with sufficient freedom Thus, polybutadiene and polyethylene have flexible chains whereas polystyrene and polymethyl methacrylate have rigid chains. The flexibility of a polymer chain depends on factors like potential energy barrier to rotation, molecular weight of the polymer, nature and size of the substituents and cross-link density etc. These are briefly described below.

The potential energy barrier is the minimum energy needed for chain molecule to undergo a conformational change via a special kind of motion i.e. internal motion. It depends mainly on the strength of intra-molecular and inter-molecular interactions. For example, isolated chains of polyacrylic acid and polymethacrylic acid should be rigid due to strong intra-molecular interaction of their COOH groups. In the condensed phase, the intra-molecular interaction of carboxylic groups lowers the potential energy barrier, the chains become flexible and assume a coiled globular structure. Conversely, chains that are flexible in the isolated state become more rigid in the condensed phase due to strong inter-molecular interaction.

The introduction of substituents containing polar groups in a polymer molecule intensifies the intra- and inter-molecular interactions. The most polar groups are the—CN,--NH$_2$,—OH groups. The distance between polar groups in a polymer chain has a profound effect on the flexibility of the chain. For example, in a copolymer of butadiene and acrylonitrile, containing 18 percentage of acrylonitrile on mole basis, the flexibility of the chain is close to that of polybutadiene. As the amount of acrylonitrile is increased in the copolymer, the flexibility decreases and the chain becomes more rigid. Polytetrafluoroethylene and polyvinylidene chloride have flexible chains even though a large number of polar groups are present in the chains. This is due to the fact that polar substituents (halogen atoms) are arranged symmetrically on a C atom. Molecular weight, in a homologous series of polymers, has a very little effect on the potential energy barrier but as the chain length increases, infinite number of conformations are possible which can change the conformation of the chain from a coiled form to a rod like shape even though there may be high energy barrier.

Bulky substituents such as phenyl groups in polystyrene make the chain less flexible and consequently rigid at insufficiently high temperature. In butadiene and styrene-butadiene copolymer (SBR), if styrene is randomly distributed in the chain and its percentage is small, the chain is flexible. As the amount of styrene is increased the chain flexibility decreases. The presence of two substituents on a carbon atom decreases the chain flexibility. For example, polymethyl methacrylate contains two substituents, e.g., —CH$_3$ and —COOCH$_3$ whereas polymethyl acrylate has only one such substituent, a —COOCH$_3$ group. Polymethyl methacrylate is more rigid than polymethyl acrylate.

The flexibility of natural rubber which has been vulcanised by the addition of 2–3% sulphur based on the rubber content is comparable to that of unvulcanised rubber. However, if rubber is highly vulcanised, a rigid network structure is formed and the mobility of the unit is decreased considerably. Consequently, the chains are said to have high-link density and become rigid.

Some common observations on the crystallinity can be listed as follows:

Linear polymers are more crystalline than their branched counter parts (HDPE is more crystalline than LDPE). *trans* 1, 4-Polyisoprene is more crystalline than *cis* form (natural rubber). Poly β-glucose (cellulose) is crystalline whereas poly α-glucose (starch) is amorphous. Stereoregular polymers are crystalline. Polarity in the chain develops crystallinity. Nylons and polyesters are crystalline.

The terms crystallinity and crystallizability are interchangeably used but they have quite different meaning. Crystallinity means the relative proportion of crystalline and amorphous regions in a polymer sample (degree of crystallinity) at a particular temperature and depends on conditions like rate of cooling, residence time, temperature of the molten polymer, heat dissipation at which the crystallisation take place. Crystallizability refers to how a polymeric substance can be obtained in the crystalline form from melt or solution. It refers to the maximum crystallinity that a polymer can achieve at a particular temperature and depends on the chemical nature of the chain, its geometrical structure, molecular weight and molecular weight distribution. Regarding the crystallizability of polymers, the following general comments can be made. Homopolymers are more easily crystallised than copolymers. Alternating copolymers are more easily crystallised than random copolymers. Polymers with bulky pendant groups are crystallised with great difficulty. For example, polyvinyl carbazole is amorphous.

Like low molecular weight substances, polymers crystallise into various morphological forms i.e. physical structures. These include single crystals to polycrystalline spherulite formations. The comparison of the size of crystallites, spherulites and single polymer crystals with those of simple molecules is shown in Table 4(a)

Table 4(a)   Structural units and their sizes

Morphological forms	minimum size Å	maximum size Å
Molecules	2–5	10^3–10^5
Crystallites	20–100	100–500
Single crystals	100	10^3–10^4
Spherulites	10^5–10^7	$> 10^7$

Depending upon the dimensions and symmetry of the simplest crystalline units and the manner of their mutual arrangement, the macroscopic physical structure in a crystallised state may consist of single crystals and when they are formed, the physical structure is built by the translation of identical unit cells. There are several types of single crystals, which vary in the mutual arrangement of units in the space and the symmetry of ultimate structures. For example, polyethylene may crystallise into orthorhombic structures. The chains of polyethylene are arranged in a planar zig-zag manner along four edges and in the centre of the orthorhombic unit cell (see Fig. 4.7).

When the linear chains with bulky side substituents are crystallised, the unit cells formed are characterised not by a planar zig-zag arrangement but by helical conformations with different loop lengths depending on the size and nature of side substituents. The various possible ordered helical structures in isotactic polymers are shown in Fig. 4.8.

Three basic single crystal structures namely lamellar, fibrillar and globular single crystals result when polymers are crystallised from dilute solutions. Also, these formations differ in dimensions and shape of unit cell and in regularity of arrangement of the chain. The formation of particular single crystal depends on

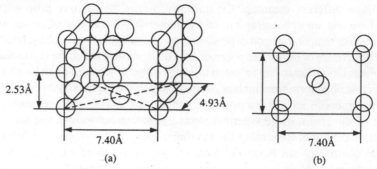

Fig. 4.7   (a) **Conformation of a polyethylene chain and (b) arrangement of chains in a crystalline cell**

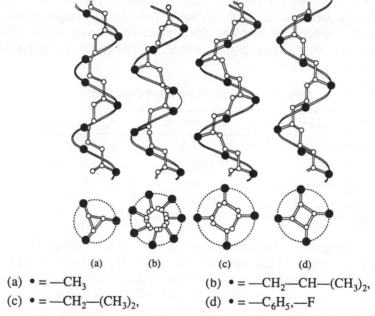

(a) • = —CH$_3$

(b) • = —CH$_2$—CH—(CH$_3$)$_2$,

(c) • = —CH$_2$—(CH$_3$)$_2$,

(d) • = —C$_6$H$_5$,—F

Fig. 4.8   The helical ordered structures in isotactic polymers

molecular characteristics of polymers as well as crystallisation conditions. The temperature of crystallisation, nature of solvent, concentration of solution and cooling rate etc. When the crystallisation is carried out in concentrated solutions or from polymeric melts and under real conditions (high viscosity, high temperatures and applied mechanical pressures) that are encountered during the processing of various polymeric articles such as films, casts and fibers, spherulite structures, unique only in polymers are formed. These structures can be seen by the naked eye or viewed as maltese-cross like formation under a polarised microscope.

Crystallisation of polymers

By crystallisation, an ordered arrangement is obtained from a disordered one which is usually present in a melt or solution. For example, when a molten polymer is cooled below its T_m, there is an inherent tendency for the randomly tangled molecules in the melt to align and form small ordered regions. Such regions are called nuclei, which are stable only at temperatures below T_m (and are disrupted above T_m by the thermal motion). Once the crystal nuclei are formed, they do show crystal growth by addition of further chains. Thus, crystallisation is an over all process consisting of nucleation and crystal growth. Nucleation may be homogeneous (i.e. where the nuclei do form randomly throughout the polymer melt) or heterogeneous (where foreign particles such as dust and impurities form part of nuclei besides the amorphous polymer molecules). Nucleation is temperature dependent. At high under-coolings, many nuclei are formed along with a large number of spherulite structures. Crystal growth can take place in one, two or three dimensions leading to variety of geometrical shapes for the crystals ranging form rods, discs and spheres, respectively. The growth takes place by the incorporation of polymer chains within the crystals, which are normally lamellar.

The change in size for a unidimensionally growing entity is generally linear with time at a given temperature of crystallisation. For example, a spherulite radius r will change with time t through the relation $r = \upsilon t$, where υ is the growth rate. This relation is valid only at initial stages of spherulite growth (i.e. until they are not very large and touch one another). The growth rate depends upon the molecular weight of the polymer (it decreases with increase in molecular weight) and temperature (often a peak is seen). These effects are shown in Fig. 4.9. The temperature dependence of υ showing a peak may be explained due to two competing effects. A decrease in temperature will increase the thermodynamic driving force for crystallisation. At the same time, decrease in temperature gives rise to increase in viscosity that offers obstacles in transporting the molecules to the point where crystal growth occurs.

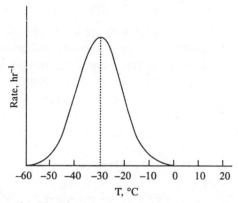

Fig. 4.9 **Effect of temperature on the crystal growth rate for natural rubber**

Kinetics of crystallisation

As discussed earlier, the crystallinity in polymers has tremendous effect on structure-performance relationship of polymeric materials. It is therefore necessary to study the kinetics of crystallisation and provide a molecular mechanism to the process. This can be done by considering some assumptions such as: (i) nucleation is homogeneous, (ii) at a given temperature the number of nuclei formed per unit time and per unit volume is a constant (say N) and (iii) there is no impingement of spherulite structures already formed. Now, if a molten polymer of mass W_0 is cooled below the crystallisation temperature, spherulites will nucleate and grow over a period of time. The total number of nuclei formed in time interval (say dt) will then be NW_0/ρ_t, where ρ_t is the density of the molten polymer.

After time t, suppose these nuclei grow into spherulites of radius, r, the mass of each spherulite will be $4/3\ \pi r^3 \rho_s$, where ρ_s is the density of spherulite.

Now since $r = vt$, so the mass of the spherulite material (dW_s) present at time t and grown in time interval dt will be

$$dW_s = 4/3\ \pi v^3 t^3 \rho_s N\ W_0/\rho_t dt \qquad (4.1)$$

where N is the number of nuclei, W_0 is the initial mass of the polymer melt, ρ_t is the density of the grown crystal at time t.

The total mass of the spherulite formed after time t from all nuclei N is,

$$W_s = \frac{4/3\pi v^3 t^3 \rho_s NW_0}{3\rho_t}\ dt \qquad (4.2)$$

or

$$\frac{W_s}{W_0} = \frac{\pi V v^3 t^4 \rho_s}{3\rho_t} \qquad (4.2a)$$

If W_t is the total mass of liquid left after time t, then $W_0 = W_s + W_t$, and

$$\frac{W_s}{W_t} = 1 - \frac{\pi V v^3 t^4 \rho_s}{3\rho_t} \qquad (4.3)$$

Equation 4.3 predicts the essential features of spherulite crystallisation and it further predicts that mass fraction of the crystals should depend initially on t^4. The above equation is derived on the assumption that the nuclei are grown independent of any impingement that may occur during the growth process. However, it is logical to expect that the growth of a nuclei may get impeded if impingement of growing nuclei is taken into consideration. In case of impingement of spherulites, the following equation can be obtained,

$$\frac{W_1}{W_0} = \exp(-kt^4) \qquad (4.4)$$

This equation is called *Avrami* equation. It can be seen that for small t, both the eqs. 4.3 and 4.4 have the same form. If the nucleation and crystal growth follow some other types, a general equation that can be written is

$$\frac{W_1}{W_0} = \exp(-kt^n) \qquad (4.5)$$

where n is known as *Avrami* coefficient, k is the constant which includes terms such as nucleation frequency, growth vector, shape factors and densities of crystalline and amorphous phases. The crystallisation can, however, be monitored by measuring change in specimen volume rather than the mass of grown spherulite. This can be done dilatometrically. Defining the volume terms as

$$V_0 = W_0/\rho_1 \tag{4.6}$$

$$V_\alpha = W_0/\rho_s \tag{4.7}$$

$$V_s = W_1/\rho_s + W_s/\rho_s \tag{4.8}$$

$$= W_0/\rho_s + (W_1(1/\rho_1 - 1/\rho_s))$$

Introducing V_0 and V_α from eqs. 4.6 and 4.7 into eq. 4.8, we get

$$V_s = V_\alpha + W_1((V_0/W_0) - (V_\alpha/W_0)) \tag{4.9}$$

or

$$\frac{W_1}{W_0} = \frac{V_t - V_\alpha}{V_0 - V_\alpha} = \exp(-zt^n) \tag{4.9a}$$

Thus, the value of *Avrami* coefficient n can be determined from the slope of log-log plot of volume terms and time.

A typical plot of $\log \dfrac{V_t - V_\alpha}{V_0 - V_\alpha}$ versus $\log t$ is shown in Fig. 4.10.

The *Avrami* equation provides a satisfactory account of polymer crystallisation rate but does not give any insight into the molecular mechanism involved in nucleation and growth. Several theories have been put forward to explain the phenomenon but none has so far been found to be perfect.

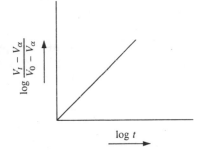

Fig. 4.10 Log volume change ratio versus log time plot

Supramolecular structure of polymers

Polymeric solids are unique. Some special solid state characteristics of these macromolecules are already discussed. The diverse physical and mechanical properties that are observed in polymers no doubt not only depend upon their chemical characteristics but also on their physical structure. Owing to the large size and presence of long chains besides smaller units, the spatial arrangement of macromolecules into separable element out of which the entire structure is built, assumes great significance. The way of chain packing, dimensions and shape of its constituent elements and the relative arrangement of elements constitute the supramolecular structure of polymers. There are three main methods for investigating the supramolecular structure. They are visual methods, interference and diffraction methods and measurements of properties, which are related directly or indirectly to the physical structure.

Visual methods

The visual methods are based upon examination of either transmitted or reflected electromagnetic radiation from the specimen. Thus, in order the above phenomenon to occur, the electromagnetic radiation must have wavelengths far smaller than the sizes of structural elements, for example of unit cells in a crystalline body. The application of these methods enables to discern the information on the shape, geometry, size and other specific features in the specimen under observation. Two common visual methods namely optical and electron microscopy are used in observing the morphological features of polymer solids.

(a) Optimal microscopy

This method is based on the formation of a magnified image of the specimen when polarised visible light is thrown on it and the incident light either gets transmitted or reflected. Since the wavelength of visible light range from 400 to 800 nm, the optimal microscopy methods can be easily applied even on the specimen having dimensions as large as several hundred microns. Another advantage with this method is that it requires to special sample preparation and thus it can be easily applied. Samples in the form of thin films or slices (from the bulk bodies) with thickness ranging from several hundred microns can conveniently be studied. The optimal microscopy technique utilising the reflected light passing through the films of the samples cast on a microscope glass slide, is used for analysing the morphological and structure related studies. The polarised light and the phase contrast microscopy methods are also employed in analysing the polymer molecular aggregates containing domains differing in refractive indices. Besides obtaining information on morphological features, the optical microscopy methods are widely used for monitoring: (i) changes in the order of molecular rearrangement during the crystallisation process, (ii) kinetics of crystallisation and (iii) effects on physical structure under the influence of various thermal and mechanical variations.

(b) Electron microscopy

In electron microscopy, electron beam emanating from cathode ray tube is made to fall on the specimen. Modern electron microscopes operate at accelerating voltage of 100–300 kV, and use electrons with wavelengths ranging from 0.03 to 0.06 Å. Thus it is possible to study the structural elements, whose dimensions range from several angstroms to few microns. Moreover, electron microscopy is a powerful technique because its resolution power is 3–20Å and magnification of image by 3×10^3 to 1×10^5 times possible. The images obtained from an electron microscope are not called photographs but as electro-micrographs. The image formed is not only due to optical phenomenon but also involves the contribution from the interaction of an electron beam with the sample. The variation in the scattered electron intensities from different regions of the sample develops contrast and with the amplification facility, images with high magnification are formed.

Unlike optical microscopy methods, electron microscopy requires stringent procedures for sample preparation. The thickness of transmitting layer is confined to 1000 Å because of strong interaction of electrons with the sample being examined.

In actual practice, the samples thinner than 100 Å are used. Thus, the method allows studying the structure even in very thin films or fractures of the polymer samples. Polymer single crystals can be easily examined under an electron microscope. As already mentioned, the contrast in electron micrographs depends on the electron scattering ability of the nuclei of various atoms present in the sample. Atoms of heavier metals scatter the electrons to a greater extent and hence metal shadowing of polymer specimens is often employed in electron microscopic methods.

Scanning electron microscopy (SEM) has been developed as a special technique in which a fine electron beam is scanned across the surface of an opaque and nonconducting polymer specimen coated with a conducting film of metals such as gold and copper. The sample is illuminated by parallel electron beams and the scattered electrons are measured by Gieger-Muller counters. The scattered electron intensities are then converted into cathode ray beam whose intensities vary with the initial intensity. This variation in the intensities develops a contrast and the image of the sample under observation is formed on the screen. The images formed have greater depth of field and excellent three-dimensional appearance. SEM method is used in the studies of (i) phase morphology of polymer blends, graft and block copolymers (ii) surface features or textures of natural and synthetic fibres and (iii) monitoring the surface or fracture failure in plastics, elastomers and composite materials under the applied mechanical stresses such as tensile, abrasion and dynamic loading etc.

Electron microscopy is also a powerful tool to investigate the unique characteristics of amorphous state in polymers. It is found that amorphous polymers are not perfectly disordered but are poorly crystalline and have some order of arrangement among its structural elements. A special technique called dark field electron microscopy has been developed to detect the ordered regions in otherwise amorphous polymers. This method is based on isolating the scattered Bragg reflexes from a specimen. In this method, the entire useful field becomes dark and the ordered regions from which the scattered electrons which produce Bragg reflexes are examined. The use of this technique has shown that like in crystalline polymers, amorphous state in polymers is characterised by the presence of some degree of order (in the form of domains) of 30–100 Å. These domains are thought to be formed by the chain folds.

Interference and Diffraction methods

These methods are based on interaction of the electromagnetic radiation with the ordered ensembles of structures as encountered in crystalline bodies of a material and following the scattering pattern by using suitable detectors. The wavelength of the radiation used is either of the same order or much larger than the characteristic size of the structural elements in the ordered bodies. The diffraction or interference pattern of x-rays from the ordered structures of a specimen provide information on wide variety of quantitative physical structure characteristics such as shape and dimensions of the entities, distance between scattering centres and the extent of order or repeat distance in a given structural arrangement. Small angle and large angle x-ray diffraction, electron diffraction and neutron diffraction are some powerful analytical techniques employed for understanding and elucidating the structure of crystalline, drawn or oriented polymers.

When monochromatic x-rays are passed onto a crystal, whose structure consists of layers in which its constituent atoms or molecules are arranged periodically with the spacing as that of the same magnitude as the wavelength of x-rays used, simple reflection and scattering of incident rays would occur. The interference coming from the reflections caused by the atoms/molecules in the same plane or from the neighbouring planes severely affects the ultimate intensity of reflected beam. Bragg, W.H. and Bragg, W.L. (father and son, both shared Nobel physics prize; the son at the age of 25) gave the condition for a constructive interference. The incident x-ray beam can be represented with all the waves in phase i.e. their electrical and magnetic vectors are same all along the path of their propagation. For some special angles, θ, the reflected x-rays coming out from particular planes separated with a periodic 'd' spacing are also in phase with each other. Under this condition the reflected waves will then add or reinforce each other, giving a net resultant wave with double of its intensity in comparison to incident beam. This phenomenon is known as constructive interference or diffraction. Fig. 4.11 shows a constructive interference pattern of x-rays at certain angles of reflections, θ from the planes stalked with spacing 'd'. Bragg and Bragg further established that the constructive interference pattern is observed whenever the phase difference of reflected x-ray beam (from successive planes) differ by an integral multiple of wave lengths. This is summarised in well known Braggs equation given as, $n\lambda = 2d \sin \theta$, where $n = 1, 2, 3, \ldots$

The diffracted x-rays are measured classically from the strong blackening of a photographic film used as detector. Modern x-ray diffractometers use photoelectric counters for measuring the intensities of diffracted x-rays. The type, size and geometry of crystal unit cells are discerned from the position and intensity of diffracted beams. There are three basic methods of x- raying that are widely used

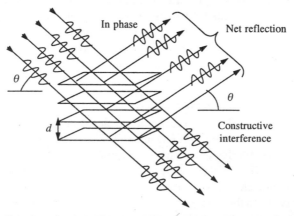

Fig. 4.11   Constructive interference (diffraction) pattern of scattered x-rays

for structural determination of ordered formations, (a) Laue method, (b) rotating crystal method and (c) Debye-Scherrer powder method. Laue method and rotating crystal methods are not employed for studying the solid polymers because both of these methods need single crystals of large dimensions for observing strong diffracted reflections for further analysis. Polymers hardly form single crystals and even if formed, their size is too small for x-rays diffraction experiments using the above two methods. Thus Debye-Scherrer powder method is of great importance and widely used for investigation of oriented polycrystalline samples. Most of the polymer crystals are polycrystalline and lattice layers are ordered in each crystallite but the crystallites themselves are not perfectly ordered. Debye-Scherrer method originally used for crystal powders suits well for polymer crystalline materials also. When monochromatic x-ray beams are incident on these ordered layers arranged along one- or two- or even three-dimensional directions, they find sufficient lattice layers for possible reflections that satisfy the Braggs condition. The multitude of orientation of layer lattices for example in a stretched polymeric fiber, generates a coaxial cones of diffracted x-rays, with a common origin at the centre of the specimen. Each spot generated by the vertex angle of each of the cones is equal to 2θ. When the photographic film is unfolded or straightened and developed, one observes the vertical cross section of the system of cone, which translates into a number of concentric rings or ellipses. Powdered pattern of diffracted x-ray from semi-crystalline isotactic polystyrene and atactic or amorphous polystyrene are shown in Fig. 4.12.

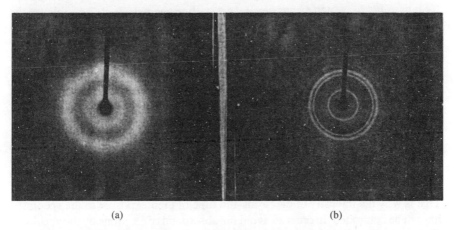

(a)   (b)

Fig. 4.12 Powder diagram of x-ray diffraction pattern for (a) amorphous or atactic polystyrene and (b) semi-crystalline isotactic polystyrene

The x-ray diffraction analysis using powder method reveals that three basic structural orders exist in polymer solids. These are small period, helical pitch and large period. These three types of periodicities have their own characteristic Bragg angle. For small period and helical pitch whose size ranges from 1–20Å, the θ values vary from several degrees to some tens of degrees. Thus, the diffraction observed from these regions is called large angle x-ray scattering. The large period, which is characterised by the presence of both ordered and disordered regions along side each other and as inseparable entities, have dimensions as large as hundred of angstroms. The diffracted angles from these large structures are low i.e. from several minutes to 1–2° and the method of diffraction is known as small angle x-ray scattering (SAXS).

The observed x-ray diffraction patterns are quantitatively processed by trial and error method. In this approach, an hypothetical model is pre-assumed of a particular arrangement in the specimen under observation and the expected diffraction patterns are calculated. Then the observed and expected calculations are compared. The comparisons are continued till a close or perfect match between expected and observed patterns are obtained. With the advent of modern computers this approach has been highly successful. The x-ray diffraction analysis has made it possible to decipher the helical conformation in deoxyribonucleic acid (DNA) and proteins, which is one of the important discoveries. The relative proportion of ordered and disordered regions that is typically found in polymer crystals is calculated from the x-rays diffraction pattern.

The orientation (full or partial) of chains especially in fibers and its changes under external effects viz. temperature, magnitude and direction of stress etc. can be easily ascertained from the characteristic changes observed in diffraction pattern. The formation of cracks or fractures in solid polymer materials are also detected from the x-ray diffraction analysis. The changes in crystallisation conditions such as crystallisation temperature, heating and cooling rates, nature of solvent and the presence of nucleating agent etc. can profoundly affect the macro conformations and packing of chains in crystals. These changes lead to 'polymorphism', which means several crystalline geometries are possible for the same polymer, when it is crystallised under different conditions. The existence of such a phenomenon can only be monitored by x-ray diffraction pattern.

Electron diffraction

The principle employed in electron diffraction is essentially same as that used in x-ray diffraction. The major advantage being that the wavelength of electrons is smaller than x-rays and hence structural information on specimens of smaller dimensions or from a part of larger domain can be obtained. However, care must be taken of the fact that electrons, when bombarded cause structural modifications, even induce fragmentation and scissioning of polymer chains producing macro free radicals, which then form macroscopic cross-linking network structures. That is why the intensity of electrons and time of their exposure to the specimen have to be controlled. In order to avoid the absorption of electrons by the specimen samples, diffraction experiments are normally conducted under high vacuum. This requirement of experimental condition, however, is very beneficial because samples in the form of fine films produced on a substrate can be scanned to elicit structural information.

Neutron diffraction

The theory of neutron diffraction essentially follows the same principle as scattering of light by macromolecular chains in dilute solutions. Neutrons are uncharged unlike x-rays and electrons. The slow neutrons produced in a nuclear reactor get scattered by the nuclei of atoms in the substance. The scattered intensities of monochromatic neutrons depend mainly on the mass of nuclei. Good contrasts are developed due to the difference in scattered neutron beam intensity emerged from nuclei of different mass. The wavelength of neutrons is dependent on their energy. Cold neutrons have wavelengths in the range of 1.6–1.8 Å. In the actual experimental practice, the systems investigated include the polymer chains containing certain amount of deuterated molecules in the medium of hydrogen containing molecules or vice versa. Since the source of neutrons is nuclear reactors, this facility is not widely accessible to users. The neutron diffraction analysis has been widely used to investigate the phase states at microscopic and macroscopic level in imperfect solid polymers especially in glassy state.

Besides the above discussed direct methods, an independent group of methods can also be used for studying the physical structure in polymer solids. These methods are called as integral methods and are based on the measurement of dependence of any physical property, which is sensitive to the structural order. These methods are based upon; (i) measurements of heat capacity and temperature induced transition points, (ii) measurements of deformational and relaxational properties, (iii) measurements of electrical permeability, dielectric loss and electric conductance, (iv) measurement of density and density related properties and their change with time and (v) special spectroscopy methods particularly infra-red dichroism. All the above mentioned methods are known as indirect methods; they measure changes in specific properties with variation in crystallisation conditions.

4.2 Thermal Behaviour

Polymeric materials differ from low molecular weight substances when they are subjected to heat. While low molecular weight substances on heating show sharp melting and boiling points, the situation is different when a polymer is subjected to heat. An amorphous polymer, for example, on heating becomes flexible above a certain temperature (usually a small range) and then melts on further heating (again over a small range of temperature). Furthermore, polymers decompose before the boiling is noticed and therefore, do not exist as vapours. Melting temperatures of most low molecular weight substances are very easy to detect since solid, crystalline materials, fuse together to become melts with low viscosities at these temperatures. The distinguishing features in two classes of substances when subjected to heat are shown in Fig. 4.13.

Amorphous polymers are known to exhibit two distinctly different mechanical properties. Some polymers like polystyrene, polyvinyl chloride, polymethyl methacrylate are hard and rigid glassy materials at room temperature whereas some others like polybutadiene, polyisoprene and polyisobutylene are soft, flexible and rubber-like. However, glassy plastic materials can exhibit rubber-like properties on heating to a certain degree and conversely, rubbery materials become glassy on cooling. Thus, all amorphous polymers undergo a change from glassy state to rubbery state and *vice versa* at a certain temperature known as glass transition temperature, T_g. Each polymer has a characteristic T_g value. By definition, glassy plastics will have T_g values above room temperature and rubber like materials will have a T_g value below room temperature.

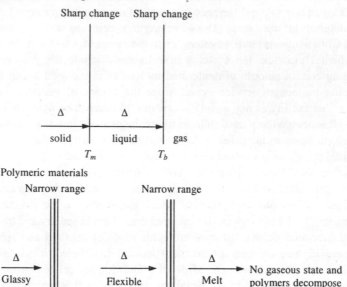

Fig. 4.13   **Distinct features of low and high molecular weight substances on heating**

The thermal transitions in a polymeric material may be described as follows. An amorphous polymer at lowest temperature behaves as a brittle glassy solid. Since it is not crystalline, it has no long range order of crystalline solids. Glassy state of a polymer can be considered as super cooled liquid state. Above a certain temperature called T_g, the polymer attains flexibility and rubber-like nature; the segmental motion sets in the constituent units, while the entire chain as a whole is at rest. In other words, it can be said that the internal Brownian motion begins but without external Brownian motion. On further heating, the polymer melts at its T_m when the motion in the whole chain begins (i.e. the external Brownian motion sets in). Fig. 4.14 depicts the effect of temperature on phase changes in amorphous and crystalline polymers.

The glass transition temperature of a polymer depends upon several factors. This can be illustrated taking some examples. Increasing inter-chain interaction as well as chain stiffness and cross-linking raises T_g value. Polyacrylonitrile because of very strong secondary bonding forces has T_g value even higher than its decomposition temperature. Higher T_g for poly α-methylstyrene than polystyrene is due to extra steric hindrance to rotation. Very high T_g values for ladder polymers e.g. polyimides are due to chain stiffness.

T_g of a polymer can be lowered with the help of plasticizer, a chemical substance deliberately added to the polymer to reduce interchain forces thus facilitating the movement of chain segments. A hard and tough polymeric material polyvinyl chloride becomes flexible at room temperature (T_g decreases) in the

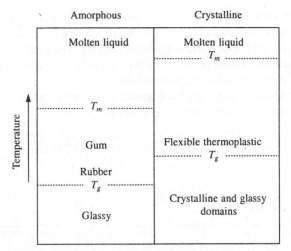

Fig. 4.14   Temperature effect on the amorphous and crystalline polymers

presence of the commonly used plasticizer dioctyl phthalate (DOP). The flexible PVC has several end uses e.g., in making rain coats etc.

Several methods can be used to determine glass transition temperature of polymers. Since T_g involves thermal transitions, any physical property when measured as a function of temperature would yield a break point at T_g. The specific volume method (dilatometery) and thermal analytical methods (DTA and DSC) are most commonly used. The dilatometric method requires the measurements of specific volume at different temperatures. A dilatometer is an all glass apparatus as shown in Fig. 4.15. The bulb of the apparatus contains the polymer sample along with a confined liquid. With the expansion or contraction of the polymer with temperature, the liquid is displaced and the quantity displaced can be measured in a graduated capillary tube. The specific volume so obtained is thus plotted against temperature. The inflection point in the specific volume–temperature curves can be used to determine T_g for most amorphous polymers. A typical plot of specific volume versus temperature is also shown in the same figure. When the polymer sample is perfectly crystalline i.e. degree of crystallinity

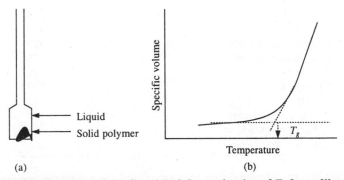

Fig. 4.15   (a) Dilatometer and (b) Graphical determination of T_g from dilatometric measurements

is hundred percent (which however is very rare in polymer solids), the plot shows a reflected Z type relation with sharp point of transition. The specific volume abruptly increases indicating direct melting. The sharpness of transition points is decreased with the presence of disordered or amorphous regions. In this case melting is preceded by glass transition.

Fig. 4.16 shows typical plots of specific volume versus temperature for amorphous, partially crystalline and crystalline polymers.

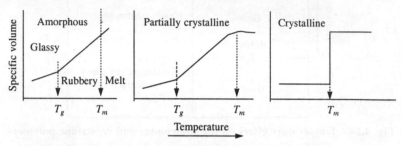

Fig. 4.16 Dilatometric data for amorphous, partially crystalline and crystalline polymers

It can be seen that, for crystalline polymers the glass transition is not detectable. Though the specific volume versus temperature plots are simplest means of determining T_g values of polymers, the data from more sophisticated methods such as DSC are needed for the final assignment of glass transition temperatures. Table 4(b) lists glass transition temperature, T_g values for some common polymers.

Table 4(b) T_g values for some selected polymers

Polymer	T_g, K
Polystyrene	373
Polyvinyl alcohol	358
Polyvinyl acetate	301
Polymethyl methacrylate (isotactic)	378
Polyvinyl chloride	354
Polypropylene (isotactic)	373
Polypropylene (atactic)	253
Cellulose triacetate	430
Cellulose acetate butyrate	320
Low density polyethylene	148
Polyacrylonitrile	378
Nylon 6, 6	330
Terylene	342
Polydimethylsiloxane	150
1, 4-*cis* Polybutadiene	165
1, 4-*trans* Polybutadiene	255

The following are some important points regarding the glass transition temperature:
(i) T_g is an inherent property of amorphous polymers originating at the onset of

segmental motion in the chain, (ii) T_g and the melting temperatures are related as follows: $T_g = 1/2\ T_m$ for regular polymers and $= 2/3\ T_m$ for amorphous polymers, (iii) Plastisizers decrease T_g (Plastisizers are low molecular weight, non-volatile substances mostly consisting of organic liquids based on alkyl di-esters. They decrease T_g because their molecules penetrate into the polymer matrix and establish polar attractive forces between them and polymer and thus decrease the inter chain attraction. This affects segmental motion and thus lowers T_g), (iv) Small amount of a monomer is often added to decrease the T_g of a polymer during its formation. This is called *internal plasticization*, (v) T_g of copolymers usually lies between the T_g values of the individual homopolymers, (vi) For block copolymers, two T_g values one corresponding to each block are seen.

From the view point of the physical structural changes in a polymer with an increase in temperature, the segmental motion in a polymer chain though plays a key role, monitoring of the changes in energy based thermodynamic parameters as function of temperature gives characteristic transition points. Below glass transition temperature, the polymer segments are rigid and do not have enough energy to undergo a rotational rearrangement and hence the material is brittle and glassy. As the sample is heated further, a small increase in volume and energy is noted until at T_g and above T_g the chains become more flexible and mobile. The nature of the material changes from brittle glassy to flexible plastic and even rubbery. Further heating provides conditions highly conducive for crystallisation because the segments and units of the polymer chain have sufficient energy and undergo rearrangement or ordering. At temperatures higher than this, the polymer melts and gets into molten state. These characteristic changes are schematically represented in Fig. 4.17. It can be seen that transition from glassy to liquid state is highly dependent on the rate of heating and cooling.

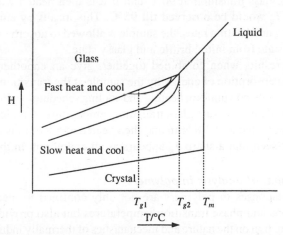

Fig. 4.17 Schematic representation of energy changes for a polymeric material with increase in temperature

The change in the heat flow as a function of temperature during the above transitions is shown in Fig. 4.18 for different polymers. The curves shown are the results based on DSC measurements. The gradual decrease followed by

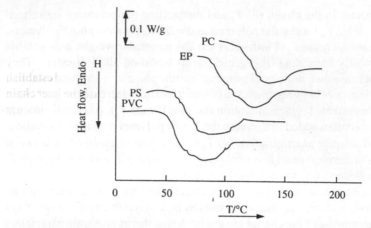

Fig. 4.18 DSC tracings for various polymers; (PC = polycarbonate, EP = epoxy-resin, PS = polystyrene and PVC = polyvinyl chloride)

sudden endothermic changes in the heat flow indicates that polymer has undergone glass transition. The value of temperature corresponding to the minimum in heat flow is taken as glass transition temperature. The value of T_g, however, depends greatly on the heating and cooling rates used in the DSC runs. If the sample is heated at very low rate of 0.1 K/min, the T_g observed is low, say 90°C and the same value is reproduced again if the sample is cooled back at the same rate. Nevertheless, if the sample is heated and cooled at much faster rate of 20 K/min, the T_g is noted at 95°C or even higher. Now imagine what happens when the heating and cooling rates are not the same. The sample cooled at a rate of 0.1 K/min undergoes glass transition at 90°C and if it is then heated with the rate of 20 K/min, no T_g would be observed till 95°C. This means by employing non-uniform cooling and heating rates, the sample is allowed to absorb more energy to reach rubbery state from initial brittle and glassy state.

The above results when combined together, give an endotherm, a distinct peak indicating absorption of energy by the sample at the transition temperature. Thus, DSC runs with non uniform heating and cooling schedules yield thermograms with well defined and accurate glass transition temperature. A typical DSC curve for polystyrene cooled at 0.2 K/min and then heated at 5 K/min is shown in Fig. 4.19. It can be seen that a sharp T_g appears as an endotherm in the figure.

Thermal methods of analysis in polymers
The thermal responses in polymers are not only confined to measure various phase transitions and phase transition temperatures but also provide wide range of useful information on the nature and mechanistics of thermally induced processes such as dehydration, degradation, curing and decomposition etc. Four basic thermal methods namely thermogravimetric analysis (TG or TGA), derivative thermal gravimetric analysis (DTG), differential thermal analysis (DTA) and differential scanning calorimetry (DSC) are widely used in analysing the temperature induced responses in polymeric materials. A brief mention of the

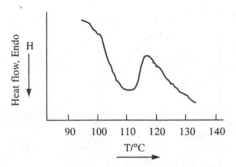

Fig. 4.19   A DSC trace of polystyrene

basics of these methods and some applications are discussed here. The reader should refer specialised literature for further details.

Thermogravimetric analysis (TGA)

This technique is based on measuring the mass (to be precise, mass loss) of a sample specimen as a function of temperature. The sample can be kept in a defined environment and the changes in the temperature are tuned to pre-programmed rate. The equipment for TGA measurement consists of four major parts (i) the electrobalance and its controller, (ii) the furnace and temperature sensor, (iii) the programmer or a computer and (iv) data acquisition device or recorder or plotter. TGA measurements are carried out mostly on solids. Samples should be ideally small, powdered and need to be evenly spread in the crucible. The crucible is usually a platinum pan. The initial mass of the sample taken ranges from 7–8 to 10–11 mg. The mass of the sample is to be measured at utmost accuracy of ± 0.001 mg with help of micro-balance. The environment of the furnace can be changed as desired for example air, nitrogen or inert atmosphere of Ar etc. by using the gas inlet and outlet chutes. Both dynamic and static modes can be employed. The results from TG measurements are represented in the form of TG curves (or traces) in which the variation of the apparent mass of the sample is plotted as a function of temperature. The mass is generally expressed in terms of mass loss ($W_0 - W_t$), where W_0 is the initial mass and W_t is the mass at a given temperature. The typical plots are usually of one, or two or three or even multi-step upturned S type curves.

In order to ascertain the steps in TGA traces, the derivative thermogravimetric (DTG) curves are frequently constructed. The DTG curve is represented by a plot of rate of mass change per pre-selected temperature interval, dm/dt as a function of temperature. The typical general TGA and DTG curves are shown in Fig. 4.20. It can be seen that DTG curve has well-defined peaks superimposing on the rapid fall in the mass loss as observed in TGA curve.

The TGA/DTG curves provide information on the nature of any process that may take place when a polymer sample is subjected to heating. The analysis of the curves helps in understanding, for example, degradation process in polymers. A proper treatment of data would provide the kinetics and mechanism of the process. Also, polymers with flexible chains need less energy and hence show low thermal resistance and melt and decompose at lower temperatures. While

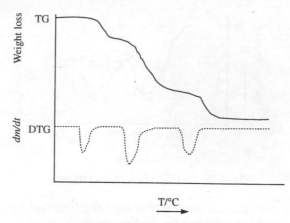

Fig. 4.20   Typical multi-step TGA and DTG curves

polymers with rigid chains absorb and withstand more energy and undergo phase transitions or decompositions at higher temperatures. Thus, thermal stability of a series of polymers of same chemical constitution with different molecular characteristics or polymers of different chemical constitution with identical molecular characteristics can be compared from this analysis.

When heated, in an inert atmosphere, polymeric materials (depending upon their mode of polymerisation and heat of polymerisation) depolymerise or carbonise. Polymers such as polymethyl methacrylate and polystyrene (whose monomers are highly reactive) have low enthalpy of polymerisation, may sequentially depolymerise back to the monomers. The linear chain hydrocarbon polymers such as polyethylene undergo the chain scission, which may occur randomly. This random scission of chain fragments produces unsaturated hydrocarbons of varying carbon chain length such as butenes, hexenes and dienes. The substitution of alkyl groups in the main chain (for example polypropylene) degrades the chain at lower temperature and while introduction of polar substituents like fluorine as in polytetrafluoroethylene, enhances the thermal stability and shifts the decomposition to higher temperature. The thermal decomposition scheme for polyvinyl chloride (PVC) is shown in Fig. 4.21.

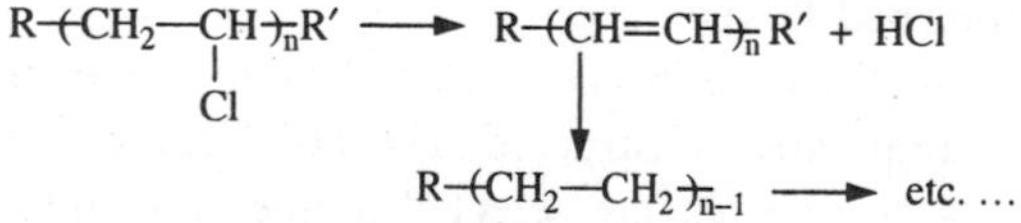

$$R\text{---}(CH_2\text{---}CH)_n\text{---}R' \longrightarrow R\text{---}(CH=CH)_n\text{---}R' + HCl$$
$$\qquad\qquad | \qquad\qquad\qquad\qquad\downarrow$$
$$\qquad\qquad Cl$$
$$\qquad\qquad R\text{---}(CH_2\text{---}CH_2)_{n-1} \longrightarrow etc. \ldots$$

Fig. 4.21   Scheme for the degradation of PVC

PVC and PAN undergo decomposition first by elimination of small molecules, formation of unsaturated links in fragments and build cross-linking among the scissioned fragments before finally degrading by complex reaction to a char. The polymers containing hydrophilic polar groups such as polyamides may absorb water and when heated loose water below 100°C before they decompose. The complex polymers such as cellulose, polyester resins and phenol-formaldehyde resins decompose through several complex reactions and finally to coke or char.

The intermediate or residual polymer fragments are highly active and hence form peroxides or other oxidised products when heated in air. The decomposition profiles of several polymers are shown in Fig. 4.22.

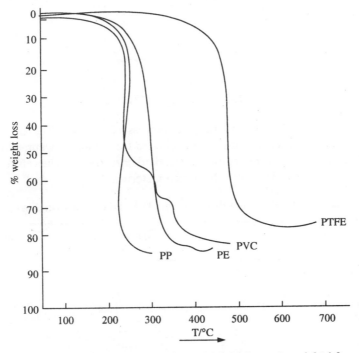

Fig. 4.22   **TGA curves for polymer samples with initial sample weight 1.0 mg, heating rate of 50 K/min (under static air conditions)**

Differential Thermal Analysis (DTA)

Whenever one needs to know the purity and stability of a material against a known reference substance, TGA measurement would not suffice. Suppose one wishes to know the melting point of a sample sharply against reference whose properties are known, it can not be done by TGA, because melting involves no mass loss. Therefore, another thermal technique, which can record the temperature difference between a sample and reference material simultaneously, is needed. This technique is known as DTA.

Based on similar lines, another important thermal method of analysis called differential scanning calorimetry (DSC) has been devised. The basic difference between DTA and DSC is that the later technique measures the difference in heat flow to a given sample and to a reference. Thus, heat is applied to the sample and to the reference material and then the temperature sensor output signal in terms of difference in heat flow of both is recorded as a function of temperature. A typical general DTA or DSC curve is shown in Fig. 4.23.

The peaks going up the base line are called as endotherms (indicating the absorption of heat by the sample) and down the base line are denoted as exotherms (heat is liberated). The value on x-axis corresponding to the peak maximum

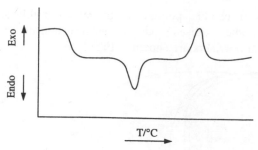

Fig. 4.23   Typical DTA/DSC trace

represents the temperature at which the given process has occurred. The temperature scale in both the instruments has to be calibrated for obtaining accurate results. Various chemical substances are used as standard or reference materials for DTA and DSC instruments. Table 4(c) lists such materials.

Table 4(c)   Calibration substances for DTA and DSC instruments

Substance		Temperature °C	Enthalpy J/g
Cyclohexane	(t)	−83	
	(m)	7	
1,2-Dichloroethane	(m)	−32	
Phenylether	(m)	30	120.41
Biphenyl	(m)	69.3	
Polystyrene	(T_g)	105	
Potassium nitrate	(t)	128	28.71
Indium	(m)	156.6	56.06
Tin	(m)	231.9	
Potassium perchlorate	(t)	300	111.18
Zinc	(m)	419.4	
Quartz	(t)	573	
Potassium chromate	(t)	665	
Barium carbonate	(t)	810	

(t) = crystal transition, (m) = melting, T_g = glass transition temperature.

There is a wide range of DTA and DSC application in the study of polymeric materials. These are broadly classified into the types, (i) physical changes and thermal transitions and (ii) study of chemical reactions such as dehydration, degradation, and curing etc. The utility of DSC method for estimating T_g of polymers is already discussed. Other important applications are mentioned here.

(i) Determination of percentage crystallinity
The integrated areas of the peaks observed in DSC/DTA measurements are used to calculate the instrument sensitivity constant, K (J/cm^2) by using the relation;
$K = \Delta H_s M_s / A_s$ where ΔH_s is the enthalpy of fusion of reference material (J/g), m_s

= mass of the reference sample (g) and A_s is the peak area (cm^2). Thus, knowing the value of instrument constant and a reference material the percentage of crystallinity can be calculated as

$$\% \text{ crystallinity} = \frac{\text{Area of the sample melting peak} \times (\% \text{ crystallinity of standard})}{\text{Area of standard melting peak}}$$

As discussed, polymeric solids do not crystallise to hundred percent crystallinity. Thus the reference material in such measurement is a polymer with known and maximum crystallinity and has chemical structure identical to the sample.

(ii) The Heat Capacity Measurements

Heat capacity is one of the most important parameters needed for adjudging the polymers especially as insulators or building materials etc. The ordinate in a DSC represents the differential rate of supply of heat energy (or heat flow). By knowing the y-axis deflection from the base line to peak point, Δy, the heating rate (dT/dt), the heat capacity of a polymer sample can be calculated as; $C_{p\,poly} = K \cdot \Delta y / (dT/dt)$, where K is the instrument sensitivity constant.

(iii) Compositional analysis of polymer blends

If two polymers having characteristic melting points and crystallinity are blended DSC and DTA techniques can be used to detect their individual presence and their amounts in the blend.

(iv) Monitoring of polymer degradation and recycling

The thermal degradation in polymers always occurs in different steps because degradation is not a simple decomposition process. Elimination of certain small molecules produced because of quick alignment of some atoms present in the repeat units is quite possible before the polymer chain gets scissioned into smaller fragments. Similarly, the small fragments show some chemical reactivity before they finally melt and decompose to char. DTA curves for the oxidative decomposition of polyvinyl chloride (PVC) and polypropylene are shown in Figs. 4.24 and 4.25, respectively.

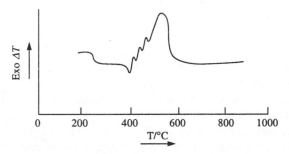

Fig. 4.24 DTA curve for polyvinyl chloride

An interesting observation can be made by comparing the two DTA curves shown in Figs. 4.24 and 4.25. PVC powder shows a small deflection at around

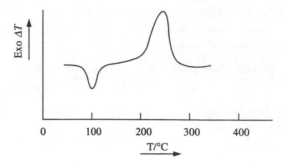

Fig. 4.25   DTA curve for polypropylene

80°C indicating its glass transition temperature and further heating shows small endotherm near 300°C followed by a large exotherm, which peaks at around 550°C. In contrast, the decomposition of polypropylene (PP) is characterised by a smooth peak. The initial single endothermic peak shows the melting followed by a large oxidation peak. The presence of several peak in DTA curves indicates that the degradation of PVC takes place in multi steps with loss of HCl and other volatile substances followed by oxidation products.

Besides the above described applications, DTA and DSC methods either individually or sometimes combined simultaneously with FTIR (Fourier transform infrared) and MS (mass spectrometry), are also used for (a) monitoring the curing of epoxy resins, (b) oxidative stability of polymers, (c) polymer blend analysis and (d) curing and thermal degradation of phenol-formaldehyde resins.

4.3   Dilute Solution

Polymer dissolution
The dissolution of polymeric substances in solvents is also different from that of low molecular weight substances. The distinguishing features of the solutions for low molecular and high molecular weight substances are shown below in Table 4(d).

Swelling of polymers
Swelling of polymers is a slow process involving the penetration of solvent molecules into the polymer matrix there by increasing its mass and volume. Swelling may be limited or unlimited. The diffusion of solvent molecules into polymer solute structures depends upon polymer – solvent interaction, which leads to the unwinding or expanding of the polymer surface and allows the penetration of the solvent molecules. Thus the polymer chain containing diffused solvent molecules is known as swollen chain. The swelling may reach a thermodynamic equilibrium. For solutes which form molecular solutions in solvents the swelling is unlimited and continues infinitely until the chains are completely separated from each other and finally acquire mobility and diffuse back into the bulk of solvent state (forming molecular solutions). For limited swelling i.e. when thermodynamic equilibrium exists between polymer – solvent systems, the polymer does not dissolve to form molecular solution.

Table 4(d)   Distinct features in dissolution behaviour of low and high molecular weight substances

Systems	Features
Solvent + low molecular weight solutes;	Immediate dissolution
	Reaches a saturation limit above which no further dissolution occurs
	Viscosity of solution is similar to that of solvent even at fairly high concentration of solute
Solvent + polymeric solute	Polymer swells before dissolution begins
	No saturation limit
	High viscosities are built up with increase in polymer concentration and may prevent further dissolution after certain limit depending upon the total molecular weight of polymer molecule
	Very dilute solutions may also be markedly viscous

Swelling can be treated thermodynamically as the phenomenon involving two processes namely mixing and expansion. The diffusion of solvent into the polymer matrix is a type of mixing phenomenon where as the expansion due to swelling is similar to an elastic deformation. For systems showing limited swelling, a parameter known as degree or amount of swelling can be defined. It is expressed as the amount of liquid sorbed by unit mass or volume of the polymer and is determined both gravimetrically and volumetrically. For example, a known mass of a polymer is kept in a solvent till equilibrium is established. The mass of the swollen polymer gel is measured after separating it from the bulk solvent. Then the amount of the swelling can be expressed as $(m - m_0)/m_0$, where m_0 is the mass of the solid polymer and m is that of swollen polymer.

Swelling and dissolution of polymers in suitable solvents constitute very important phenomenon in solution chemistry and applications of polymers. These two successive process are influenced by several factors viz. chemical nature of polymer and solvent, molecular weight of polymer, chain flexibility of polymer, packing density in solid polymer, phase state and temperature etc.

The nature of polymer/solvent system is often described in terms of polymer-polymer, polymer-solvent and solvent-solvent interactions. The molecular weight effect on swelling degree for given polymer-solvent systems is an important factor in fractionation of polydisperse polymers into fine fractions. High is the molecular weight, finite the swelling would be while the low molecular weight polymers swell infinitely. The flexibility i.e. the ability of the polymer chains to undergo conformational changes due to special kind of motion called internal rotation, has remarkable influence on the degree of swelling in polymers. Non-polar polymers which as a rule are more flexible usually form solutions in non-polar solvents. Rigid or less flexible chains offer problem in swelling and dissolution because lot of energy is required to unwind the polymer-polymer segmental contacts. This makes the dissolution of rigid polymer chains thermodynamically less favourable. Amorphous polymers containing bulky polar substituents readily swell in polar solvents but do not form solutions at room temperature. Similarly crystalline polymers with close packing between the chain segments do not easily dissolve and need high temperatures so that the crystalline order is broken to facilitate the individual segment-solvent contacts. Thus, crystalline polymers usually form solutions above their melting temperature. Similarly, more the cross-linking among the chains, poorer would be their solubility and highly cross-linked polymers do not show swelling even above melting temperatures.

Dimensions of polymer molecules in solution

The shape of low molecular weight solute molecules in solution can be considered as spherical and for this reason, their molecular size is expressed in terms of collision diameter. However, it is not always so in case of polymers, except for polymers with flexible chains which remain in solution as coils. Also, polymers even with most rigid chains do not remain in fully extended form. So one uses the terms, full contour length and extended length in describing the polymer chain in solutions. A full contour length can be treated as the product of length of the monomer (b) and the number of monomers (N) where as the extended length can be treated as $Nb \sin \theta/2$, where θ is the bond angle between the chain links.

A polymer molecule may be considered to exist as a thin long chain as shown (see Fig. 4.26).

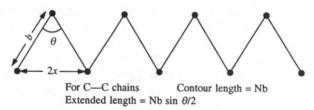

For C—C chains   Contour length = Nb
Extended length = Nb sin θ/2

Fig. 4.26   The polymer molecule showing extended length

However, in reality a polymer molecule exists as a random coil with large number of conformations that arise as a result of segmental motion and internal rotation among the units. Chain segments constituting a polymer molecule (one segment usually consists of a few monomer units) are capable of rotating around fixed angles at links between successive segments. The segmental conformations are governed by random flight probability. The polymer molecule thus assumes innumerable conformations. The probability is that two chain ends would be separated by a distance r_0 (which is extremely small as compared to the size of the polymer molecule). The parameter r_0 (which is often expressed as root mean square end-to-end distance $(\bar{r}_0^2)^{1/2}$, thus gives an idea of the size of the polymer coil.

In branched polymers, which contain several chain ends, the root mean square end-to-end distance does not provide correct information on the size of the polymer coil. In such cases, a new term called *radius of gyration* (s_0) is considered which is also expressed as root mean square value $(\bar{s}_0^2)^{1/2}$ and is defined as the average distance between molecular coil's centre of gravity and the chain ends.

The end-to-end distance for linear polymers and radius of gyration for branched polymers are shown in the Fig. 4.27.

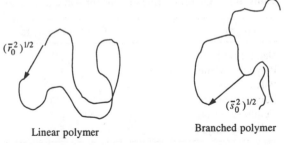

Linear polymer   Branched polymer

Fig. 4.27   The dimensions of polymer coils in solution

For linear polymers, the size of polymer coil can be expressed by any of the above two quantites. These are related to each other as $(\bar{r}_0^2)^{1/2} = 6(\bar{s}_0^2)^{1/2}$

When a polymer is dissolved in a solvent, different types of interactions exist. For example, one is the cohesive force between chain segments of the polymer molecule, which tightens the polymer coil. Also, strong interaction between chain segment and solvent would operate and loosen the polymer coil. Basically, these two opposing forces would then determine the actual size of the polymer coil. The polymer molecules are thus more and more expanded as the polymer-solvent interaction strengthens (quality of the solvent goes on improving). The end-to-end distance thus increases. When the polymer coil has no interaction with the solvent molecules; it acquires what are known as unperturbed dimensions. The solvents in which polymer coil acquires unperturbed dimensions are called theta or *Flory* solvents. The temperature at which the polymer molecules are just held in solution or are held at the threshold of precipitation is called theta temperature (or *Flory* temperature).

The average dimension of the polymer coil in real solutions is therefore different from the unperturbed dimension. In good solvents, for example, the root mean square end-to-end distance, $(\bar{r}_0^2)^{1/2}$ can be written as, $(\bar{r}^2)^{1/2} = \alpha(\bar{r}_0^2)^{1/2}$. The end-to-end distance of a polymer coil in any solvent compared with the same in theta solvents provides the value of the chain expansion factor, α. In very dilute solutions, isolated polymer coils do exist which are independent of their neighbouring molecules. In moderately concentrated solutions the polymer coils may interpenetrate and entangle with each other. In more concentrated solutions, viscoelastic swollen gels do form as the coils are aggregated. The determination of the rms end-to-end distance in theta solvents should therefore provide the size of the polymer chain identical to that can be simply calculated for an undissolved polymer coil. Dilute solution viscosity measurements are used for determining the rms end-to-end distance for a polymer coil in theta solvents. The ratio between the calculated rms end-to-end distance to the experimental value gives an idea about the stiffness of the chain. For flexible polymer chains, the ratio is close to 1.0 and while rigid chains with hindered rotation among its segments as well as units (rigid or stiff chains), are characterised by a ratio more than 1.0. The values of chain stiffness ratio for some polymers are shown in Table 4(e)

Table 4(e)   Chain stiffness ratios for some polymers

Polymer	Chain stiffness ratio
Natural rubber	1.1
Polyisobutylene	1.9
Polystyrene	2.4
Polyacrylonitrile	3.0
Cellulose nitrate	4.5
Cellulose butyrate	5.0

Besides chain stiffness ratio, another important parameter known as chain expansion factor can also be estimated from the calculated and measured dimensions of polymer coils. For example, the root mean square end-to-end distance of a polymer coil in a θ-solvent, $(\bar{r}_\theta^2)^{1/2}$ and the same dimension for the polymer coil in a non theta solvent, $(\bar{r}_0^2)^{1/2}$ can be related by; $(\bar{r}_\theta^2)^{1/2} = \alpha(\bar{r}_0^2)^{1/2}$, and thus the ratio between the two is given as α, which provides the value of chain expansion factor. The utility of the value of α is that based on its magnitude the solvent quality is decided for a given polymer. For example, a value for α greater than unity indicates better quality of solvent (good solvent). Value less than unity is observed for poor solvents and a value of unity is noted in θ-solvents.

Thermodynamics of polymer dissolution

Thermodynamic principles governing the dissolution of low molecular weight substances and polymers are essentially the same. The dissolution of a solute involves a change of state and, therefore, depends on the changes in entropy and enthalpy of the system. Whether or not a polymer dissolves (mixing of polymer with solvent) in a solvent can be examined from the well known Gibbs relationship, $\Delta G_{mix} = \Delta H_{max} - T\Delta S_{mix}$. For the thermodynamic feasibility of the dissolution of polymer in the solvent, the free energy change of the solution (mixing) i.e. ΔG_{mix}. should be negative. The entropy change, ΔS_{mix} for any dissolution process is always positive because of the increased randomness of the molecules in solution. However, the magnitude of entropy change for the mixing of a macromolecule in a liquid is far less as compared to the mixing of a low molecular weight substance in a liquid. Since ΔS_{mix} is positive, (though small in magnitude) and T is always positive, the quantity $T\Delta S_{mix}$ will always be positive and will favour the dissolution of polymer. Now ΔH_{mix} can be either positive or negative. For a negative value of ΔH_{mix}, the terms, $\Delta H_{mix} - T\Delta S_{mix} = \Delta G_{mix}$ will be large negative and therefore the solute solubility will be very high. A negative ΔH_{mix} means the existence of specific interactions such as hydrogen bonding between the polymer solute and solvent molecules. If ΔH_{mix} to be positive, the solution is possible only if $\Delta H_{mix} < T\Delta S_{mix}$ (i.e. ΔG_{mix} is negative). A positive ΔH_{mix} means that the polymer and solvent 'prefer their own company' i.e. these are in lower energy state. Thus when ΔH_{mix} is small, the sign of ΔG_{mix} is governed by ΔS_{mix}. Hildebrand, however, gave the following empirical equation for the estimation of ΔH_{mix}:

$$\Delta H_{mix} = V_s \phi_s \phi_p (\delta_s - \delta_p)^2 \qquad (4.10)$$

where ϕ_s and ϕ_p are the volume fractions of solvent and polymer, respectively. V_s is the molar volume of the solvent and δ_s and δ_p are the solubility parameters for solvent and polymers, respectively. The solubility parameters are estimated as the square root values of cohesive energy density (CED), which is a measure of the magnitude of inter-molecular attraction. Further, the cohesive energy density is defined in terms of the molar energy of vaporisation, ΔE_{ev} by the relation;

$$\delta = (CED)^{1/2} = (\Delta E_{ev})^{1/2}/V \qquad (4.11)$$

By convention, these values are always positive and irrespective of their magnitude; the term $(\delta_s - \delta_p)^2$ will also be positive. Thus, ΔH_{mix} will always be positive. Now as discussed before $T\Delta S_{mix}$ is very small in magnitude; for the condition favouring polymer solubility, i.e. $\Delta H_{mix} < T\Delta S_{mix}$, a minimum positive value of ΔH_{mix} can be upset by matching the polymer and solvent such that their solubility parameters are as close as possible. To a rough approximation a polymer and solvent differing in solubility parameters by $(\delta_1 - \delta_2) < 0.5$ will form the solution.

Solubility parameter of solvents can be experimentally determined easily from the values of molar internal energy change on vaporisation. Solubility parameter for polymers can not be obtained by this method as they decompose long before reaching vaporisation temperature. But δ values for polymers can be indirectly determined. The basis to do this is that solvent which shows maximum swelling of a lightly cross-linked polymer has the same solubility parameter value as polymer. Thus degree of swelling for a given polymer is determined in various solvents to obtain the solubility parameter of the polymer. In table 4(f) are shown solubility parameter values for some polymers.

Table 4(f)   Solubility parameters of some polymers

Polymer	Solubility parameter H
Polystyrene	9.1
Polyethylene	7.9
Polypropylene	9.2
PVC	9.7
PMMA	9.5
PET	10.7
Nylon 6, 6	13.6

Hilderbrand's solubility parameter approach is simple and in common practice even today is taken as guide to select a solvent for a polymer. Since the solubility parameters can have only positive values, the ΔH_{mix} as defined by below given equation 4.12 is obtained always positive (i.e. endothermic mixing). *Flory* introduced a term which eventually known as Flory interaction parameter χ (which can take positive or negative values), to explain the dissolution process in systems (both exothermic and endothermic) of solute-solvent pair systems. The enthalpy of mixing is defined as

$$\Delta H_{mix} = \delta_s \delta_p \chi RT \qquad (4.12)$$

where $\chi = Z\,\Delta w\,RT$ and Z is the number of nearest neighbour (or coordination number) and Δw will yield large χ (which is dimensionless) and thus large positive values to ΔH_{mix} and would not favour dissolution (or phase separation would result). Small positive or negative values of χ would explain the spontaneity of dissolution process. Table 4(g) lists interaction parameters for some polymer-solvent pairs.

Table 4(g)   Flory interaction parameter values for some polymer-solvent systems

Polymer	Solvent	Temperature °C	Flory parameter, χ
Polyvinyl acetate	n-Hexane	100	1.93
	Benzene	100	0.37
	Chloroform	100	0.11
HDPE	n-Decane	150	0.18
LDPE	n-Dodecane	120	0.18
Natural rubber	n-Hexane	25	0.53
	n-Heptane	25	0.50
	n-Octane	25	0.48
	Benzene	25	0.46
	Toluene	25	0.35

Thermodynamics of Solution

Before we attempt to develop relations for explaining the thermodynamic aspects of polymer solutions, it is highly essential to consider the following thermodynamic principles derived for solutions of simple molecules.

Solutions of simple molecules

A simple example for a solution of small molecules would be a binary system consisting components 1 and 2 and formed by mixing n_1 and n_2 moles (N_1 and N_2 molecules) at constant pressure, P and temperature, T. The total free energy of such simple systems is given as

$$G_{12} = n_1 G_1^0 + n_2 G_2^0 + n_1\,RT\ln(p_1/p_1^0) + n_2\,RT\ln(p_2/p_2^0)$$

where G_1^0 and G_2^0 are the free energies of pure components, p_1^0 and p_2^0 are the vapour pressures of pure components and p_1 and p_2 the partial vapour pressures of pure components in solution, respectively.

Raoult's law defines the partial vapour pressure of the pure components in ideal solutions as

$$p_1 = x_1 p_1^0 \qquad (4.14)$$

$$p_2 = x_2 p_2^0 \qquad (4.15)$$

where x_1 and x_2 are the mole fractions of the respective components such that $x_1 + x_2 = 1$ and $x_1 = (1 - x_2) = n_1/(n_1 + n_2)$. Thus

$$G_{12} = n_1 G_1^0 + n_2 G_2^0 + n_1 RT \ln x_1 + n_2 RT \ln x_2 \qquad (4.16)$$

or

$$G_{12} - (n_1 G_1^0 + n_2 G_2^0) = n_1 RT \ln x_1 + n_2 RT \ln x_2 \qquad (4.16a)$$

or

$$\Delta G_{mix} = n_1 RT \ln x_1 + n_2 RT \ln x_2 \qquad (4.16.b)$$

or

$$\Delta G_{mix} = RT \sum n_i \ln x_i \qquad (4.16c)$$

Differentiating eq. 4.16b with respect to temperature, we can write

$$\partial \Delta G_{mix}/\partial T = n_1 RT \ln x_1 + n_2 RT \ln x_2 \qquad (4.17)$$

$$-\partial \Delta G_{mix}/\partial T = \Delta S_{mix} = -n_1 RT \ln x_1 - n_2 RT \ln x_2 \qquad (4.18)$$

Combining eqs. 4.16b and 4.18, we get

$$\Delta G_{mix}^{ideal} = \Delta S_{mix}^{ideal} \qquad (4.19)$$

or

$$\Delta H_{mix}^{ideal} = 0$$

For non-ideal or real solutions, however, eqs. 4.14 and 4.15 should be modified to give allowance to the nonideality by introducing a term for activity coefficient. Then

$$p_1 = x_1 \gamma_1 p_1^0 \qquad (4.21)$$

$$p_2 = x_2 \gamma_2 p_2^0 \qquad (4.22)$$

Now for real solutions the free energy, entropy and enthalpy of mixing can be expressed in terms of

$$\Delta G_{mix} = RT \sum n_i \ln x_i + RT \sum n_i \ln \gamma_i \qquad (4.23)$$

$$\Delta S_{mix} = -R \sum n_i \ln x_i - R \sum n_i \ln \gamma_i - RT \sum n_i (\partial \ln \gamma_i/\partial T) \qquad (4.24)$$

$$\Delta H_{mix} = RT^2 \sum n_i (\partial \ln \gamma_i/\partial T) \qquad (4.25)$$

Defining an excess function as the difference between the change in it upon mixing over its value in ideal state, we can write

$$G^E = \Delta G_{mix} - \Delta S_{mix}^{ideal} = -RT \sum n_i \ln \gamma_i \qquad (4.26)$$

$$S^E = \Delta S_{mix} - \Delta S_{mix}^{ideal} = -R \sum n_i \ln \gamma_i - RT \sum n_i (\partial \ln \gamma_i/\partial T) \qquad (4.27)$$

$$H^E = \Delta H_{\text{mix}} - \Delta H_{\text{mix}}^{\text{ideal}} = RT^2 \, \Sigma \, n_i \, (\partial \ln \gamma_i / \partial T) \qquad (4.28)$$

The utility of excess functions is manyfold in explaining the solution behaviour in simple and complex systems. It can be seen from eqs. 4.26 to 4.28 that the excess functions namely G^E, S^E, and H^E are zero for ideal solutions. In case of nonideal solutions, the values deviate from zero either by a positive or negative magnitude. Similarly one expects either positive or negative deviation from Raoult's law when applied to real solutions;

$G^E > 0$ positive deviation from Raoult's law

$G^E < 0$ negative deviations from Raoult's law

Thus solutions showing positive deviation from Raoult's law undergo immiscibility or phase separation only if $G^E = \Delta G_{\text{mix}}^{\text{ideal}}$. Taking into consideration the well known Gibbs relation, $G^E = H^E - TS^E$, the following can be written for regular as well as for non ideal solutions;

$$S^E = 0, \; G^E = H^E \qquad (4.29)$$

Mixtures containing components of same size and polarity show such behaviour. For example for the system benzene + CCl_4, $G^E = 82$ J mol^{-1}, $H^E = 115$ J mol^{-1} and $S^E = 0.11$ J mol^{-1} K^{-1} at 298.15 K. When $S^E > 0$, positive deviation is expected from Raoult's law and since $S^E > 0$, then $G^E > 0$, $\Delta G_{\text{mix}}^{\text{ideal}} < 0$. Thus, no phase separation occurs when $G^E > \Delta G_{\text{mix}}^{\text{ideal}}$. For $S^E < 0$, negative deviation is expected from Raoult's law and since $S^E < 0$, mixtures have more orderliness than their respective pure components (due to hydrogen bond formation or any other specific interaction etc.). For $S^E < 0$, $H^E < 0$, so also $G^E < 0$, i.e. all the three functions are negative (e.g. benzene + tetrahydrofuran systems with $G^E = -131$ J mol^{-1} $H^E = -271.1$ J mol^{-1} and $S^E = -0.47$ J mol^{-1} K^{-1}).

By applying statistical thermodynamics and assuming certain models the relations for the fundamental thermodynamic functions of mixing can further be refined. For example, using Boltzmann relation, the entropy of mixing can be expressed as

$$\Delta S_{\text{mix}} = k \ln \Omega \qquad (4.30)$$

where Ω is the number of distinguishable arrangements of N_1 and N_2 molecules in $N_1 + N_2$ sites in a lattice (see Fig. 4.28) and defined as

$$\Omega = (N_1 + N_2)! / N_1! \, N_2! \qquad (4.31)$$

$$\Delta S_{\text{mix}} = k \ln (N_1 + N_2)! / N_1! \, N_2! \qquad (4.32)$$

Introducing Stirling's approximation $N! = (N \ln N) - N$, we get

$$\Delta S_{\text{mix}} = - k \, [N_1 \ln \{N_1/(N_1 + N_2)\} + N_2 \ln (N_2/(N_1 + N_2))] \qquad (4.33)$$

For expressing the number of molecules N in terms of either the moles n or mole fraction (X), we can define $n_1 = N_1/N_A$, $n_2 = N_2/N_A$, $X_1 = n_1/(n_1 + n_2)$ and $X_2 = n_2/(n_1 + n_2)$ and also writing the term $kN_A = R$, one obtains

$$\Delta S_{mix} = - R \ [n_1 \ \ln X_1 + n_2 \ \ln X_2] \qquad (4.34)$$

since for ideal solutions $\Delta H_{mix} = 0$, the free energy of mixing can be written as

$$\Delta G_{mix} = T \Delta S_{mix} = RT \ [n_1 \ \ln X_1 + n_2 \ \ln X_2] \qquad (4.35)$$

Thermodynamics of polymer solutions

Polymer solutions are mostly nonideal. At concentration above a few per cent, the deviations from ideality are so large that the Hilderbrand's law is of very little significance for predicting the thermodynamic properties of polymer solutions. The heat of mixing is zero and the entropy of the mixing is simply given by the ideal combinatorial entropy that results from the mixing of two types of solid spheres of the same size for ideal solutions. For athermal solutions the heat of mixing is zero but the entropy of mixing is larger than the ideal combinatorial entropy by an excess entropy of mixing. But in theta solution, the heat of mixing just compensates the excess entropy of mixing at a certain temperature. The deviations from ideality are mainly attributed to the small values for the entropy of mixing. Also, the simple lattice model for calculation of ΔS_{mix} given by eq. 4.34 is not only insufficient but also invalid as the polymer solute and low molecular weight solvent differ in size by several decades of magnitude. Flory and Huggings independently proposed the calculation procedures for the estimation of ΔG_{mix}, ΔS_{mix} and ΔH_{mix}.

Flory Huggins theory

The theory takes into consideration the modified lattice model as shown in Fig. 4.28. It is assumed that each space in the lattice is occupied by either one solvent molecule or one chain segment. Also polymer chains are divided into segments based on their ease of internal rotation. These segments are known as thermodynamic or Kuhn segments. A freely jointed chain is characterised by segments of equal size and identical internal rotation. Also, each site occupied by a polymer segment has two adjacent polymer sites so that there is a continuous path of polymer segments. The solution is assumed to be concentrated enough that the occupied lattice sites are randomly distributed as shown in Fig. 4.28b. A small entropy of mixing for polymer solutions as compared to larger values for low molecular weight solutes in solvents, can be qualitatively understood here. There will be fewer ways in which the same number of lattice sites can be occupied by polymer segments.

Now, if there are N_1 solvent molecules and N_2 polymer molecules and each polymer molecule is made up of x segments, then the total number of lattice sites will be $[N_1 + xN_2]$. The calculation of configurational entropy of mixing in this case is more complicated than for the solutions of low molecular weight solutes. Flory and Huggins in 1942 independently derived a simple equation:

$$\Delta S_{mix} = - k[N_1 \ \ln \ \{N_1/ \ (N_1 + xN_2)\} + N_2 \ \ln \ \{x \ N_2/(N_1 + xN_2)\}] \quad (4.36)$$

$$\Delta S_{mix} = - R[n_1 \ \ln \ \phi_s + n_2 \ \ln \ \phi_p], \qquad (4.37)$$

where $\qquad \phi_s = N_1/(N_1 + xN_2) \ \text{and} \ \phi_p = xN_2/(N_1 + xN_2) \qquad (4.38)$

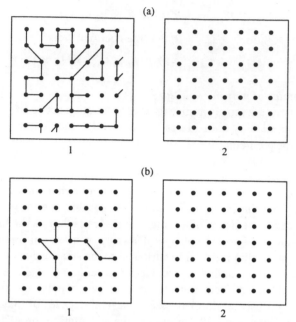

Fig. 4.28 **Model of quasicrystalline lattice (a) at low concentration (b) at high solvent concentration. 1—polymer units connected into segments and 2—free polymer units**

It should be noticed that the entropy contribution that may be present as a result of specific interaction between polymer and solvent molecules is not considered here. The Flory-Huggins theory also allows the simple calculations of entropy of mixing. It assumes that ΔH_{mix} is not equal to zero because the polymer-solvent interaction energies in solution are different from the polymer-polymer and solvent-solvent in the pure unmixed state. Considering uncharged polymer and solvents, the interaction energy is primarily governed by the three types of contacts viz. solvent-solvent, solvent-polymer and polymer-polymer. Let these be represented by w_{ss}, w_{sp} and w_{pp}. Now for every two polymer-solvent contacts taking place in solution, one solvent-solvent and one polymer-polymer contact of the pure components are broken. Thus the energy difference per polymer-solvent contact, Δw, between the mixed and unmixed states will be

$$\Delta w = w_{sp} - 1/2[w_{ss} + w_{pp}] \qquad (4.39)$$

Using the liquid lattice model, it can be shown that the number of such contacts, p, is given by

$$p = x N_2 z \, \phi_s \qquad (4.40)$$

Here z is the coordination number of lattice. The enthalpy of mixing, ΔH_{mix} is simply the product of the energy per contact and number of contacts.

$$\Delta H_{\text{mix}} = x N_2 \, z \phi_p \, \Delta w \qquad (4.41)$$

but $\qquad\qquad x N_2 \phi_s = N_1 \phi_p$

so
$$\Delta H_{\text{mix}} = N_1 z \phi_p \Delta w \tag{4.42}$$

Flory and Huggins theory introduced a dimensional quantity known as interaction parameter, χ, which represents the interaction energy per solvent molecule divided by kT

$$\chi = z \Delta w / kT \tag{4.43}$$

$$\Delta H_{\text{mix}} = \chi N_1 z \phi_p kT = n_1 \phi_p \chi RT \tag{4.44}$$

Now knowing entropy and enthalpy contribution, the free energy of mixing can be obtained as

$$\Delta G_{\text{mix}} = RT [n_1 \ln \phi_s + n_2 \ln \phi_p + n_1 \phi_p \chi] \tag{4.45}$$

This equation becomes similar to that of ideal solutions when $x = 1$ and $\chi = 0$.

The free energy contribution arises from two entropy factors; one due to the contact between polymer segments and solvent molecules and the other due to conformations of polymer molecules. Thus, the overall free energy of mixing

$$\Delta G_{\text{mix}} = \Delta G_{\text{conform}} + \Delta G_{\text{contact}} \tag{4.46}$$

These two contributions are

$$\Delta G_{\text{conform}} = RT[\ln(1 - \phi_p) + (1 - 1/n) \phi_p] \tag{4.47}$$

$$\Delta G_{\text{contact}} = RT \chi \phi_p^2 \tag{4.48}$$

$$\Delta G_{\text{mix}} = RT[\ln (1 - \phi_p) + (1 - 1/n)\phi_p + \chi \phi_p^2] \tag{4.49}$$

when χ exceeds some critical value χ_c, the system will separate into two phases. It can be derived mathematically

$$\chi_c = 1/2 \, n + 1/n^{1/2} + 1/2 \tag{4.50}$$

Although Flory-Huggins theory is an important improvement over ideal solution theory, it has certain serious drawbacks to fully account dilute polymer solutions. The theory suffers from its limitation because of the following:

(i) the occupation of lattice sites is purely statistical and ignores the tendency of polymer molecules to form isolated coils specially in dilute solutions and (ii) does not consider specific interactions between the polymer and solvent.

Determination of Flory interaction parameter from osmotic pressure measurements

The reduced osmotic pressure, π/C, is related to molecular weight of polymers by

$$\pi/C = [RT/M](1 + A_2 C) \tag{4.51}$$

where A_2 is the second virial coefficient. The magnitude of A_2 is related directly to the extent of deviation from ideality. Thus, we can derive a relation between A_2 and Flory interaction parameter χ. The osmotic pressure π is related to the free energy term by

$$\pi V_s = -\Delta G_{\text{mix}} \tag{4.52}$$

where V_s is the partial molar volume of the solvent

$$-\pi V_s = RT[\ln \phi_s + (1 - 1/n)\phi_p + \chi\phi_p^2]$$

or $\qquad -\pi V_s / RT = \ln \phi_s + (1 - 1/n)\phi_p + \chi\phi_p^2$

$$= \ln(1 - \phi_p) + (1 - 1/n)\phi_p + \chi\phi_p^2$$

For very dilute solutions $\phi_p \ll 1$ and

$$\ln(1 - \phi_p) = -[\phi_p + \phi_p^2/2 + \phi_p^3/3 + \ldots] \tag{4.55}$$

$$= -\phi_p - \phi_p^2/2$$

and we get $\qquad \pi V_s / RT = -\phi_p - \phi_p^2/2 + \phi_p + \phi_p/n + \chi\phi_p^2 \tag{4.56}$

$$= -[\phi_p/n + \phi_p^2(1/2 - \chi)]$$

Now, since

$$\phi_p = n_p V_p / V_s = n_p \rho_p \tag{4.57}$$

$$= C(V_s/M)\,\rho_p \text{ (as } n_p = CV_s/M),$$

also $\qquad \phi_p = n_p V_p / V_s = CV_p/M = C/\rho_p \text{ (} \rho_p = \text{density of polymer)} \tag{4.58}$

Thus $\quad \pi V_s / RT = CV_s / M + C^2(1/2 - \chi)/\rho_p^2 \tag{4.59}$

$$\pi / CRT = 1/M + (1/2 - \chi)C/V_s\rho_p^2 \tag{4.60}$$

$$\pi / C = RT(1/M + ((1/2 - \chi)CV_s\rho_p^2)) \tag{4.61}$$

$$= RT(1/M + A_2 C)$$

where $\qquad A_2 = (1/2 - \chi)V_s\,\rho_p^2$

It becomes clear that when $\chi = 0.5$, the second virial coefficient A_2 is zero. Thus, osmotic pressure measurements can also be utilised to estimate the interaction parameter, χ and (θ solvent/temperature) for a given polymer-solvent system. Second virial coefficient (A_2) is a measure of polymer-solvent interaction which depends on several factors such as (i) the temperature (increase in temperature leads to a decreasing value of A_2), (ii) the molecular weight and molecular weight distribution of polymer and (iii) the tacticity of polymer.

Phase Separation

Figure 4.29 shows the plots between total free energy of mixing against volume fraction of solute ϕ_2 (for convenience can be taken as x_2). At high temperature, the free energy of mixtures is less than the sum of the free energies of the

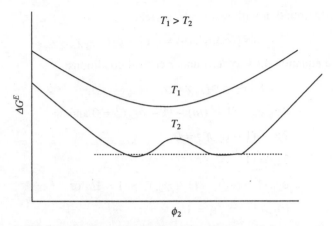

Fig. 4.29 **Schematic representation of variation of free energy of mixing with the composition of solute**

constituting components and thus; ΔG_{mix} is negative at all compositions and hence a homogeneous solution is always formed. At lower temperatures (see Fig. 4.29), ΔG_{mix} is positive in the composition range ϕ_2' and ϕ_2'' and negative in the ranges $0 - \phi_2'$ and $\phi_2'' - 1$. Thus, the occurrence of phase separation in the composition range ϕ_2' and ϕ_2'' and a homogeneous solution in rest of the composition is quite evident. The phase diagram for a typical polymer is thus expected to have the feature shown in Fig. 4.30, where both the upper critical solution temperature (UCST) and lower critical solution temperature (LCST) boundaries of solubility and phase separation are depicted.

The inflexion observed in free energy versus composition plot can be mathematically formulated as

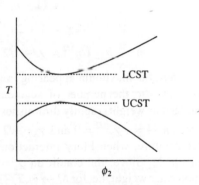

Fig. 4.30 **Phase temperatures in polymer solutions**

$$(\partial^2 \Delta G_{mix}/\partial \phi_2^2) = 0 \text{ and } (\partial^3 \Delta G_{mix}/\partial \phi_2^3) = 0 \qquad (4.62)$$

also, the chemical potential of the solvent in the solution μ_1 can be related to ΔG_{mix} $\mu_1 = (\partial G_{mix}/\partial n_1)$ when temperature T and number of moles of solution n_2 are constant. The critical conditions for miscibility can be written in terms of chemical potential as

$$(\partial \Delta \mu_1/\partial \phi_2) = 0 \text{ and } (\partial^2 \Delta \mu_2/\partial \phi_2^2) = 0 \qquad (4.63)$$

$$1/RT \, (\partial \Delta \mu_1/\partial \phi_2) = \partial/\partial \phi_2 \, [\ln \phi_1 + \phi_2(1 - 1/n) + \phi_2^2 \, \chi] \qquad (4.64)$$

$$= -(1 - \phi_2)^{-2} + \chi$$

Further differentiation with respect to ϕ yields

$$1/RT(\partial\Delta\mu_1/\partial\phi_2) = -(1 - \phi_2)^{-2} + \chi \qquad (4.65)$$

Setting the equation 4.64 to zero under critical conditions:

$$T = T_c, \; \phi_2 = \phi_{2,c}, \; \chi = \chi_c$$

$$2\phi_{2,c}\chi_c + (1 - 1/n) - (1 - \phi_{2,c})^{-1} = 0 \text{ and} \qquad (4.66)$$

$$2\chi_c - (1 - \phi_{2,c})^{-2} = 0 \qquad (4.67)$$

thus, $\qquad 2\chi_c = (1 - \phi_{2,c})^{-2} \qquad (4.68)$

So, $\qquad \phi_{2,c}/(1 - \phi_{2,c})^2 - (1 - \phi_{2,c})^{-1} + 1 = 1/n \text{ or} \qquad (4.69)$

$$[\phi_{2,c}/(1 - \phi_{2,c})]^2 = 1/n \qquad (4.69a)$$

or $\qquad \phi_{2,c} = (1 + n^{1/2})^{-1} \qquad (4.69b)$

For $r \rightarrow \infty$

$$\phi_{2,c} = 1/n^{1/2} \qquad (4.70)$$

$$2\chi_c = (1 - \phi_{2,c})^{-2} \qquad (4.71)$$

$$= (1 - n^{1/2})^{-2}$$

$$2\chi_c = 1 + 2n^{1/2} \text{ or} \qquad (4.72)$$

$$1/n^{1/2} = \chi_c - 1/2 \qquad (4.73)$$

Thus, the maximum appearing in free energy plot at critical composition χ_c^2 is related to the number of segments, n. An increase in 'n' (i.e. increase in molecular weight) rapidly diminishes χ_c^2. Furthermore, under critical condition when $n \rightarrow \infty$, $\chi_c^{-1/2} = 0$ and $\chi_c = 1/2$. Thus, for polymer-solvent pair at a given temperature, when Flory interaction parameter $\chi > 1/2$, the polymer will be insoluble. Since, the condition $\chi_c = 1/2$ is valid for a polymer with infinite molecular weight, i.e. for $M \rightarrow \infty$, $T_c \rightarrow \theta$. This implies that ϕ-temperature equals to CST for polymer with infinite molecular weight.

Viscosity of polymer solutions and size of the polymer coil

Generally, as would be expected the viscosity of polymer solutions increase with increase in concentration and decrease in temperature. However, the opposite behaviour is shown in few cases. For example, the viscosities of dilute aqueous solutions of polymers with ionisable group increases on further dilution (this is called a polyelectrolyte behaviour). Such an effect as discused in details elsewhere (chapter 8) is primarily due to the increased repulsion between like charges on the same chain, which is further enhanced due to greater ionisation in more dilute systems. A very high viscosity of phenol-formaldehyde liquid resin in solvent at elevated temperature can be explained due to the formation of cross-links.

The viscosities of motor oils can be effectively controlled to a constant value over wide temperature range, using polymers, for the purpose of all weather lubricants. While the viscosity of the oil will decrease at higher temperatures, as is the case with most simple organic liquids, the presence of dissolved polymer will have opposite effect. As the polymer coil will expand with increase in temperature (bestowing the oil with better solvent power), viscosity of solution is increased rather than usually expected decrease in viscosity with rise in temperature.

The viscosity of molten polymer also varies with molecular weight. Thus, molecular weight dependence of viscosity provides highly useful information in the processing of polymers. The viscosity becomes very high above the threshold molecular weight making the processing very difficult, and therefore, often polymers with extremely high molecular weight are not easily processible. *Fox* gave following equation relating viscosity and molecular weight (as applicable at low shear above a threshold molecular weight (or degree of polymerisation, DP);

$$\log \eta = K + 3.4 \log DP \tag{4.74}$$

For most polymers this DP is around 600. A log-log plot between melt viscosity and molecular weight for polymer is shown in Fig. 4.31.

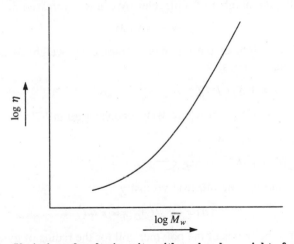

Fig. 4.31 **Variation of melt viscosity with molecular weight of polymers**

Intrinsic viscosity and dimension of a polymer coil

Contrary to low molecular weight substances which do not show significant increase in the viscosity of their solution even at fairly high concentrations, polymeric materials markedly show enhanced viscosity in very dilute solutions. Such an increase in viscosity of liquids in the presence of polymers can be explained as follows. When a liquid flows through a capillary, it moves in several layers with different velocity. The dissolved polymer molecules show translational motion of both the random (Brownian) and directional (diffusion and sedimentation) types due to large size. The single polymer molecule occupies position in different layers of solvents (moving with different velocity) which results in rotational motion of the molecules. These translational and rotational motions of polymer molecules with respect to the static phase produce friction. This leads to high viscosity even in dilute polymer solutions.

A polymer molecule has been considered to remain as curled coil in solution (i) from which the solvent molecules can pass (free draining model) or (ii) which takes up a few solvent molecules to form swollen impermeable gel. In both these cases the polymer molecules can be treated according to hard sphere model. For colloidal dispersions with non-interacting dispersed particles, the Einstein equation relates the specific viscosity η_{sp} to the volume fraction, ϕ as

$$\eta_{sp} = 2.5\phi \tag{4.75}$$

A polymer solution can thus behave similar if the total number of polymer molecules N are present in a given volume V, then ϕ is defined as

$$\phi = \text{Volume of one molecule} \times N/V \tag{4.76}$$

If the polymer molecule is considered as a spherical coil of radius r, then the volume of a polymer molecule is $4/3\,\pi r^3$, thus

$$\phi = 4/3\pi r^3/V \quad \text{or} \tag{4.77}$$

$$\eta_{sp} \propto Nr^3/V \tag{4.77a}$$

Conventionally, the number of molecules, N can be expressed as

$$N = WN_A/M \tag{4.78}$$

where W is the weight of the polymer of molecular weight M and N_A is the Avogadro's number. Thus,

$$\eta_{sp} \propto WN_A r^3/MV \tag{4.79}$$

$$\propto CN_A r^3/M \text{ where } W/V = \text{concentration 'C'}$$

$$\eta_{sp}/C \propto N_A r^3/M \quad \text{or} \tag{4.80}$$

$$\eta_{sp}/C \propto r^3/M \tag{4.80a}$$

or for dilute solutions, the intrinsic viscosity

$$[\eta] = (\eta_{sp}/C)_{C=0} \propto r^3/M \tag{4.81}$$

Now, r which is the radius of the polymer coil (or the radius of gyration) can be related to the root mean square end-to-end distance $(\bar{r}^2)^{1/2}$ (i.e. r = radius of gyration $(\bar{s}^2)^{1/2}$ and $(\bar{r}^2)^{1/2} = 6(\bar{s}^2)^{1/2}$.

The intrinsic viscosity thus can be shown as

$$[\eta] = K(\bar{r}^2)^{3/2}/M \tag{4.82}$$

where K is proportionality constant, the product of intrinsic viscosity and molecular weight of the polymer is directly proportional to the 3/2 power of the size of the polymer coil, i.e.

$$[\eta]M \propto (\bar{r}^2)^{3/2} \tag{4.83}$$

The end-to-end distance of a polymer coil in a solvent is related to the similar value in θ-conditions; $(\bar{r}_\theta^2)^{1/2}$ as $(\bar{r}^2)^{1/2} = \alpha(\bar{r}_\theta^2)^{1/2}$, where α is Flory's chain expansion factor. Thus,

$$[\eta] = K\alpha^3 [(\bar{r}_\theta^2)/M]^{1/2} \times M^{1/2} \tag{4.84}$$

The ratio $(\bar{r}_\theta^2)/M$ remains constant for a given polymer [as increase in M increases $(\bar{r}_\theta^2)$]. So

$$[\eta] = K' \, \alpha^3 M^{1/2} \tag{4.85}$$

Under θ-conditions, the expansion factor $\alpha = 1$, i.e. $(\bar{r}^2)^{1/2} = (\bar{r}_\theta^2)^{1/2}$

$$[\eta] = K' \, M^{1/2} \tag{4.86}$$

This is the well known Mark-Houwink equation. The exponent value of 0.5 has been verified for number of polymer solutions in theta solvents. K' was erroneously treated as an universal constant with a value of 2.84×10^{21}. The ratio of intrinsic viscosity in a good solvent to that of in theta solvent, $[\eta]_\theta$ can be written as $[\eta]/[\eta]_\theta = \alpha^3$.

The viscosity of dilute polymer solutions thus depends on several factors like molecular weight of polymer, nature of solvent, polymer concentration and temperature. The concentration dependence of polymer viscosity can be represented by several equations as proposed by Huggins (4.87), Kraemer (4.88), Martin (4.89) and Schulz-Blaschke (4.90). These equations are written as,

$$\eta_{sp}/C = [\eta] \{ 1 + k_H [\eta]C \} \tag{4.87}$$

$$(\ln \eta_r)/C = [\eta] \{ 1 - k_k [\eta]C \} \tag{4.88}$$

$$(\ln \eta_{sp}/C) = \ln [\eta] + k_M [\eta]C \tag{4.89}$$

$$\eta_{sp}/C = [\eta] \{ 1 + k_{SB}\eta_{sp} \} \tag{4.90}$$

Thus, the intrinsic viscosity needed for determining the molecular weight of a given polymer can be estimated by respective plots of viscosity functions of polymer solutions. A typical Huggins plot showing the dependence of reduced viscosity on concentration is shown in Fig. 4.32. Thus, the dilute solution viscometry

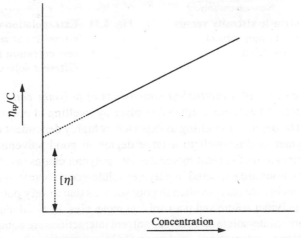

Fig. 4.32 **Typical Huggins plot of reduced viscosity Vs concentration**

has proved to be an important technique employed for obtaining viscosity average molecular weight of polymers.

Determination of intrinsic viscosity in theta conditions

Theta solvents are known for limited polymers. A slight change in temperature may transform a theta solvent into good or poor solvent making the estimation of intrinsic viscosity less reliable. Thus, correct value of intrinsic viscosity corresponding to nearly θ-conditions has to be found. There are two ways to obtain the correct $[\eta]_\theta$ from the viscosity functions data as obtained from common solvents. In one such method, intrinsic viscosities of a given polymer are determined in a solvent/nonsolvent mixture for different compositions. The compositions of solvent and nonsolvent are varied such that polymer gets precipitated at particular proportion. The intrinsic viscosity obtained in a composition close to precipitation of polymer solute is then determined (see Fig. 4.33). This value is supposed to be identical with $[\eta]_\theta$. Another way to obtain intrinsic viscosity in theta conditions is to measure its value for a given polymer in different good solvents and then extrapolating the respective reduced viscosity plots beyond the zero value of concentration i.e. to the left of the abscissa to a common intersection point where all the plots converge. Then the value of intercept from the concentration base line to the common intersection point gives the $[\eta]_\theta$. Such a procedure is depicted in Fig. 4.34.

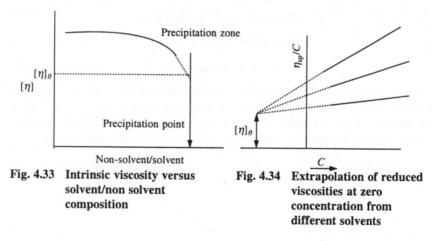

Fig. 4.33   Intrinsic viscosity versus solvent/non solvent composition

Fig. 4.34   Extrapolation of reduced viscosities at zero concentration from different solvents

Intrinsic viscosity and unperturbed dimensions of polymer coil

The dissolution of a polymer solute takes place by swelling of polymer chains in the solvent. The degree of swelling as expected, is highly dependent upon solvent quality. Polymer chains swell to a large degree in good solvents because of excellent permeation of solvent molecules into polymer chains (as strong solute-solvent interactions are expected in polymer solute-good solvent systems) while the polymer chains are least swollen in poor solvents due to very poor interaction between the polymer solute and poor solvent molecules. In θ-solvents, however, where the pure solute-solute and solvent-solvent interactions are equally balanced by solute-solvent interactions, polymer chain, as explained earlier acquire unperturbed dimensions. Thus hydrodynamic measurements can be of great utilitiy in evaluating unperturbed state of polymer chain or θ-condition. The relation between intrinsic viscosity and the size of an unperturbed coil has been theoretically shown to be

$$[\eta]_\theta = \phi_0 \frac{(\overline{r}_\theta^2)^{1/2}}{M} \tag{4.91}$$

and the same relation in non ideal (or non-theta) condition becomes

$$[\eta] = \phi_0 \frac{(\overline{r}_0^2)^{1/2}}{M} \times \alpha \tag{4.92}$$

where parameter ϕ is an universal constant for all polymers equal to 2.84×10^{21}. The disadvantage of above relation is that it requires the accurate determination of viscosity function in θ-solvents. Finding a θ-solvent in a useful and measurable temperature is highly difficult. Hence, attempts have been made to develop relations useful for computing the unperturbed dimensions from the measurement of viscosity functions in solvents under normal conditions. Out of several such attempts, the relation proposed by *Stockmayer* and *Fixman* has been found to be more popular. The relation is given as

$$[\eta] = K_\theta M^{1/2} + 0.51 \, \phi_0 BM \tag{4.93}$$

where K_θ is the constant determined in θ-solvent, M the molecular weight of polymer and B the quantity related to Flory-Huggin's interaction parameter χ by the relation

$$B = V_{sp}^2 (1 - 2\chi)/V_1 N_A \tag{4.94}$$

Where V_{sp} is the specific volume of a polymer, V_1 the molar volume of the solvent and N_A the Avogadro's number.

Rearrangement of eq. 4.93 gives

$$[\eta]/M^{1/2} = K_\theta + 0.51 \, \phi_0 BM^{1/2} \tag{4.95}$$

Thus, knowing the molecular weights of a series of fractions of a polymer and their corresponding intrinsic viscosities in a single solvent at a selected temperature, the constants K_θ and B can be evaluated from the intercept and slope of $[\eta]/M^{1/2}$ versus $M^{1/2}$ plots. Then the root mean square end-to-end distance in θ-condition $(\overline{r}_\theta^2)^{1/2}$ for a polymer chain can be calculated from the relation

$$K_\theta = \phi \left(\frac{\overline{r}_\theta^2}{M} \right)^{3/2} \tag{4.96}$$

Determination of θ-temperature/solvent
Several methods can be used to determine the theta temperature for a given polymer-solvent system.

(a) Phase equilibrium method
In this method, the maximum critical precipitation temperature T_c for a polymer

of different molecular weight is determined by solubility method. The plot of $1/T_c$ and $M^{1/2}$ is linear with an intercept equal to $1/\theta$.

(b) Second virial coefficient (A_2) as obtained from osmometry measurements
From membrane osmometry, one obtains the reduced pressure, π/C for a given concentration of polymer solute dissolved in suitable solvents. Then, a plot of π/C versus concentration is constructed to yield an intercept and slope values. The magnitude of slope depends highly on the quality of solvent. Such a plot for solutions of polymer solute in different solvents is shown in Fig. 4.35. The solvent in which the slope is zero ($A_2 = 0$) corresponds to θ-solvent at that temperature.

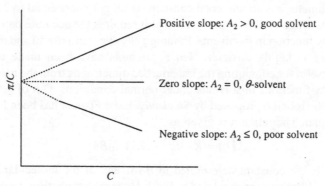

Fig. 4.35   Reduced osmotic pressure versus concentration plot for polymer solution in different solvents

(c) From viscosity-molecular weight relationship
The dependence of intrinsic viscosity $[\eta]$ on molar mass of polymer chains is given by the empirical Kuhn-Mark-Houwink-Sakurada equation as

$$[\eta] = K_\eta M^\alpha \tag{4.97}$$

This equation is valid for macromolecules with wide range of molar masses. In the above equation K_η and α are polymer solute-solvent system specific constants that depend on the constitution, configuration and molar mass distribution of the polymer as well as on the quality of solvent and temperature. The exponent α value is dependent on the shape of the polymer molecules and segmented distribution. It is generally predicted by theory that the value of $\alpha = 0$ for spheres, $\alpha = 0.5$ when the polymer chain acquires unperturbed dimensions (in θ-solvents) $\alpha = 0.764$ for perturbed coils (in good solvents) and $\alpha = 2$ for infinitely thin rigid rods. Thus, from the measurements of viscosity functions for several fractions of the same polymer with different molar masses and in wide variety of solvents, and using eq. 4.97, the value of α for each polymer solute-solvent system is determined. Also, the system in which α is close to 0.5 can be regarded to exist in θ-conditions.

(d) Titration (cloud appearance) method
Polymer solutions of different concentrations are titrated with a non-solvent till

the first appearance of turbidity. The log [nonsolvent] versus log [polymer] plot on extrapolation to 100% turbidity gives the composition of solvent/non-solvent mixture corresponding to θ-conditions.

Table 4(h) gives θ-conditions i.e. θ-solvents and θ-temperatures for various polymers.

Table 4(h) θ-Conditions for various polymers

Polymer	Solvent	Theta temp., K
Polyethylene	Biphenyl	398
Polypropylene	Isoamyl acetate	307
Polystyrene	Cycohexane	308
Polystyrene	Decalin	300
PMMA	4-Heptanone	307
PVC	Benzyl alcohol	428
Polyacrylic acid	Dioxane	297

4.4   Rheological Behaviour

Rheology is the science of deformation and flow of material (*rheos* in Greek means flow; the term was coined by Bingham who is known as father of rheology). The basic aspects of material deformation involve melt flow and hence are important from the view point of polymer processing. Low molecular weight solids and liquids exhibit the characteristic flow behaviour according to Hooke's equation (elasticity of solids) or Newtonian equation (viscosity of liquids). A polymeric material is often visualised as viscoelastic (that can act as both solids and liquids).

Hooke's equation

The elastic component is dominant in solids and hence their mechanical properties are described by Hooke's law, which states that 'applied stress' (S) is proportional to the resultant strain γ but is independent of the rate of the strain ($d\gamma/dt$). Many metals, rubbery materials etc. display elastic behaviour at load (stress) levels below elastic limit.

The behaviour of an ideal elastic solid can be represented by the elongation versus time plot as shown in Fig. 4.36. The application of load at a certain point of time t elongates the material by a certain length, which remains same as long as the load is retained. Releasing the load say after time t_2 leads the material to restore its original length instantaneously. Such a time independent behaviour is called elastic deformation (or elasticity) and is shown by the linear stress-strain plot. The elastic behaviour of solids can be

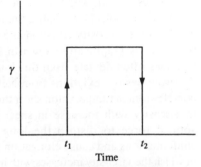

Fig. 4.36   Elongation versus time plot

compared with a spring that elongates instantaneously on applying load but recoils to its original shape on unloading.

Fig. 4.37 shows the typical elastic deformation and Hooke's spring model.

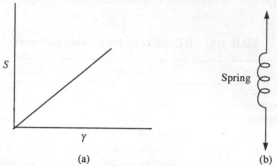

(a)   (b)

**Fig. 4.37   (a) Behaviour of elastic solid under strain and
(b) Hooke's spring model**

The stress, strain and modulus are defined as: stress S = force/area, strain γ = change in length/original length, modulus elasticity E = stress/strain = S/γ. The E values in psi (pound/inch^2) for some materials are: glass = (10×10^6), nylon 6 = (6×10^5) and polyethylene = (4×10^4).

Newton's equation
Like in liquid, the elongation versus time plot obeys Newton's law which states that the applied stress S is proportional to the rate of strain $d\gamma/dt$ but independent of strain γ (the strain rate is sometimes written as $\dot{\gamma}$). Thus, $S = \eta \cdot d\gamma/dt$, where η is proportionality constant and equals the viscosity of liquid. The viscosities (in Pa.s) of various materials for example in liquid state are; water (10^{-3}), glycerin (10^0), polymer melt (10^2-10^6) and glass (10^{21}). The variation of γ and t and S and $\dot{\gamma}$ are shown in Fig. 4.38. The application of load at a certain point of time t_1 to a viscous liquid (say honey, coal tar etc.) which then starts elongating and continues to do so until the force is released at time t_2 when the flow stops but the material does not recover its original length. Such a permanent and time dependent deformation is called as viscous flow. This behaviour can be compared to the movement of a piston inside a cylinder filled with a viscous liquid (the liquid does not allow the piston to move fast).

Many low molecular weight systems display laminar flow behaviour as described by Newton's viscosity equation and hence are called as Newtonian fluids. Such behaviour is graphically shown in Fig. 4.39. Polymer melts and concentrated solutions often deviate from this linear behaviour.

Two types of deviations from Newton's flow are commonly observed. Some non-Newtonian fluids exhibit shear thinning. Shear thinning is a reversible decrease in viscosity with increase in shear rate that results from the tendency of the applied force to disturb the long chains from their favoured equilibrium conformations and cause elongation in the direction of shear. The opposite effect in which the viscosity increases with increasing shear rate is called shear thickening. This phenomenon is very rare in polymer solutions.

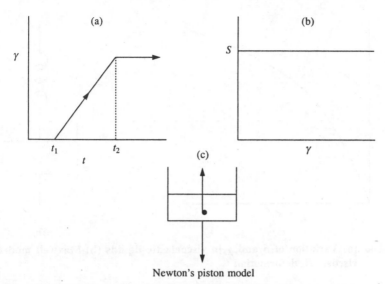

Fig. 4.38 Variation of (a) strain versus time, (b) stress versus strain and (c) schematic representation of Newton's model

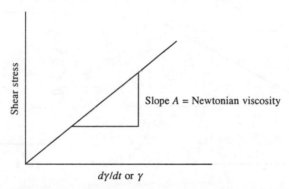

Fig. 4.39 Newtonian variation of shear stress as a function of shear rate

Viscoelasticity is the property of both a solid and liquid and since polymeric materials are viscoelastic solids, combination of the above Hooke's model of elasticity and Newton's model of viscosity can be used to demonstrate the deformation resulting from the application of stress. Maxwell put forth a combination of the above two models in series to explain viscoelasticity. He assumed that the contributions of both the spring and dash pot to strain were additive and that the application of stress would cause instantaneous elongation of the spring followed by a slow response of the piston in the dash pot.

According to Maxwell's model for viscoelastic deformation the shear rate is given as $\gamma_{total} = \gamma_{elastic} + \gamma_{viscous}$. Maxwell model assumes the following dependence of S on γ (see Fig. 4.40).

Another model to explain the viscoelasticity proposed at the same time is called Voigt-Kelvin model (see Fig. 4.41) which combines a spring and dash pot in parallel. In this model, the applied stress is shared between the spring and the

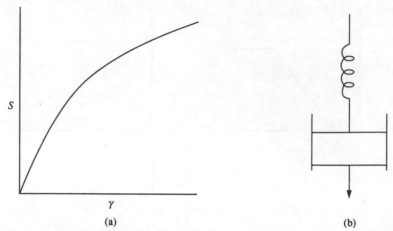

(a)   (b)

Fig. 4.40 (a) **Variation of** S **and** γ **in viscoelastic liquids (b) Maxwell model for viscoelastic deformation**

dash pot and thus the elastic response is retarded by the viscous resistance of the liquid in the dash pot.

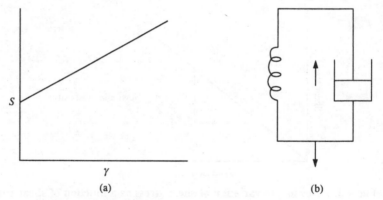

(a)   (b)

Fig. 4.41 (a) **Stress-strain relation and** (b) **Voigt-Kelvin model for representing viscoelasticity**

Stress-strain behaviour in polymers

High E values means rigid materials since stress will be very large and elongation small. Based on typical stress-strain behaviour, polymers are classified as: (a) soft and weak: low E, low yield point and moderate time dependent elongation, Poisson ratio (contraction/elongation) ~ 0.5, similar to liquids, e.g. polyisobutylene, (b) hard and brittle: high E, poorly defined yield point, little elongation before failure, Poisson ratio ~ 0.3, e.g. polyethylene, (c) soft and tough: low E, high elongation, a well defined yield point, Poisson ratio ~ 0.5–0.6 e.g. plasticized PVC, (d) hard and strong: high E, high yield strength e.g. rigid PVC, (e) hard and tough: moderate elongation prior to the yield point followed by nonrecoverable elongation e.g. ABS resins. In all the cases, the behaviour prior to yield point is Hookean. The reversible recoverable elongation prior to the yield point is called elastic range. A graphical sketch on the stress-strain behaviour in above mentioned classes of polymers is shown in Fig. 4.42.

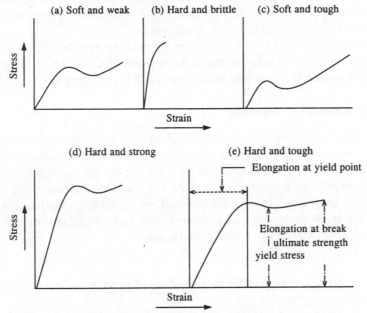

Fig. 4.42 (a–e) Stress-strain behaviour in various polymers based on their viscoelastic characteristics

The stress-strain behaviour of typical classes of polymers is shown in Fig. 4.43. The typical rubbers, plastics and fibers show a varied response to high stresses. A reinforced or crystallizable rubber exhibits a relatively low value of breaking stress, but high elongation. A ductile plastic such as polyethylene exhibits yielding, drawing and at high elongations, some strengthening due to orientation. A brittle plastic such as polystyrene does not yield much and breaks at high clongation.

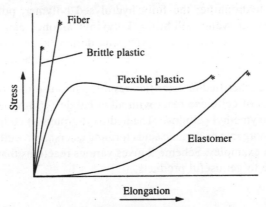

Fig. 4.43 The Stress-elongation behaviour of typical classes of polymers (*—point at which failure occurs)

A fibre exhibits the highest strength, high stiffness and low elongation. In all the three cases, the shape and position of the stress-strain curve are determined by the balance struck between cohesion and segmental mobility at the testing rate and temperature concerned. As with the low stress behaviour discussed above, behaviour is sensitive to these last two parameters.

Thus a polymer that is brittle at room temperature may become ductile on raising the temperature and lowering the testing rate. Similarly a normally ductile polymer becomes brittle at low temperatures and high testing rates.

4.5   Chemical Behaviour

Various transformations and chemical reactions can be carried out by polymers but the number of such reactions is very limited unlike low molecular weight substances. Polymer degradation and cross-linking are the examples involving reactions in polymer backbone in solid state. Polymers, like low molecular weight substances may react in selected solvents, provided the reaction sites are accessible in polymer molecules. Some of the typical reaction of polymers involving functional groups are described below:

Class-I
Polyvinyl alcohol is obtained by the hydrolysis of polyvinyl acetate as shown in scheme-1.

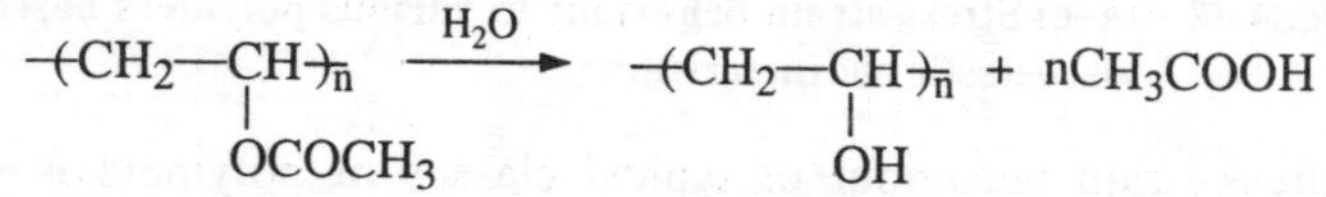

$$-(CH_2-CH)_{\overline{n}} \xrightarrow{H_2O} -(CH_2-CH)_{\overline{n}} + nCH_3COOH$$

$$OCOCH_3 \qquad\qquad OH$$

Polyvinyl acetate   Polyvinyl alcohol

Scheme   1

No direct method for the polymerisation of the vinyl alcohol monomer is possible (as this monomer is not stable). Polyvinyl acetate is water insoluble and its hydrolysis can be controlled to a definite degree. Increased hydrolysis makes the polymer more hydrophilic; the fully hydrolysed polymer, polyvinyl alcohol possesses excellent water solubility. Polyvinyl alcohol can be sulfated or phosphorylated to obtain polyelectroytes with desired ionic content.

Class-II
Cellulose can be etherified or esterified resulting into useful products. The free hydroxyl groups of cellulose react with alkyl halide forming ethers e.g. ethyl cellulose, carboxymethyl cellulose. The hydroxyl groups of cellulose react with acid groups of inorganic or organic acids forming esters. Nitrocellulose, cellulose acetate are such examples. Scheme-2 gives various reactions that can be carried out on cellulose to get useful products.

Class -III
Polyacrylates and polyacrylamide are hydrolysed (partially or completely) to yield polymers with characteristics properties (see scheme-3)

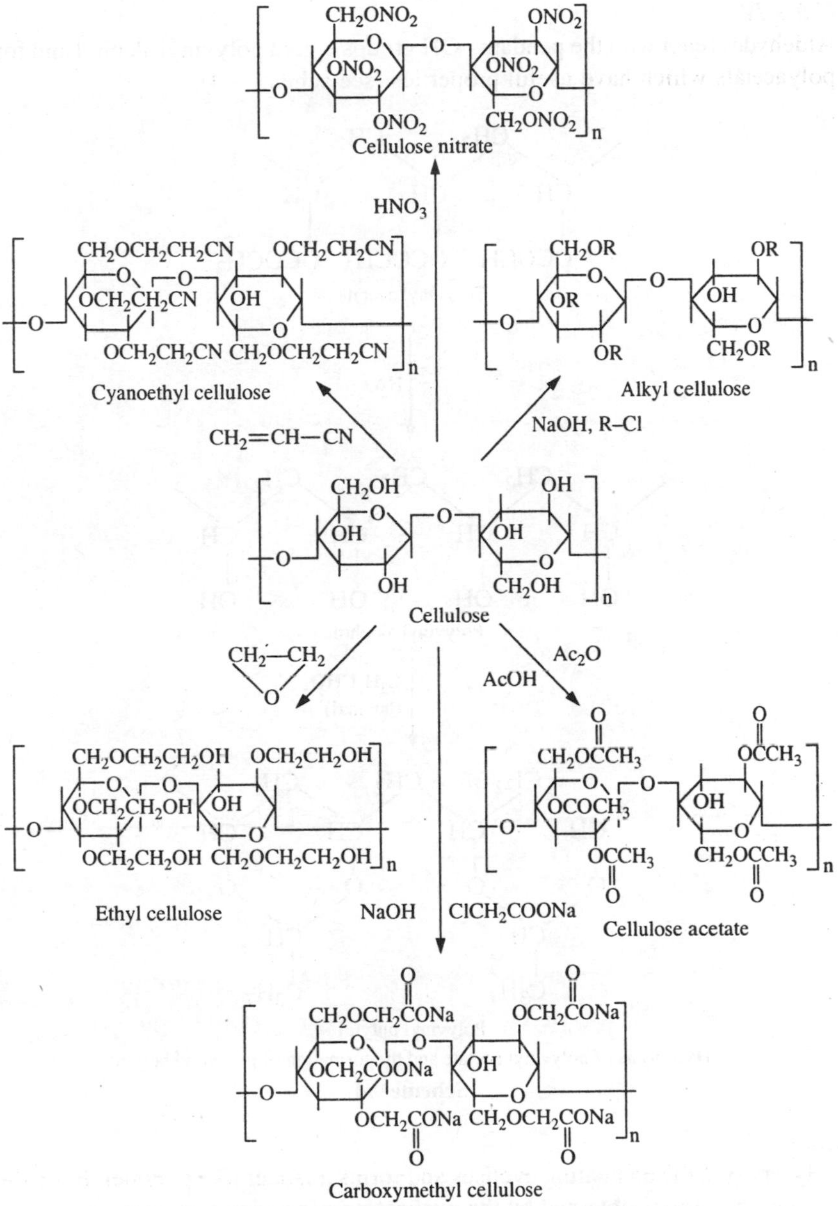

Cellulose nitrate

HNO₃

Cyanoethyl cellulose

CH₂=CH—CN

Alkyl cellulose

NaOH, R–Cl

Cellulose

CH₂—CH₂ / O

Ac₂O
AcOH

Ethyl cellulose

NaOH | ClCH₂COONa

Cellulose acetate

Carboxymethyl cellulose

Scheme 2   Reaction of cellulose giving useful products

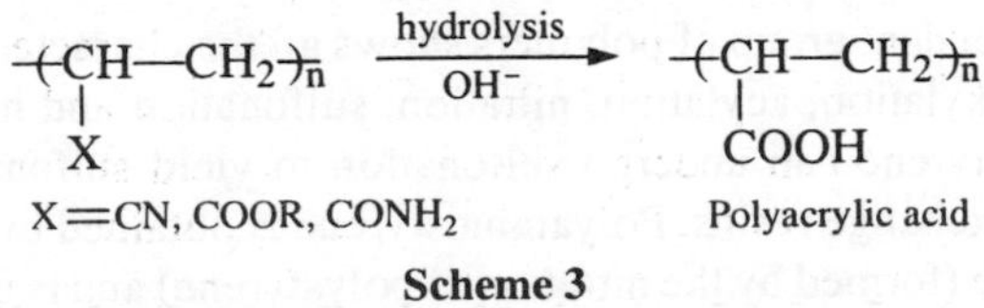

$$+CH-CH_2+_n \xrightarrow[OH^-]{hydrolysis} +CH-CH_2+_n$$

X

X = CN, COOR, CONH₂

COOH

Polyacrylic acid

Scheme 3

Class-IV

Aldehydes react with the pendant –OH groups e.g. in polyvinyl alcohol and form polyacetals which have useful properties (see scheme - 4).

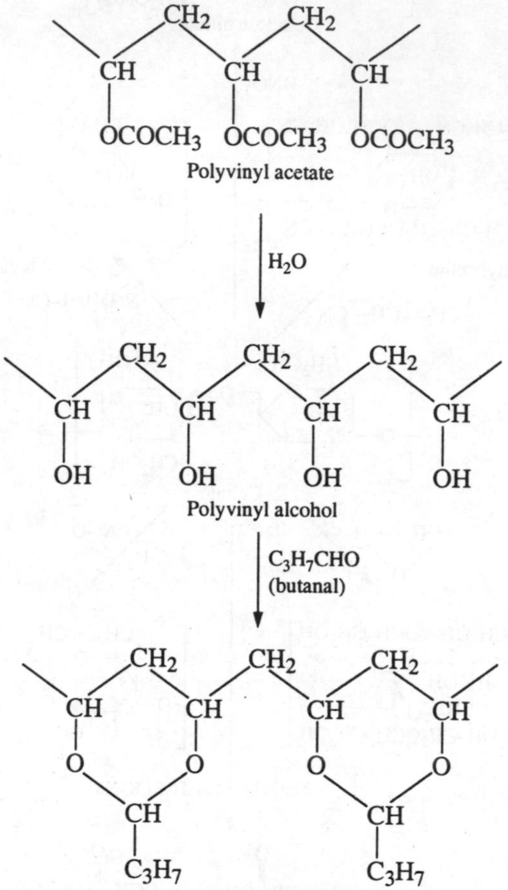

Polyvinyl acetate

H_2O

Polyvinyl alcohol

C_3H_7CHO
(butanal)

Polyvinyl butyral

Hydrolysis of polyvinyl acetate and the formation of polyvinyl butyral

Scheme   4

Class -V

Polyacrylonitrile on heating cyclises and forms a ladder like polymer. It is called carbon fiber and is obtained by the cyclization of the pendant nitrile group. The formation of carbon fiber is depicted in scheme-5.

Class -VI

The aromatic pendant group of polymers shows all the characteristic reactions of benzene e.g. alkylation, acylation, nitration, sulfonation and halogenation. For example, polystyrene can undergo sulfonation to yield sulfonated polystyrene useful for ion exchange resins. Polyaminostyrene is obtained by the reduction of polynitrostyrene (formed by the nitration of polystyrene) and is useful for making

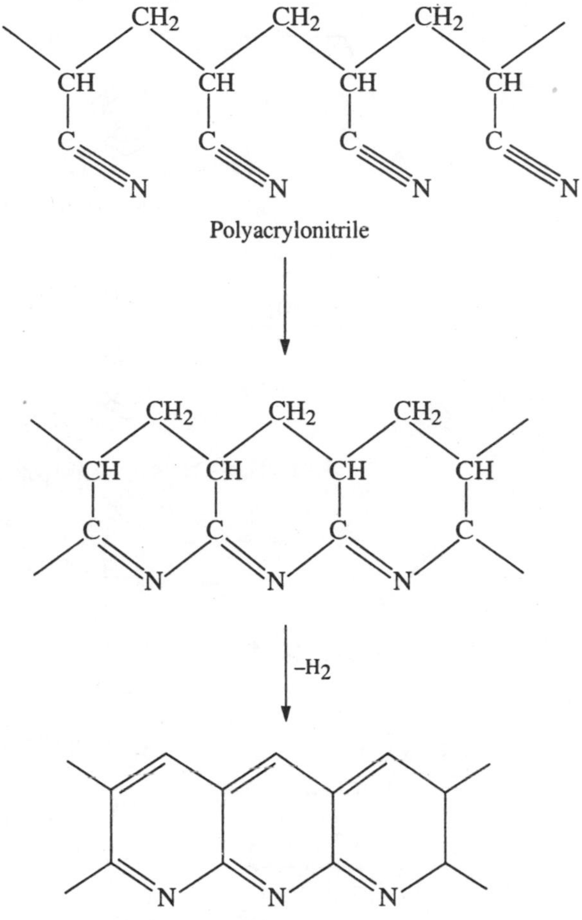

Polyacrylonitrile

$-H_2$

Polyquinizarine (Carbon fibre)

Cyclization and dehydrogenation of polyacrylonitrile

Scheme 5

polymeric dyes. The various possible chemical reactions of polystyrene are shown in scheme-6.

Some other well known examples involving chemical reactions are the formation of styrene-maleic acid copolymers by the hydrolysis of styrene-maleic anhydride copolymer, hydrogenation of styrene-butadiene block copolymer to styrene-butane block copolymers, formation of polyimides by heating polyamides and transformation of natural rubber into useful derivatives. These all are illustrated in schemes-7-10 respectively.

Cross-linking

Cross-linking means the formation of bonds between different chains of linear polymers leading to three dimensional network structures. Though the cross-linking can result due to secondary bonds (e.g. H-bonding), chemical cross-

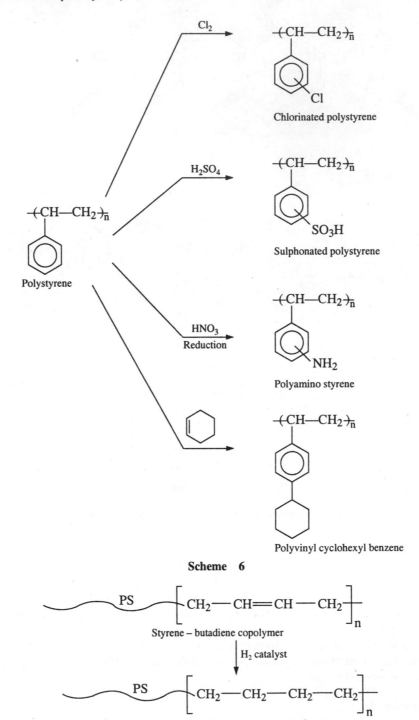

Chlorinated polystyrene

Sulphonated polystyrene

Polyamino styrene

Polyvinyl cyclohexyl benzene

Scheme   6

Styrene – butadiene copolymer

H₂ catalyst

PS – hydrogenated PB block copolymer

Hydrogenation of styrene-butadiene block copolymer

Scheme   7

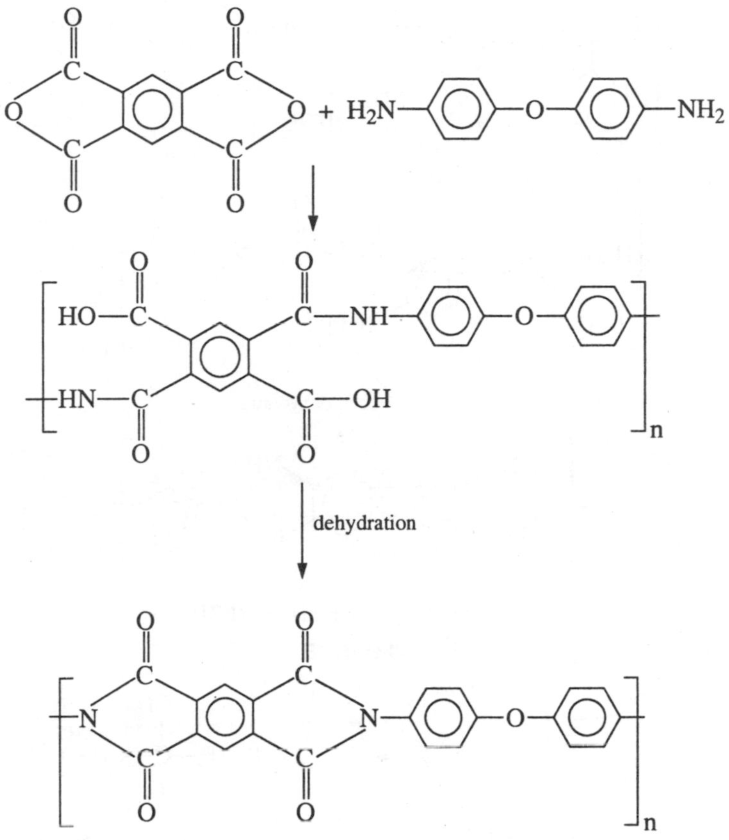

Scheme 8

linking involves covalent linkages, which is by nature irreversible. A chemically cross-linked polymer becomes a thermosett which does not melt on heating and does not dissolve in any solvent. A typical network formed due to cross linking of linear polymer chains is shown in Fig. 4.44.

Cross-linking in an elastomer (Vulcanisation)

Addition of some reagents develops cross-linking in linear polymers. The best known example of such cross-linking is of vulcanisation—a process which was discovered by *Charles Goodyear* in 1839. Mastication of natural rubber with sulphur develops cross-links thus improving processing and utility of rubber. Although the mechanism of vulcanisation appears quite complex and still not known in detail. The cross-linking as earlier thought is not an auto-accelerated process. It is not initiated by free radicals but by organic bases and acids, thus it is probably an ionic (cationic) process. The sulphur added as a cross-linking agent, itself is a multimer of 8 atoms i.e. cyclooctasulphur S_8. The probability is that the S_8 dissociates into $S_m^{\oplus}$ and S_n ions in the initial stages of vulcanisation. The sulphur cation $S_m^{\oplus}$ reacts with a diene molecule of rubber to give I. This cation adds to another polydiene molecule to give II and III. The allylic carbonium

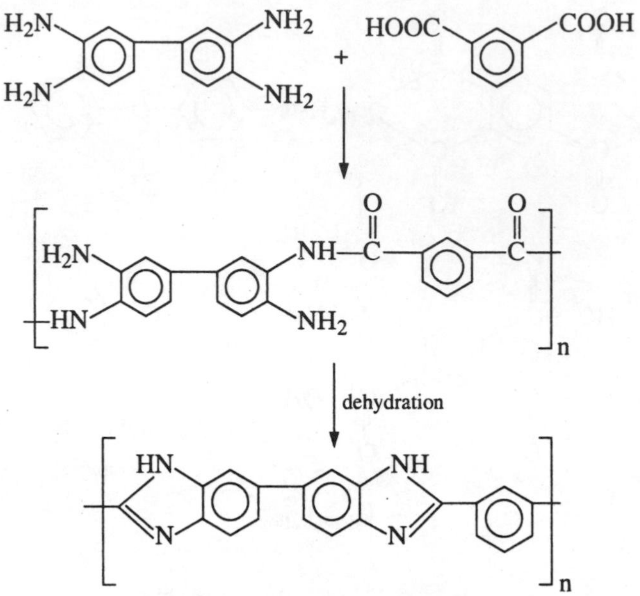

Aromatic polybenzimidazole (PBI)

Scheme   9

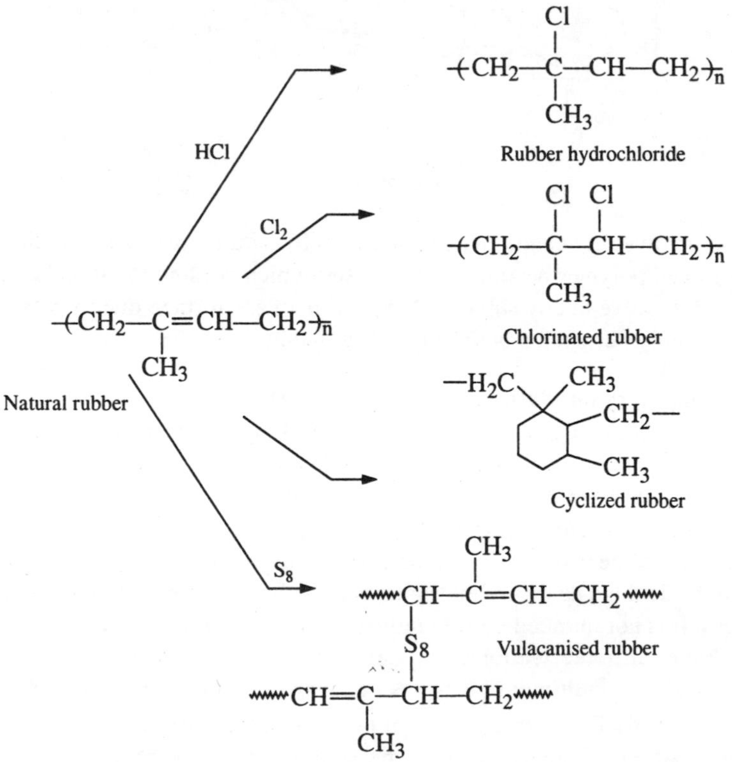

Scheme   10

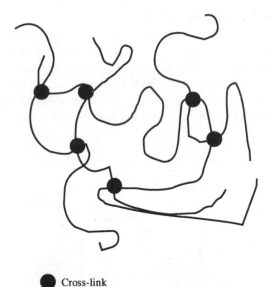

● Cross-link

Fig. 4.44   A typical cross-linked network

cation III reacts with S_8 to form IV and this adds to polydiene to form V via a sulphur bridge. Form V also has a cation, which undergoes a transfer reaction to polydiene, which regenerates III, and the cycle or curing will go on. The scheme-11 as given below depicts various steps during the vulcanisation process.

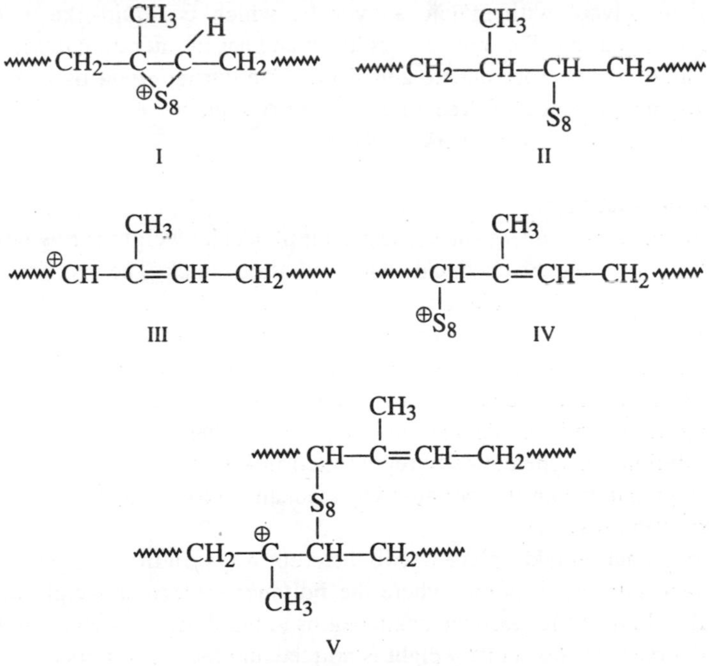

$$\text{wwww CH}_2\text{—}\underset{\underset{\oplus S_8}{|}}{\overset{\overset{CH_3}{|}}{C}}\text{—}\underset{\overset{H}{\diagup}}{C}\text{—CH}_2\text{ wwww}$$

I

$$\text{wwww CH}_2\text{—CH—}\underset{\underset{S_8}{|}}{\overset{\overset{CH_3}{|}}{CH}}\text{—CH}_2\text{ wwww}$$

II

$$\text{wwww}\overset{\oplus}{C}H\text{—}\overset{\overset{CH_3}{|}}{C}\text{=CH—CH}_2\text{ wwww}$$

III

$$\text{wwww CH—}\underset{\underset{\oplus S_8}{|}}{\overset{\overset{CH_3}{|}}{C}}\text{=CH—CH}_2\text{ wwww}$$

IV

$$\text{wwww CH—}\underset{\underset{S_8}{|}}{\overset{\overset{CH_3}{|}}{C}}\text{=CH—CH}_2\text{ wwww}$$
$$\text{wwww CH}_2\text{—}\underset{\underset{CH_3}{|}}{\overset{\oplus}{C}}\text{—CH—CH}_2\text{ wwww}$$

V

Vulcanisation of natural rubber

Scheme   11

Cross-linking of thermoplasts (Curing)

Low molecular weight polymers usually those containing either reactive functional groups or double bonds in their molecule can be cross-linked to obtain a three dimensional polymer. This process is called curing (see scheme-12). The low molecular weight polymer, which is liquid-like, is called prepolymer. Substances, which develop cross-links, should have at least bifunctionality and are called curing agents. An important example of curing reaction is described below.

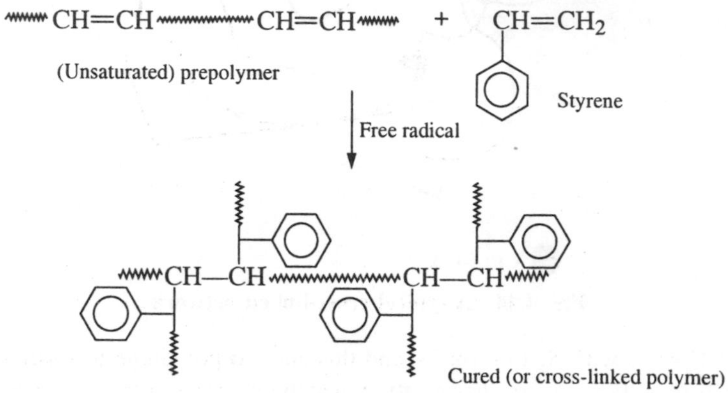

Scheme 12

An unsaturated polyester resin made as prepolymer by the polycondensation of ethylene glycol with maleic anhydride, which is liquid-like and yields polymer upon curing. This polyester is linear and has unsaturation in the polymer chain which comes from maleic anhydride. The curing agent used is styrene, which in the presence of free radical develops chemical cross-links in the prepolymer producing a network structure.

Polymer degradation

Polymeric materials undergo decrease in their molecular weight during fabrication and/or long term use, by chemical and/or radiation attack. The loss of molecular weight is mostly undesirable (however desirable at times) and is called polymer degradation. The degradation is caused by the scission of the backbone by certain agents like heat, light and other ionizing radiations, water and bacteria etc. Degradation in polymer chains is described under several titles depending on the factor/agent that induce degradation in a polymer chain. Several possible types of degradation are schematically represented in Fig. 4.45.

The degradation can also be caused by mechanical force, high energy radiation and ultrasound wave etc.

The degradation takes place in two different ways. In the first chain scission takes place at random points where the polymer molecular weight decreases drastically. In the other case the chain begins to break from its ends. In this case a slow decrease in molecular weight is noticed and the monomers are liberated. This type of degradation is called unzipping and is reverse of polymerisation. It is therefore also called *depolymerisation*. Polymer degradation can be studied

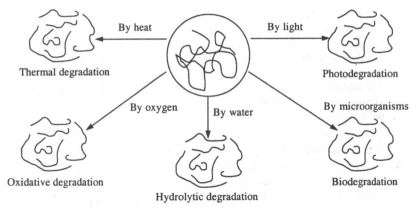

By heat

By light

Thermal degradation

Photodegradation

By oxygen

By water

By microorganisms

Oxidative degradation

Biodegradation

Hydrolytic degradation

Fig. 4.45   Types of polymer degradation

using thermal methods of analysis, spectroscopy and by chromatography. Since the degradation is accompanied by a decrease in molecular weight, routine determination of intrinsic viscosity would serve a better way to investigate polymer degradation. Agents like antioxidants, photo stabilisers are added to the polymers to prevent the unwanted degradation. Thermal degradation of polymers is practically important and it may follow either unzipping or random scission. Since many polymers contain C–C bonds in the main chain (backbone), the thermal stability of a polymer would thus depend on the stability of C–C bond. Several factors influence the C–C bond stability. Introduction of methyl substituents in the polymer decrease the bond stability and it is for this reason Polyolefins show the following order in thermal stability; polypropylene > polyisobutylene. For the same reason the following trend in the thermal stability of polymers can be understood: polyethylene > polystyrene > poly α-methylstyrene.

High thermal stability of teflon is the exception to this rule. Introduction of aromatic ring in the polymer main chain increases thermal stability of polymers. Polycarbonates, polyphenylenes, aromatic polyamides are fairly stable to high temperature. Branching or introduction of oxygen atom in the polymer chain decreases the thermal stability of polymers. Polyethylene oxide is degradable while polyethylene is stable. In some cases, the polymer chain remains intact but the pendant groups are eliminated. For example, polyvinyl chloride degrades at around 200°C with the elimination of HCl.

Numerical exercises

1. Estimate the contour length and root mean square end-to-end distance of polypropylene chain with DP = 5000. The end-to-end distance between carbon atoms is 0.126 nm.
   Solution — The contour length = 2 × 0.126 × 5000 = 1260 nm, the root mean square end-to-end distance = $r = nl^{1/2} = 0.126 \times (2 \times 5000)^{1/2} = 0.126 \times 100 = 12.6$ nm.
2. The enthalpy change for the polymerisation of methyl methacrylate monomer is

-57 kJ mol^{-1}. If the ceiling temperature for this monomer is 434 K, what would be the value of entropy change in polymerisation?

Solution At ceiling temperature the equilibrium between monomer and polymer is established and thus free energy change is 0. Since $\Delta G = \Delta H - T\Delta S$, $O = -57 - 434\ \Delta S$ or $\Delta S = -(57/434)$ kJ K^{-1} mol^{-1}. $= 131.3$ JK^{-1} mol^{-1}.

3. The intrinsic viscosity of a polymer is measured in THF and in CHCl$_3$. The intrinsic viscosities are in the ratio $[\eta]_{THF}/[\eta]_{CHCl_3} = 1.7$. Calculate the ratio of the mean-square end-to-end distances.

   Solution Intrinsic viscosity, $[\eta]$ is proportional to r^3, hence $[(\bar{r}^2)^{1/2}$ in THF]$/$ $[(\bar{r}^2)^{1/2}$ in CHCl$_3] = \{[\eta]$ in THF$/[\eta]$ in CHCl$_3\}^{1/3} = \{1.7\}^{1/3} = 1.2$.

4. Two samples of polystyrene with mol. mass 1×10^5 and 1×10^6 g mol^{-1} in a θ-solvent have intrinsic viscosities of 0.26 and 0.83 dl g^{-1}, respectively. A third sample of the same polymer in similar condition has intrinsic viscosity of 0.50 dl g^{-1}. Calculate the mol. mass of the third sample.

   Solution $[\eta] = KM^{1/2}$, for polymer solvent pair in θ-conditions. Thus $0.26 = K (1 \times 10^5)^{1/2}$ and $0.83 = K (1 \times 10^6)^{1/2}$. Solve the equations to get a K value of 8×10^{-4} dlg^{-1}. Then for the third sample, we can write $0.50 = (8 \times 10^{-4})M^{1/2}$ and its molar mass, M is 3.6×10^5.

5. The intrinsic viscosity for a polyethylene oxide sample in aqueous potassium sulphate under θ-conditions ($K = 130 \times 10^{-3}$) is 0.82 dl g^{-1}. For this sample in water ($K = 16.6 \times 10^{-3}$, $\alpha = 0.82$), calculate the chain expansion factor and rms end-to-end distance of the macromolecular coil.

   Hint Calculate M value from equation, $[\eta]_\theta = KM^{0.5} = 130 \times 10^{-3} M^{0.5} = 0.82$. Now calculate $[\eta]_{water}$ from equation (using calculated value of M).$[\eta]_{water} = KM^\alpha = 16.6 \times 10^{-3} M^{0.82}$. Evaluate chain expansion factor α from $\alpha^3 = [\eta]/[\eta]_\theta$. Now obtain n or degree of polymerisation $= M/44$ and $N = 2 \times n$. Calculate $(\bar{r}_0^2)^{1/2} = b[2N]^{1/2}$ (here $b = $ C—C bond length, 1.54Å). Calculate rms end-to-end distance of PEO in water from the relation $(\bar{r}_0^2)^{1/2} \times \alpha$.

6. Polyethylene oxide (PEO) in water and 0.1 M K$_2$SO$_{4(aq)}$ at 25°C has the following Mark-Houwink constants, K and α;

	$K \times 10^3$	α
Water	16.6	0.82
Aq. K$_2$SO$_4$	130	0.5

   Calculate the chain expansion factor for a PEO sample of mol. mass 50,000.

   Solution $\alpha = 0.5$ for aq. K$_2$SO$_4$ since it is a θ-solvent. Hence, $[\eta]_\theta = KM^\alpha = 130 \times 10^{-3} \times (50,000)^{0.5} = 29.1$ dl g^{-1}. For water, $[\eta] = K M^\alpha = 16.6 \times 10^{-3} \times (50,000)^{0.82} = 118.4$ dl g^{-1}.
   Chain expansion factor, $\alpha = ([\eta]/[\eta]_\theta)^{1/3} = (118.376/29.069)^{1/3} = (4.072)^{1/3} = 1.597$.

7. A PMMA sample in acetone at 30°C gave intrinsic viscosity of 0.565 dl g^{-1}. Calculate the mol. mass of the polymer when $K = 6.2 \times 10^{-5}$ dl g^{-1} and $\alpha = 0.72$. Also calculate rms end-to-end distance if under θ conditions the intrinsic viscosity of the polymer is given by relation $[\eta] = 4.8 \times 10^{-4} M^{0.5}$.

   Solution $[\eta] = KM^\alpha = 0.565 = (6.2 \times 10^{-5}) M^{0.72}$ or $M = 3, 16,000$ g mol^{-1}. Now, $[\eta]_\theta = 4.8 \times 10^{-4} \times (3, 16,000)^{0.5} = 0.269$ dl g^{-1}. The chain expansion factor $\alpha = [([\eta]/[\eta]_\theta)]^{1/3} = (0.565/0.269)^{1/3} = 1.280$. The repeat unit in PMMA has a mol. mass of 100. Thus, $n = 3160$ and $N = 6320$. Hence $(\bar{r}^2)^{1/2} = b\sqrt{2N} = 1.54 \sqrt{12640} = 173$Å. In acetone, $(\bar{r}^2)^{1/2} = \alpha(\bar{r}^2)^{1/2} = 1.279 \times 173 = 221$ Å.

8. Calculate (i) the contour length and (ii) the extended chain length and (iii) rms end to end distance for polyethylene of mol. weight 1,00,000.

Solution The polymer is made up of —$(CH_2$—$CH_2)_n$— repeat unit. Thus C—C bond length = b = 1.54 Å and C—C—C bond angle θ = 109.5°, mol. wt. of repeat unit = 28 and thus

Contour length = $2 \times n \times b$ = $2 \times (100000/28) \times 1.54$ = 11000 Å

$$\text{Extended length} = 2 \times n \times b \sin \theta/2$$
$$= 2 \times (100000/28) \times 1.54 \sin (54.75)$$
$$= 8983 \text{ Å}$$

rms end-to-end distance for freely jointed chain = $b\sqrt{2N}$ = $1.54\sqrt{2} \times 100000/28$ = 130 and rms end-to-end distance with bond angle restriction = $b\sqrt{2N}$.

Suggested Questions

1. Describe briefly some factors that affect the solubility of polymers. Define:
   (a) Good solvent
   (b) Poor solvent
   (c) θ-Solvent
   (d) Selective solvent
   (e) Hildebrand's regular solvent
2. Indicate how solvent "power" (good solvent versus poor solvent) influences:
   (a) The intrinsic viscosity of the polymer sample
   (b) The molar mass of a polymer sample as determined by membrane osmometry
3. Write units of
   (a) Intrinsic viscosity
   (b) Solubility parameter
   (c) Partial specific volume
   (d) Second virial coefficient
   (e) Modulus of elasticity
4. (i) What is the value of free energy change for mixing at the θ-temperature? (ii) What is Hildebrand's solubility parameter? How is it important in dissolution of polymers in solvents? (iii) Ethyl benzene is soluble in methanol but not polystyrene, why? (v) Addition polymers are normally not crystalline, why?
5. Explain the terms:
   (a) Rheopectic
   (b) Viscoelastic
   (c) Thixotropic
   (d) Isotropic
   (e) Microgels
   (f) Newtonian and non-Newtonian law
6. Write notes on
   (a) Solubility parameter
   (b) Dimensions of macromolecular coil
   (c) Polymer chain flexibility
7. Explain the terms
   (a) Chemical heterogeneity
   (b) Polymer chain flexibility
   (c) Radius of gyration
   (d) Ideal and regular solutions
8. What are polyelectrolytes? Explain viscosity behaviour of a polyelectrolyte in water and aqueous salt solution.
9. Describe the dissolution process of polymers in solvents. Discuss thermodynamics of polymer dissolution. How do the following factors influence the polymer solubility?
   (i) Increase in molecular weight   (ii) Crosslinking   (iii) Crystallinity
10. Describe the Flory-Huggins theory of polymer solutions. Obtain the necessary derivation.

11. Comment on the limitations of Hildebrand solubility parameter equation. How is the solubility parameter related to the heat of mixing?

12. Explain the terms Flory's interaction parameter and give its importance. Describe the method which can be used to determine its value for:
    (a) soluble polymers   (b) cross-linked polymers

13. Answer the following:
    (a) How does the crystallinity of polymers influence the film and fiber properties?
    (b) How does the tacticity influence the solution and solid state properties of polymers?

14. Explain the differences in (a) Plasticization and vulcanisation (b) Newtonian and non-Newtonian flow (c) Amorphous and crystalline polymers (d) Maxwell and Voigt model of viscoelsaticity

15. What effect do the crystalline polymers have on the solid state and solution properties? Picturise the 'Fringed micelle model' of the crystalline-amorphous structure of polymers.

16. (a) Draw schematically the stress-strain curves for different plastics with varying hardness and toughness (b) Draw schematically the stress-strain curves for plastics, fibres and elastomers

17. Define: (a) θ-Temperature (b) Ceiling temperature (c) Glass transitiion temperature (d) Softening temperature (e) Critical solution temperature

18. Expalin the term polymer chain flexibility. What factors influence it? How is the chain flexibility related to end-to-end distance?

19. Comment on the following:
    (i) The polymers of ω-amino acids are termed as 'nylon n', where n is the number of consecutive carbon atoms in the chain (general formula $-[NH-CO-(CH_2)_{n-1}]-$. The polymers are crystalline and do not dissolve in either water or hexane. They, however, absorb solvent when immersed in each liquid.

    (ii) Polyvinyl alcohol is made by the hydrolysis of polyvinyl acetate because the monomer vinyl alcohol is unstable. Hydrolysis can be controlled anywhere from 0–100%. Polyvinyl acetate is insoluble in water, polymer with increased hydrolysed character becomes more and more soluble and 100% hydrolysed polyvinyl acetate, i.e. polyvinyl alcohol is very soluble in water. Explain its solubility behaviour.

    (iii) Cellulose and starch, both are polymers of glucose, yet cellulose does not dissolve in water whereas starch is soluble. Comment why methyl cellulose is more water soluble than cellulose.

    (iv) Crystalline polymers can be dissolved only above their T_m in the presence of strongly interacting solvent.

20. The chemical reactions shown by polymers are few as compared to those by low molecular weight substances. Why?

21. What are various chemical transformations in polymers? Describe in detail polymer degradation, cross-linking and stabilisation.

22. How can the following polymer transformations be carried out?
    (a) Polyvinyl acetate to polyvinyl alcohol (b) Styrene-isoprene block polymer to styrene (ethylene/propylene) block copolymer (c) Polyacrylonitrile to carbon fiber (d) Cellulose to acetate rayon. Mention the usefulness of these transformations.

Suggested Further Readings

Crystalline Behaviour of Polymers
Alexander, L.E., *X-ray Diffraction Methods in Polymer Science*, New York: John Wiley, 1969.

Balta-Calleja, F.J., and C.G. Vonk, X-Ray *Scattering of Synthetic Polymers*, New York: Elsevier, 1989.

Geil, P.H., *Polymer Single Crystals*, New York: John Wiley, 1963.

Glatter, O., and O. Kratky, *Small Angle X-Ray Scattering*, New York: Academic Press, 1982.

Hemsley, D.A., *The Light Microscopy of Synthetic Polymers*, New York: Oxford Univ. Press, 1985.

Kakudo, M., and N. Kasai, *X-ray Diffraction by Polymers*, Amsterdam: Elsevier 1972.

Keinath, S.E., R.L. Miller, and J.K. Rieke, (eds.), *Order in the Amorphous State of Polymers*, New York: Plenum, 1987.

Mendelkern, L., *Crystallisation of Polymers*, New York: McGraw-Hill, 1964.

Rhodes, G., *Crystallography Made Crystal Clear. A Guide for users of Macromolecular Models*, New York: Academic Press, 1993.

Sandman, D.J. (ed.) *Crystallographically Oriented Polymers*, Washington, D.C.: ACS, 1987.

Sawyer, L.C., and D.T. Grubb, *Polymer Microscopy*, New York: Chapman and Hall, 1987.

Sharples, A., *Introduction to Polymer Crystallisation*, London, Edward Arnold, 1966.

Tadakoro, H., *Structure of Crystalline Polymers*, New York: Wiley-Interscience, 1979.

Ward, I.M. (ed.), *Developments in Oriented Polymers*, 2 Parts, New York: Elsevier Applied Science, 1982 and 1987.

Woodward, A.E., *Understanding Polymer Morphology*, Munich: Hanser, 1995.

Wunderlich, B., *Marcomolecular Physics*, (3 vols.), New York: Academic Press, 1980.

Thermal Behaviour of Polymers:

Mathot, V.B.F., (ed.), *Calorimetry and Thermal Analysis of Polymers*, Munich: Hanser, 1994.

Turi, E.A., (ed.), *Thermal Characterisation of Polymeric Materials*, 2nd ed., New York: Academic Press, 1982.

Wunderlich, B., *Macromolecular Physics*, Vol. 3, Crystal Melting, New York: Academic Press, 1980.

Solution Behaviour of Polymers:

Barton, A.F.M., *CRC Handbook of Solubility Parameters and Other Cohesion Parameters*, Boca Raton, Fla.: CRC Press, 1983.

Bohdanecky, M., and J. Kovar, *Viscosity of Polymer Solutions*, Amsterdam: Elasevier, 1982.

Flory, P.J., *Principles of Polymer Chemistry*, Ithaca, New York: Cornell Univ. Press, 1953.

Fujita, H., *Polymer Solutions*, Amsterdam: Elsevier, 1990.

Jannık, G., and J. des Cloizeaux, *Polymers in Solutions*, Oxford: Oxford Univ. Press, 1990.

Kurata, M., *Thermodynamics of Polymer Solutions*, New York: Gordon and Breach, 1982.

Morawetz, H., (ed), *Macromolecules in Solution*, 2nd ed., New York: Wiley-Increascience, 1975.

Tanford, C., *Physical Chemistry of Macromolecules*, New York: Wiley-Interscience, 1961.

Tompa, H., *Polymer Solutions*, New York, Academic Press, 1956.

Yamakawa, H., *Modern Theory of Polymer Solutions*, New York: Harper and Row, 1971.

Rheological Behaviour of Polymers:

Aklonis, J.J., W.J. MacKnight, and M. Shen, *Introduction to Polymer Viscoelasticity*, 2nd ed., New York: Wiley-Interscience, 1983.

Alfrey, T., *Mechanical Behaviour of High Polymers*, New York: Interscience, 1948.

Brydson, J.A., *Flow Properties of Polymer Melts*, Lancaster Penn.: Technomic, 1981.

Cogswell, F.N., *Polymer Melt Rheology,* New York: Halsted-Wiley, 1981.

Deanin, R.D., *Polymers: Structure, Properties and Applications,* Boston: Cahners, 1972.

Ferry, J.D., *Viscoelastic Properties of Poymers,* 3d ed., New York: John Wiley, 1980.

Mashelkar, R.A., A.S. Mujumdar, and R. Kamal, (eds.), *Transport Phenomenon in Polymeric Systems,* Englewood Cliffs, New Jersey: Prentice Hall, 1989.

Severs, E-T, *Rheology of Polymers,* New York: Reinhold, 1962.

Chemical Behaviour of Polymers

Cassidy, P.E. *Thermally Stable Polymers,* New York: Dekker, 1980.

Guven. O, *Crosslinking and Scission in Polymers,* Dordrecht: Kluwer, 1990.

Grassie, N., and G. Scott, *Polymer Degradation and Stabilisation,* Cambridge: Cambridge Univ. Press, 1988.

Mark, J.E., B. Erman, and F.R. Eirich, (eds.), *Science and Technology of Rubber,* 2nd ed., San Diego: Academic Press, 1994.

Moore, J.A. (Ed.) *Chemical Reactions of Polymers,* New York, Wiley-Interscience, 1964.

5

Polymer Technology

5.1 Physical Properties Versus Application

After the synthesis, identification and characterisation in terms of evaluation of
molecular weight and molecular weight distribution, crystallinity, glass transition
temperature and the determination of other physical properties (in solid state and
in solution), the polymers are then processed into a material which is in useful
form for a direct application. The processing of polymers is not straight forward.
Usually, polymers with very high molecular weight offer problems in processing
due to enormous increase in melt viscosity (see Fig. 5.1).

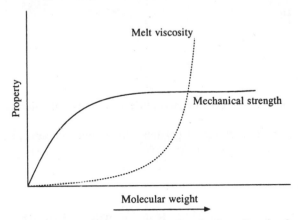

Fig. 5.1 The mechanical properties and melt viscosity of polymers as
a function of molecular weight

Depending upon various applications most polymeric materials can be classified
into major groups such as plastics, fibers, elastomers, coating materials and
adhesives. The materials belonging to these different groups vary in many properties
especially in solid state properties. One of the most important physical properties
i.e. mechanical properties differ to a great extent and also play a decisive role in
classifying the industrial polymers. The typical responses of rubbers, plastics
and fibers to large stresses are depicted in Fig. 5.2. A reinforced or crystallizable

rubber exhibits a relatively low value of breaking stress, but high elongation. A ductile plastic such as polyethylene exhibits yielding, drawing and at high elongation some strengthening due to orientation. A brittle plastic such as polystyrene does not yield much and breaks at a low elongation; a fiber exhibits the highest strength, high stiffness and low elongation. In all these cases, the form of the final curves i.e. the shape and position are determined by the balance struck between cohesion and segmental mobility at the respective and concerned temperature and testing rate. The strain behaviour under applied low stress is very sensitive to the temperature and testing rate. A polymer that is brittle at room temperature may thus become ductile on raising the temperature or lowering the testing rate. Similarly, a normally ductile polymer becomes brittle at low temperatures and high testing rates.

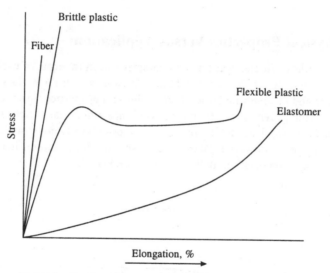

Fig. 5.2 **Stress-elongation behavior of typical polymer**

The characteristic features of plastics, fibers, elastomers and adhesives are discussed in the following manner.

5.2 Plastics

The word plastics is commonly used for a material that usually consists of organic polymers. The term plastic is nowadays refers to all thermoplastic polymers. These are the macromolecular compounds which, within a more or less wide temperature range, pass over from the solid state into a plastic mouldable state without having undergone chemical changes. This process is reversible. The thermoplastic behavior of polymers is due to the absence of cross-linking. Examples of plastics include polymeric hydrocarbons (e.g polyethylene, polypropylene, polyisobutylene etc.), vinyl polymers (e.g. polystyrene, polyvinyl chloride, polyvinyl acetate, polyvinyl alcohol, polyvinyl butyral, etc.), acrylics (polyacrylamide, polyacrylates, etc.). Many fluorine containing polymers (e.g.

teflon, polychlorotrifluorethylene and polyvinyl fluoride etc.) polyacrylonitrile and polymethyl methacrylate are well known examples of plastics.

The synthesis, properties and applications of these substances have already been described in previous chapters. Plastics are divided into two types, thermosetting and thermoplastic materials. The difference between these two is that thermoplastics become fluid upon heating above a certain temperature called heat distortion temperature and on cooling they again restore elastic or solid state. The process can be repeated many times without any effects on its chemical structure. Thermosetts can be heated to fluid state only once and the reverse process is not possible, because of the change in their chemical structure upon heating. Such plastics can not be re-softened because of cross-linking that takes place during the course of first heating. Some of the examples of thermoplastics are polyamides, polyesters, polyolefins and polystyrenes. Cured unsaturated polyesters, condensation products of urea-formaldehyde(UF) and melamine-formaldehyde(MF), phenol-formaldehyde(PF) and epoxy resins etc. are representative of thermosetts. Table 5(a) describes the typical properties of various thermoplastic materials.

Table 5(a)   Physical properties of some common thermoplastics

Property	PS	LDPE	HDPE	PP	PMMA	PTFE
Elongation %	1.2–2.5	90–800	20–1000	200–700	2–10	200–400
Tensile strength $(10^3$ psi$)$	5–12	0.6–2.3	3.1–5.5	4.3–5.5	7–11	2–5
Impact strength (ft.lb/inch of notch)	0.25–0.4	16	0.5–2.0	0.5–2.0	0.3–2.0	3.0
Tensile Modulus $(10^5$ psi$)$	4–6	0.2–0.4	0.6–1.8	1.6–2.3	4.5	0.58
Burning Rate	slow	very slow	very slow	slow	slow	none
Effect of alkalis	attacks	resistant	resistant	resistant	attacks	very resistant to acids
Effect of organic solvents	soluble	resistant (below 80°C)	resistant (below 80°C)	resistant (below 80°C)	soluble	very resistant
Clarity	transparent	opaque	opaque	opaque	transparent	opaque
Specific gravity	1.4–1.09	0.91–0.92	0.94–0.96	0.90–0.91	1.17–1.2	2.14–2.2

Thermosett materials are also often called as resins. The term resin was applied originally to naturally occurring amorphous organic solids such as amber, copal, dammar, shellac and rosin etc., but is now commonly used for synthetic polymeric materials similar to natural resins in physical properties and used for surface coatings. Resins differ from linear polymers in having cross-linked structures that produce three dimensional network. Thermosetting resins change irreversibly on heating from a fusible and soluble material to infusible and insoluble mass; such a conversion results into a thermally stable network. These materials are used for variety of purposes. Some common physical properties of thermosetting resins are listed in Table 5(b). A few examples of commercially important thermosetts are (i) phenolic resins, (ii) amine resins, (iii) epoxy resins (iv) unsaturated polyester resins (v) polyurethane resins (vi) silicone resins, and (vii) alkyd resins etc.

Table 5(b)   Physical properties of some common thermosetting (cross-linked) resins

Property	PF	Polyester	Epoxy	silicone
Elongation (%)	1–1.5	5	3–6	100–300
Tensile modulus (10^5 psi)	7.5–10	3–6.5	3.5	0.1
Impact strength (ft.lb/inch of notch)	0.2–0.35	0.2–0.4	0.2–1	—
Burning rate	very slow	burns	slow	self-extinguishing
Effect of organic solvents	resistant	attacks	resistant	attacks
Specific gravity	1.25–1.3	1.1–1.45	1.1–1.4	1–1.5

All resins are attacked by strong acids or alkalis and they are transparent in appearance. However, reinforced polymers are opaque and have modified properties.

5.3   Fibers

Fibers by definition are macromolecular substances, whose molecules have length to diameter ratio of at least 100 : 1. To qualify as a useful textile material, a polymer has to meet the following requirements, (i) high degree of polymerisation, (ii) high degree of orientation of molecules relative to longitudinal axis, (iii) high melting (or softening) point, (iv) fair elongation to break, (v) elasticity and (vi) chemical stability. Different polymer materials are used as fibers. Fibers can be broadly classified as natural, semi-synthetic and synthetic types. As expected natural fibers are derived from naturally available sources. Semi-synthetic fibers are produced by modifying the otherwise naturally occurring material. Synthetic fibers are, however, produced exclusively by the laboratory process. Cotton, silk and wool are the well-known natural fibers. Rayons are cellulose based semi-synthetic fibers. Among the synthetic (man made) fibers, the most common are nylons, polyesters and acrylic fibers. These are described below.

Natural Fibers

Cotton

Cotton is a vegetable fiber and is essentially made of cellulose, which is a linear polymer of β-glucose. Cellulose has a sheet like structure and is the principal structural component of plants. The strong intermolecular hydrogen bonding between hydroxyl groups gives cellulose a high degree of crystallinity and thus strength to cotton. For the same reason, cellulose despite containing a large number of hydroxyl groups is insoluble in water. However, the water uptake is high enough, and therefore, cotton clothes take long time to dry up. Fibers derived from cellulose are called rayons. The viscose rayon, cupraammonium rayon and acetate rayon are the derivatized celluloses which are used as fibers. These useful fibrous materials are called cellulosic (semi-synthetic) fibers. An early process for producing regenerated cellulose fiber involved treating cellulose with solution of ammonical copper sulfate and the resulting solution was then forced through spinnerets into an acid bath to produce cellulose as long filaments.

An alternative method of regeneration is to dissolve cellulose in a solution of sodium hydroxide and carbon disulphide. The resulting solution is called viscose and it is then forced through a spinneret into an acid solution. This form of cellulose is also called regenerated cellulose and can be further processed into fibrous form. It is known as viscose rayon and is a major commercial textile fiber. Cellulose acetate (acetate rayon) can be prepared by heating cellulose with acetic anhydride and acetic acid in the presence of sulfuric acid. It is also used in making films and lacquers in addition to fibers.

Wool

Wool is an animal protein fiber. Like all other proteins, wool is also made up of α-amino acids. The amino acids are joined end-to-end in the form of peptide chains making long chain molecules. These polypeptide chains are joined through cross-linkages which give wool many characteristic structural and tensile properties. Polypeptide chains in wool occur normally in folded form and these chains can be unfolded on stretching the fiber. The folded form is β-keratin. Two sulfur containing α-amino acids found in wool as well as in other keratins are cysteine and methionine. Many of the physical and chemical properties of wool are due to the disulfide bonds formed by covalent cross-linking of the cysteine molecules. Any chemical that weakens or destroys disulfide bonds may cause structural break-down of the fiber. Many other acidic and basic side chains are also present in the polypeptide chain of wool. The presence of these side chains causes the formation of "salt-linkage" which are responsible for the amphoteric nature of wool, i.e. it behaves both as acid and base. The peptide chains are held together also through hydrogen bonds which are responsible for the elastic properties of individual wool fiber.

Dry wool fiber swells up when placed in water. The degree of swelling depends on factors such as the temperature and strain on the fiber. In cold or luke-warm water, the fiber swells to about 10% in diameter, but on drying, the fiber usually returns to its original size. This recovery is thought to be due to reversibility of hydrogen bonding transformation. This explains the important property of elasticity of wool. Fibers dried in extended state do not recover to their original size. When wool is heated in dry air at 100°C to 105°C over a long period, it loses moisture and the fibers become harsh, if returned to moist air, the fiber rapidly reabsorbs moisture and regains its softness. Wool, when burnt emits a characteristic odor like that of burning of hair or feather, probably due to the presence of nitrogen and sulphur in the keratin. The vapors coming off during burning have an alkaline reaction and probably contain ammonia and sulfur compounds. Although it burns, wool is fire resistant. When removed from the flame, wool fiber assumes a knot or charred globule on its end but does not continue to burn. Oxidizing agents like hydrogen peroxide, $KMnO_4$, $K_2Cr_2O_7$, attack cysteine disulfide bonds, resulting in decreased sulfur content, strength and fiber weight. The main use of oxidizing agent is in bleaching. H_2O_2 is the most common bleaching agent for wool and imparts permanent white colour. When wool fiber is exposed to light in the presence of water, reaction between water and cysteine linkage takes place. This photochemical reaction is responsible for the yellowing of woollen fabrics. So white woollen garments should not be dried in bright light. Wool is more hygroscopic than other textile fibers. This characteristic feature confers a number of advantages on wool as a clothing and furnishing fabric.

Silk
Silk is a natural protein fiber produced by the insect, silk worm. It has all the desirable qualities of fiber such as softness, strength, elongation, etc. Besides, silk has a bright lusture. It is lighter than other natural fibers and has the specific gravity of 1.25–1.27. Silk is an excellent insulator of heat, hence it is warm in winter and cool in summer. Depending upon the atmospheric conditions, silk can absorb about 20–25% moisture without becoming wet, which makes it comfortable to wear. Silk is a polypeptide made up of only four amino acids, glycine, alanine, serine and tyrosine. Glycine and alanine are present in 2 : 1 ratio and cover more than 75% mole percent of the silk. The fiber can be made to form a continuous filament about 300–700 metre long. Silk is noted for its strength, toughness and smooth soft feel. It is a poor conductor of heat and electricity. After proper cooking, the cocoons are deflossed mechanically to get the continuous filament. Several cocoons are fed per end of the reeling machine to produce silk yarn of a particular thickness. These yarns are then twisted or doubled and used for fabrication to get the bright silk cloth.

Synthetic Fibers
The first synthetic fiber was produced by Carothers by the reaction of hexamethylene glycol and adipic acid. The molten polyester thus obtained gave a long filament like material that could be stretched several times of its original length. Since then, there has been a phenomenal growth in the production of synthetic fibers. In 1950, production of synthetic fibers was 0.73% of the total fibers while in 1980 it rose to 37% and presently it is estimated around 60%. The increasing trend will still continue. Polyesters, nylons and acrylic fibers are produced in major quantities. An important development in the synthetic fiber

industry is the production of staples of short length that can be blended with natural animal or other synthetic fibers. The advantage of such composite or blend fibers is that the properties of individual components complement each other. For example, a blend of polyester-cotton (67: 33) is superior and exhibits more comfort than that of its individual components. Many drawbacks of synthetic fibers such as low moisture absorption, accumulation of static charges, easy soiling, metallic lusture, and lack of ideal warmth and comfort can be overcome by blending them with other fibers.

Synthetic fibers are produced from highly pure monomers under strictly controlled reaction conditions. The following steps are generally involved for the manufacture of fibers; (i) conversion of monomer into a polymer, (ii) polymer to granules or chips, (iii) pressing the melt or dope of the polymer through spinning jet to form filaments, (iv) development of morphological fine structures and (v) modification of filament to suit the end use. Synthetic fibers are produced in a batch or continuous process. Some of such fibers are described below:

Polyester (polyethylene terephthalate)
Polyethylene terephthalate is produced by the condensation of ethylene glycol (EG) and dimethyl terephthalate (DMT) or terephthalic acid (TPA). The condensation of DMT and EG is carried out in two steps namely trans-esterification and polycondensation. The catalyst for *trans*-esterification process consists of acetates of sodium, manganese, zinc or cobalt or a mixture there of. Antimony trioxide or acetate is used as a polycondensation catalyst. The polymer melt is spun into ribbons or bands that are granulated. The granules are dried under reduced pressure till the moisture content reaches to 0.05% since polyester is hydrolysed in the presence of traces of water. The dried chips are melted and the molten mass is spun into filaments. The appropriate spin finish is applied and the filaments or tows are collected in a drum.

The technology of staple manufacture is slightly different from that of filaments. Many filaments are brought together to form a thick tow. The tow is passed through water to remove the spin finish, squeezed, heated and stretched before crimping, drying, heat set and cut into stable fibers of desired length. The stretched tow is used for yarn spinning. The staples are pressed into bales and marketed. The polyester staples are directly spun into filaments or blended with cotton before spinning into fibers.

Polyamides (Nylons)
Among the large number of polyamides (nylons) only two, i.e. nylon 6 and nylon 6, 6 could get commercial importance. Nylon 6, 6 is synthesised by the condensation of hexamethylene diadipate salt (HD Salt). The salt is produced by the condensation of equimolar amount of hexamethylene diamine and adipic acid. The salt is relatively insoluble in methanol and crystallises out on cooling. The crystals are separated, purified and dissolved in water to get a 50–60% solution. Acetic acid (5%) is then added as viscosity stabiliser. The solution is heated in an autoclave when a pressure of 15–20 kg / cm^2 builds up. The pressure is slowly released and heating is continued to remove all water by distillation.

The autoclave is evacuated to facilitate the removal of water from highly viscous molten polyamide mass. The melt is spun into filament or fibers by conventional methods. The filaments are cold drawn 3–4 times of their original length to improve the mechanical properties.

Polyamides (nylon 6)

6-aminocaproic acid or caprolactam is the raw material for the production of nylon 6. Caprolactam, water and acetic acid are charged to an autoclave preheated to 250°C under pressure ranging from 12–15 kg/cm². The steam is slowly released, nitrogen is flushed and vacuum is applied to remove last traces of water. The reaction never reaches to a state of completion. The melt is extruded and granulated in water. The granules are extracted with hot water to remove 10% low molecular weight compounds, containing unreacted monomer and oligomers, in a counter current manner. After the extraction, granules are washed with fresh water and vacuum dried. They are then spun at 270–280°C. Nylon 6 is invariably spun into filaments because its staple fibers could not get any commercial importance.

Acrylic Fibers

World wide demand for the substitute of natural wool led to the invention of acrylic fibers. It was in 1940, when Du Pont produced first acrylic fiber that was difficult to dye. After 1955, cationic dyeable fibers were produced. Copolymers of acrylonitrile with other monomers were synthesised in the presence of few radical catalysts such as potassium persulfate. Presently available acrylic fibers are the copolymers of acrylonitrile and other vinyl monomers. In such polymers the amount of acrylonitrile is not less than 85% and rest 12% may be constituted by one or two monomers.

The most common monomers used with acrylonitrile are vinyl acetate, methyl acrylate. Such monomers reduce glass transition temperature of polyacrylonitrile from 104°C to 80–90°C, thereby facilitating the dyeing at boil. The acidic comonomers such as acrylic acid, allylsulfonic acid, and itaconic acid impart cationic dyeability to acrylic fibers. The fibers containing basic comonomers such as vinyl pyridine, or ethylene imine are dyeable with acid dyes. The commercially available fibers are used as substitute for wool because of their wool like feel and inertness to chemicals, bacteria and resistance to humidity etc.

Acrylic polymers do not melt before decomposition and are therefore not melt spun. The acrylic polymers are wet spun. In such a process, polymer is dissolved in a suitable solvent such as dimethylformamide, dimethylacetamide, dimethylsulphoxide. The polymer dope is carefully prepared to avoid jet formation and forced through spinnerets having 5,000 to 70,000 holes. The fiber is passed through a series of coagulating baths to remove the solvent. The coagulating baths are filled with dilute aqueous solutions of the solvent. The major fraction of the solvent diffuses out from the fibers and the residual solvents are removed by washing with water during the process of hot stretching. The stretched tow is crimped and heat set (annealed by steam and cut) into staple fibers.

Polypropylene

Stereoregular, high molecular weight, fiber forming polypropylene is synthesised using a Ziegler-Natta catalyst by anionic mechanism. Both batch and continuous polymerisation plants are under operation to make polypropylene. The reactor is filled with monomer, diluent and catalyst. The polymer is produced by heating the mixture. The heat of reaction is controlled by the addition of fresh propylene in order to produce a homogeneous product. The slurry thus obtained is passed into the flask, which is maintained at relatively low pressure wherein monomer and diluents are flushed out. The catalyst is deactivated by the addition of alcohol and is removed from the main polymer. The polymer is centrifuged, washed and dried. Paraffins, cycloparaffins or their mixtures are used as the inert medium. The filaments are produced from the polymer by melt spinning at temperatures of 100–150°C higher than its melting point because it exhibits high melt viscosity. The spinneret jet holes are also larger (higher length to diameter ratio) than those used for PET or nylon spinning. Polypropylene fibers have excellent properties and are used for various purpose. Since the fiber has low moisture absorption, wet and dry strengths are the same. It has lowest fiber density of all the commercial fibers. Polypropylene is extensively used in ropes, netting, industrial fabrics and carpets.

The main problem of polypropylene fiber is its low softening point (160°C). Fabrics having more than 30% PP can not be exposed to temperatures higher than 125–130°C as they become harsh due to shrinkage. However, PP fiber is chemically inert to most chemicals, acids, alkalis and oxidising agents. In the Table 5(c) are given some physical properties of common fibers.

Table 5(c)   **Summary of some characteristics of synthetic fibers**

Polyesters	Polyamides	Acrylics
Acidic at room temperature. Disintegr ate on boiling in strong alkali solutions but resistant to bleaches and organic solvents. Have good resistance to sun light on glass support but prolonged exposure results into a decrease in strength. Can be dyed using dispersed and cationic dyes	Inert to alkalis. Dissolve in strong mineral acids, are bleach resistant, have little or no effect of most organic solvents. They are soluble in phenols. The fibers show marked loss in the physical strength on prolonged exposure to sun light. Can be dyed with dispersed and acid dyes	Have good resistance to weak alkalis. They have excellent resistance to sun light, to bleaches and common solvents. Can be dyed using cationic, dispersed, acid and vat dyes

Properties of synthetic fibers
Synthetic fibers are characterised by high strength, superior abrasion resistance, good crease recovery, low density and resistance to moth and bacteria. Synthetic fibers are very durable compared to natural fibers such as cotton, viscose and wool. The main problems with synthetic fibers are the low moisture absorption, development of static charges, easy soiling, poor comfort and loss of aesthetic properties during their use.

Most of the synthetic fibers melt before burning. Synthetic fibers recover well from crosses formed during wear. However, pleats and creases set in the fibers at high temperatures remain almost throughout the life of the garments. Synthetic fibers have good resistance to chemicals. Polyester and acrylic fibers are attacked by hot alkaline solutions. The stability against the light and UV radiation is better than those of natural fibers. Synthetic fibers have a tendency to shrink in hot water and are usually heat set during the process of manufacturing to avoid the shrinkage during their use. After heat setting, nylons and polyesters do not shrink. Acrylic fiber is heat set in the presence of moisture or preferably by steam.

Synthetic fibers are produced in different deniers as staples and yarns. The denier of a fiber measures its size and is defined as the weight in grams of 9000 m of the fiber. The size of the fiber is also expressed in many other measures, but the denier is the most common of all. The size of the fiber measured in deniers is proportional to its density and to its cross sectional area. One measures the tensile strength of the fiber from deniers when the later is expressed as weight per unit length. The yarns are directly used for weaving the garments whereas staples of different cut lengths are blended with other staple fibers and the blend is spun into yarns before the processing, weaving or knitting. Variable cut lengths of staples are common in acrylic fibers. Fibers with different cross sections are produced by changing the spinnerets having different types of holes. Trilobal, multilobal yarns of nylon, polyester are used for various purposes and exhibit different types of lusture and comfort depending upon the dimension of the cross section. Some important properties of synthetic fibers are listed in Table 5(d).

Table 5(d)   Physical properties of some typical fibers

Fiber	Tenacity g /denier	Tensile strength (10^3 psi)	Elongation %
Cotton	2.1–6.3	42–125	3–10
Viscose rayon	0.7–3.2	28–47	15–30
Acetate rayon	1.2–1.4	20–24	25–45
Wool	1.0–1.7	17–28	20–50
Silk	2.8–5.2	45–83	13–31
Polyester	2.4–7.0	39–106	12–55
Polyamide (Nylon 6,6)	2.3–6.0	40–106	25–65
Polyacrylonitrile	2.3–2.6	32–39	20–28
Polyurethane	0.7–0.9	11–14	400–625
Polypropylene	3–4	35–47	80–100
Glass		30–50	0.5–2.0

5.4   Elastomers

Natural rubber is a polymer of isoprene that has pronounced elasticity (high extension and low modulus of elasticity). In other words, a rubber-like state is a state in which a polymeric material is capable of undergoing very great recoverable deformations that occur under the influence of very small loads. Elastomer is a general name given to synthetic polymers with rubber-like elasticity. The formation of rubber from its monomer isoprene is as follows;

$$CH_2{=}\underset{\underset{CH_3}{|}}{C}{-}CH{=}CH_2 \longrightarrow {-}\!\!\left[CH_2{-}\underset{\underset{CH_3}{|}}{C}{=}CH{-}CH_2\right]_{\!n}$$

Isoprene   Polyisoprene
                                 (natural rubber)

The characteristic property of rubbers or elastic materials is that they possess little or no crystallinity. The extension and contraction is only possible if the polymer is in amorphous state and at an operating temperature above its glass transition temperature. Under these conditions, segments of polymer chains are able to move so that the material can extend or contract rapidly. However, chains must not slip past each other on stretching, otherwise the material will not regain its original shape. Slipping of chains (creep) can be prevented by introducing cross-links into the material, a process known as vulcanisation. An elastomer can thus be described as an amorphous polymer, which is above its T_g, and contains cross-links to prevent slipping of chains. Table 5(e), (f) and (g) describe some important properties of elastomers.

Table 5(e) Molecular weight dependence of physical properties of natural rubber

Mol. wt. range	Appearance	Solubility in benzene (1%)
Upto 700	Thick liquid	Quickly dissolves without swelling
700–7,000	Sticky elastic	Soluble without swelling
7,000–35,000	Tough and elastic	Soluble with weak swelling
35,000–2,00,000	Very tough and high elastic	Soluble with strong swelling

Table 5(f) Typical properties of elastomers

Elastomer type	Tensile strength (10^3 psi)	Elongation (%)
Natural rubber	210	700
Styrene-butadiene rubber	28	800
Acrylonitrile-butadiene rubber	42	600
Thiokol	21	300
Neoprene	245	800
Butyl rubber	210	1000
Polyurethane elastomer	350	600

The properties of natural rubber drastically change upon hardening. Even the appearance and physical states are immediately imminent. Natural or raw rubber is in latex form (dispersion) and the vulcanised forms of rubbers are however rigid and in glassy state. A contrast of different properties for natural raw and vulcanised rubber is shown in Table 5(h).

Table 5(g)   Upper and lower use temperatures of typical elastomers

Elastomers	Use temperature limit (°C)	
	Upper	Lower
Natural rubber (polyisoprene)	80	−50
SBR [butadiene-co-styrene (25%)]	110	−50
Nitrile rubber [butadiene-co-acrylonitrile (25%)]	120	−50
Butyl rubber [isobutylene-co-isoprene (5%)]	100	−50
Fluorinated rubbers	230	−40
Polysulfide (thiokol)	80	−50
neoprene (polychloroprene)	100	−50

Table 5(h)   Raw versus vulcanised rubber

Property	Raw rubber	Vulcanised rubber
Tensile strength (10^3 psi)	0.2	2.0
Elongation at break (%)	1,200	800
Rapidity of retraction	good	very good
Water absorption	large	small
Swelling in organic Solvents	infinite (soluble)	large, but limited
Tackiness	marked	slight
Useful temperature Range	10 to 60°C	−40 to +100°C
Chemical resistance	very poor	much better
Elasticity	very high (300 to 1,000 %)	Low, depending on degree of vulcanization

The chains of a stretched elastomer revert to the highly coiled state on the release of tension. This is due to the fact that highly coiled polymer system has a higher degree of disorder, i.e. higher entropy than a stretched oriented sample. Thus, the elastic behaviour is a direct consequence of the tendency of the system to assume spontaneously a state of maximum entropy. Since free energy, enthalpy and entropy are related by well known Gibbs equation, $\Delta G = \Delta H - T\Delta S$, a stretched rubber band immediately held to the lips is warm.

Several elastomers have been commercialised. These include styrene-butadiene rubber, butyl rubber (copolymer of isoprene and isobutylene), neoprene (polychloroprene), nitrile rubber (copolymer of acrylonitrile and butadiene), thiokols (polysulfide), silicone rubbers and polyurethane rubbers etc. The synthesis and application of these rubbery materials and the production of the monomers used for making them are already described in Chapter 2. Some newly commercialised elastomers include ethylene/propylene copolymers, polypropylene oxide, copolymer of epichlorohydrin with ethylene oxide, copolymers of ethylene and vinyl acetate.

5.5   Adhesives

Adhesives are the substances, often polymeric (natural or synthetic) that are used to glue two surfaces together. Plant exudates have been known in use as adhesives by Egyptians for about 6000 years for bonding ceramic vessels. Starch and sugar, casein (from milk) and glue from animals and fish have been used for 4000 years. The adhesives often work on the principle that they form either primary covalent bond or interact with the surface through physical forces (secondary bonds). There are several types of adhesives. Solvent based adhesives are dissolved in solvent and the solvent is then allowed to evaporate in the presence of surface to be glued. The evaporation of the solvent that leads to the formation of a thick or thin solid coating. Latex adhesives (which are directly used as the dispersion of the polymer made by emulsion polymerization) must have low T_g. It gives a flow and good surface contact on evaporation of water from a water based latex. Pressure sensitive adhesives are viscous polymer melts at room temperature. The adhesives flow on applying pressure and thereafter on releasing the pressure the high viscosity of the polymer provides the adhesion. Hot melt adhesive works on similar principle except here the material flows on applying heat. Important and highly effective of all the adhesives are reactive adhesives which are low molecular weight liquid polymers and solidify due to the cross-linking. Cyanoacrylates, phenolic resins, silicones, epoxy polymers and unsaturated polyesters are some examples of this class.

5.6   Polymer Additives

Only few polymers like PMMA, polystyrene, gums and some fiber forming materials etc., are used as pure polymers (*virgin polymers*). Many plastics, for example formaldehyde resins (PF, UF and melamine-formaldehyde resins) which are highly useful materials would have been of no commercial value in the absence of substances like wood flour, cotton and other textile fibers, glass wool etc. Most polymers often contain a variety of substances which are added during their preparation or processing. Chemical additives are often used in plastics to produce some desired characteristics. For instance, antioxidants protect a polymer from chemical degradation by oxygen or ozone; similarly, ultraviolet stabilizers protect against weathering. Plasticizers make a polymer more flexible, lubricants reduce problems with friction, and pigments add colour to polymeric materials. The other important additives namely flame retardants, antistatics and antibiocides etc. are also added for specific functions.

Many plastics are manufactured as composites. This involves a system where reinforcements (usually fibers made of glass or carbon) are added to a plastic resin matrix. Composites have strength and stability comparable to that of metals but generally with less weight. Plastic foams, which are composites of plastic and gas, offer bulkiness with less weight. Additives are thus essential functional ingredients of polymers and their proportion in the material should be optimum in order to obtain high quality materials. Table 5(i) lists various types of additives along with the suitable examples. Each one plays a definite and distinctive role.

Table 5(i)   Some polymer additives

Function of additives	Examples
1. Fillers	Cellulosic and lignin products, carbon black, silica products and silicates, metals, metallic oxides and carbonates
2. Plasticizers	Paraffin oils, n-dialkyl phthalate, tricresyl phosphate and camphor etc.
3. Antioxidants	Phenol derivatives, e.g. di (t-butyl) p-cresol
4. Heat Stabilizers	Organic phosphides, toxic lead, barium and cadmium salts
5. UV Stabilizers	Benzotriazoles, phenyl salicylate
6. Flame Retardants	Halogen and phosphorus compounds, antimony oxide
7. Colorants	Inorganic pigments (iron oxides, lead chromate, zinc chromate, prussian blue, ultramarine blue, carbon black), organic pigments (phthalocyanines, and dyes of the class triphenylmethane, anthraquinone etc)
8. Curing Agents	Benzothiazoles (e.g. 2-mercaptobenzothiazole), zinc butyl xanthate

Fillers

Fillers are often used in substantial amount. A filler is an inert material added to a plastic to modify its strength and working properties or sometimes to lower its cost. Besides reducing the cost, a filler helps in improving physical properties viz. high heat resistance, high mechanical strength, low moisture absorption, good electrical characteristics. Fillers are supposed to be available abundantly at cheaper cost. These should be compatible with the polymer and other additives and should not have abrasive or chemical action on the mould. Cellulosic products (wood, paper, fibers viz. cotton and jute), carbon, inorganic materials such as silica (sand, quartz etc.), silicates (asbestos, clay, mica, talc etc.), glass, metals, metal oxides (alumina, titania, magnesia and zinc oxide etc.) and even synthetic fibers (nylons, polyesters and orlon) are the commonly used fillers. Fibrous fillers greatly enhance the mechanical strength in the polymers. Such a fibrous reinforcement is much more effective than spherical fillers. The effect of a fibrous filler depends on the fiber length and the interfacial bond between it and the continuous resin matrix.

Plasticizers

Many synthetic polymers and cellulose derivatives are often obtained as white powders or horny tough materials as a result of which they do not flow to a significant level on heating or under pressure. This makes their processing difficult. Such thermoplastic materials are thus modified by adding some substances called *plasticizers*. Plasticizers are non-volatile liquids, decrease T_g of polymers and thus impart flexibility. Tricresyl phosphate, dialkyl phthalates, camphor, glycerol, paraffin oils etc. are well known plasticizers. Since plasticizers are essentially non-volatile liquids, they would be compatible with the polymer if the solubility parameter of the two matches (as discussed in Chapter 4). Thus the difference in the solubility parameter of plasticizer and polymer should not be more than 2. Dioctyl phthalate (DOP) is widely used for plasticizing PVC (whose solubility parameter is 9.66 H). It is interesting that a plasticizer when used in small amounts often functions as anti-plasticizer i.e. increases the hardness and decreases the elongation of polymers. Structures of some commonly used plasticizers are shown below;

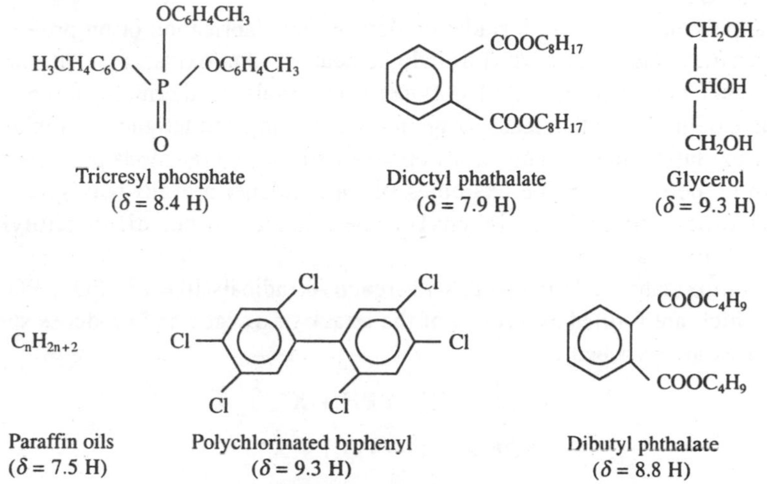

Tricresyl phosphate
(δ = 8.4 H)

Dioctyl phathalate
(δ = 7.9 H)

Glycerol
(δ = 9.3 H)

Paraffin oils
(δ = 7.5 H)

Polychlorinated biphenyl
(δ = 9.3 H)

Dibutyl phthalate
(δ = 8.8 H)

Fig. 5.3 Structures of some common plasticizers

Plasticizers function by reducing intermolecular forces between the polymer chains (and decreasing the glass transition temperature of polymer taken) and thus makes it flexible at lower temperature. Camphor has long been used as plasticizer in the manufacture of celluloid from cellulose nitrate. PVC and cellulose esters are often plasticized by dialkyl phthalates. The development and applications of plasticizers have been associated with the toxicity. The use of highly toxic polychlorinated biphenyls (PCB) has been discontinued. Also, plasticizers are often leached out after a prolonged use of the polymeric material which as a result becomes brittle. One usually notices the inner surface of buckets used in the bath rooms becoming greasy and sticky after water storage. This is essentially due to the leaching out of the plasticizer. Also, it is a common experience that plastic buckets get cracked and broken after their life time. The flexibility of a polymer can also be increased by internal plasticisation. For example, flexible PVC can be obtained if vinyl chloride monomer is copolymerized with small amount of vinyl acetate.

UV Stabilizers and Absorbers

A part of UV radiation in the wave length range 280–400 nm reaches earth's surface. This is associated with the energy of 100–72 kcal and is strong enough to cleave covalent bonds and thus causes yellowing and embrittlement of organic polymers. UV stabilizers protect the material from preferentially harmful high energy UV radiation emanating from sun. An UV absorber usually dissipates the absorbed energy by transferring it to its surroundings as heat or by remitting it at higher wave length (or harmless low energy) radiation. Carbon black can function as UV absorber but is restricted to articles where colour is not a criterion of choice. Many commercial UV absorbers have alkoxy groups at the 4th positioned carbon of the phenyl group. Phenyl salicylate, 2-hydroxy-4-octoxy-benzophenone, α-hydroxy-3, 5di-alkyl benzotriazole are commonly used UV-stabilisers.

Antioxidants

Several polymers undergo degradation during their fabrication or on prolonged use when they come in contact with air, light, heat, etc. Antioxidants are substances that protect the deterioration of polymeric materials by thermal-, photo- and oxidative degradation that leads to ageing, weathering and fatigue. Antioxidants oppose oxidation and in many cases oppose undesirable reactions promoted by oxygen or peroxides. Some examples of antioxidants are di(t-butyl)p-cresol, phenyl β-naphthylamine, diphenyl-p-phenylenediamine, diβ-naphthyl-p-phenylenediamine.

An antioxidant (XH) interacts with reactive radicals like P*, PO*, POO*, OH*, which are formed as a result of the attack of oxygen and produces stable substances as described below.

$$P* + XH \rightarrow PH + X*$$

$$PO* + XH \rightarrow POH + X*$$

$$POO* + XH \rightarrow POOH + X*$$

$$OH* + XH \rightarrow HOH + X*$$

A typical behaviour of a polymer with or without antioxidant is shown in Fig. 5.4.

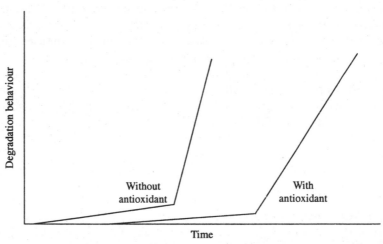

Fig. 5.4   Comparative behaviour of two samples of a polymer with and without antioxidant

Flame Retardants

Most organic polymers like other organic compounds have the tendency to burn at elevated temperatures. While burning, the polymeric materials would melt (thermoplastics) or decompose giving smoke and even toxic gases like HCl,

HCN and CO etc. Since polymers are often used as building materals and many household items are made of polymers, it is thus important that they have good fire resistance. Flame retardants are chemicals added to a polymer to minimise fire hazards. Inorganic phosphates like ammonium phosphate have been used as flame retardants in wood, paper and cotton. Tricresyl phosphate, 2-chlorethyl phosphate, $(ClCH_2CH_2O)_3 P = O$, antimony trioxide (Sb_2O_3), aluminium oxide trihydrate $(Al_2O_3\ 3H_2O)$, mixture of borax and boric acid are few examples of flame retardants.

Colorants

Colouring gives the material a better look and therefore most polymeric materials are sold as coloured articles. The polymers do not have the colour of their own but are intentionally coloured for aesthetic appearance. Several dyes and pigments are used to impart colours to polymeric products. Inorganic pigments (Rutile, TiO_2 (for white), lead chromate (for yellow), ferriferrocyanides or prussian blue (for blue), Carbon black (for black) and organic pigments, phthalocyanines (for blue) are common. Dyes are often more compatible with polymers and are therefore used to get transparent coloured products. Acid and basic dyes such as rhodamine red, victoria blue, anthraquinone dyes such as flavanthrone yellow, dioxazine dyes such as carbazole violet etc. provide yellow to red range of colours.

Curing agents

The use of curing agents began with the accidental discovery of vulcanisation of hevea rubber with sulphur by Charles Goodyear in 1838. Curing of phenol-formaldehyde resins is another early example. The low mol. wt. substances used to cross-link plastics (usually liquid prepolymers) to obtain useful resins are called curing agents. These substances chemically react with the polymer through functional groups and develop cross-links.

Shaping and Finishing

The techniques used for shaping and finishing plastics depend on three factors: time, temperature, and flow (or deformation). Many of the processes are cyclic in nature, although some fall into the categories of continuous or semi-continuous operation. One of the most widely used operations is extrusion. The extruder is a device that pumps a plastic through a desired die or shape. Extrusion products, such as pipes, have a regularly shaped cross section. The extruder itself also serves as the means to carry out other operations, such as blow moulding and injection moulding. In extrusion blow moulding, the extruder fills the mould with a tube, which is then cut off and clamped to form a hollow shape called a parison. This hot, molten parison is then blown like a balloon and forced against the walls of the mould to form the desired shape. In injection moulding, one or more extruders are used with reciprocating screws that move forward to inject the melt and then retract to take on new molten material to continue the process. In injection blow moulding, which is used in making bottles for carbonated beverages, the parison is first injection moulded and then reheated and blown.

Compression moulding uses pressure to force the plastic into a given shape. Another process, transfer moulding, is a hybrid of injection and compression moulding: The molten plastic is forced by a ram into a mould. Other finishing processes include calendering, in which plastic sheets are formed, and sheet forming, in which the plastic sheets are formed into a desired shape. Some plastics, particularly those with very high temperature resistance, require special fabrication procedures. For example, polytetrafluoroethylene has such a high melt viscosity that it is first pressed into shape and then sintered—exposed to extremely high temperatures that bond it into a cohesive mass without melting it. Some polyamides are processed by a similar process.

Health and Environmental Hazards

Because plastics are relatively inert, they do not normally present health hazards to the maker or user. However, some monomers used in the manufacture of plastics have been shown to cause cancer. Similarly, benzene, which is an important raw material for the synthesis of nylon, is a carcinogen. The problems involved in the manufacture of plastics are parallel to those of the chemical industry in general. Most synthetic plastics are not environmentally degradable; unlike wood, paper, natural fibers, or even metal and glass, they do not rot or otherwise break down over time. (Some degradable plastics have been developed, but none has proven compatible with the conditions required for most sanitary land-fills). Thus, there is an environmental problem associated with the disposal of plastics. Recycling has emerged as the most practical method to deal with this problem, especially with products such as the polyethylene terephlalate bottles used for carbonated beverages, where the process is fairly straight forward. More complex solutions are being developed for handling the mixed plastic scrap that constitutes a highly visible, albeit relatively small, part of the problem of solid waste disposal.

5.7   Polymer Processing

Polymeric materials are needed for a variety of uses in the form of rods, tubes, sheets, foam, coatings, adhesives and in other different moulded forms. It is therefore not surprising that several moulding and fabrication techniques are used in polymer processing to obtain a finished product. The polymer processing methods however depend on the nature of the material i.e. thermoplastics, thermosetts, fibers etc. Fibers are processed in altogether different way. In the foregoing discussion the processing techniques used for thermoplastics, thermosetts and fibers are described. The polymeric material is often first mixed with necessary additives, other polymers (polyblends or polymer alloys) or other inorganic or fibrous materials (composites or reinforced plastics). This process is called compounding. The compounded polymer is then subjected to various processing methods that yield films, sheets, tubes or the finished moulded articles.

5.7.1   Casting

It is one of the cheapest and simplest processes by which polymers can be shaped into a desired solid form. Both the thermoplastics and thermosetts can be moulded by this process. In case of thermoplasts, the solid polymer powder is compounded with other additives such as filler, curing agent, colorants, antioxidants and plasticizers and the compounded mixture is dissolved or swollen or gelled in a suitable solvent. Then the mixture is poured into dies or moulds. When thermosetts are to be casted, usual practice is that the polymerisation mixture at prepolymer stage is separated from the reaction vessel and then mixed with curing agents, plasticizers and other additives, which impart color and smoothness. Then the whole mixture is poured into a mould or a die. The mould or die is prepared out of plaster of paris, lead or even glass and designed into a given shape with a particular cross section and depth. The different designs of the mould or die enable one to get sheets, tubes, rods or even thick films.

The die loaded with the compounded mixture of a thermosett or thermoplast is then subjected to heating at a programmed schedule, to allow the complete curing of poured semi-solid mixture into a well shaped solid form. Once the polymerisation or curing is complete, the die is taken out of the oven and cooled. The inner lining of the die or mould is often coated with mould releasing agents. This facilitates easy separation of articles from the mould cavity. The casting process has following advantages: (i) the mould costs are low, (ii) good surface finish of the product can be obtained, (iii) the process needs to be conducted only at ordinary conditions of temperature and pressure, (iv) the whole process is economically and energetically cheap, (v) large parts with thin to thicker cross sections can be shaped or solidifed. However, major disadvantages of the casting process are, (i) only limited shapes can be obtained, (ii) some of the thermoplasts and thermosetts cannot be casted, (iii) the method is uneconomical when large volume production is desired, because this process is rarely automated.

Casting process can be divided into three types: (i) die casting, (ii) rotational casting or slush moulding and (iii) film casting.

Die Casting

This process is a simple one and as discussed above involves the transfer of the polymer (thermoplasts in compounded and solution form and thermosetts at prepolymer stage) plus added curative agents, catalysts and other additives to the cavity of the pre-designed and pre-shaped die. Then, the die is subjected to further reaction conditions. The reaction conditions mainly involve the change in temperature, because casting process seldom involves application of pressure or vacuum. However, in the case of certain thermoplast and complex parts, casting has to be done either under applied pressure or in vacuum. To make die casting energetically and economically viable, following criterion should be adopted. Free flowing plastics with low surface tension and viscosity can be casted into intricate shapes easily. The low viscosity of the precast plastic mixture is highly preferred because bubble free articles can be obtained. However, to get casted articles with better physical properties viz. mechanical strength, and transparency etc., the precast mixtures with high viscosity are highly desirable. But handling of high viscosity mixture is difficult and expensive. Moreover, air can be trapped into highly viscous mixtures causing voids in the final product. The selection temperature of casting must be based on the following simpler considerations. The curing or further polymerisation reaction is usually exothermic and the heat evolved is dependent upon the shape and thickness of casted article. When the thin articles with the large exposed surface area in relation to the total volume of the plastic taken are to be casted, the heat of the process is dissipated rapidly and thus the cast temperature is not very high. Thus, the thin sections of the plastics can be casted at room temperatures. When low viscous and free flowing mixtures are used, one can obtain almost crack free casts of thinner dimensions. The amount of heat liberated is high when highly viscous and thick sheets are to be cured during casting process. Thus the proper heating rate has to be applied for getting bubble- and crack free articles under this condition. In such cases, heat sinks or heat pipes are used for absorption or transfer of heat from one side to another side of the die. Different liquids and bulk graphites can be used as heat sinks. Heat pipes consists of long tubes with a working fluid flowing through it. This fluid can be vaporized by application of external heat and thus generated heat can be transferred from one part to another. Similarly, the liquid can adsorb the heat generated during the curing process and eliminate the excess heat in the die.

Formation of air entrapments and bubbles is the chief drawback of casting process. So extreme care must be taken from calendering stage to final removal of hardened plastics from the mould. All the ingredients used are to be properly deaerated and their mixing has to be carefully controlled during calendering to keep the amount of air entrapped to minimum. The die should be carefully designed to avoid sharp corners and provision of adequate number of air releasing outlets must be made for its free circulation. High speed mixers and vortexing of material has to be avoided. Application of desired level of vacuum during mixing and pouring eliminates the entrapment of air completely.

Die casting is widely employed in the formation of thin or thick sheets, rods and other shapes from acrylic, epoxy, polyester, urethane and phenolic polymer materials. The cast acrylic products in the form of rods, tubes, lenses, spheres and other intricate shapes are available with high transparency, in wide range of colors, rigidity, moderate impact strength, good thermal and chemical resistance and excellent weatherability. From 1930's, acrylic products are being die cast. The casting of acrylic products has been highly preferred because bulk and solvent free radical initiated polymerisation of acrylic monomers is possible in the cavities of moulds made of either plaster of paris, reinforced plastic, wood or even concrete moulds, which have adequate heat transfer and surface release characteristics. Transparent, long and thick acrylic sheets are obtained by allowing the polymerisation of monomers mixed with free radical catalysts for few hours by heating in a die made up of two thick glass or metal plates with rubber gaskets fitted in between. Once the polymerisation is complete, the die is cooled gradually under controlled cooling rates and sheets with better optical characteristics are separated. The casting process just described is known as batch casting.

This method of casting can be made continuous by taking solutions of already formed thermoplasts or monomer syrups with catalysts added on and the mix is then rolled through continuous conveyer belts passing through a controlled temperature environment adequate for polymerisation process. Though the continuous casting process is economically cheaper, batch casting yields sheets of better quality.

Rotational Casting

For obtaining articles such as toys, hollow balls, rain boots and doll heads etc., a specialised apparatus based on rotational casting procedure is used. The apparatus is designed such that the mould which is usually a hollow tube with a desired shape can be rotated simultaneously along its breadth and length. The mould is filled either with a fine powder of a thermoplast or plastisol (a dispersion of thermoplastic polymer in a liquid plasticizer) with additives such as stabilizers, coloring agents and antioxidants. etc. and closed. The uniform distribution of suspension mixture is ensured by initial heating and rotation. The mould is then chilled after a specified time period when it is still under rotation. By this way the thermoplast is melted first and cooled to a solid into a desired shape. Instead of a thermoplastic, if one takes a prepolymer of thermosetting resin and curative and stabilizing agents mixture, the curing is done at elevated temperatures and chilling step is avoided. In both the cases the conveyor belt is taken through an oven (kept at appropriate temperature range) and excess liquid plasticizer is drained out of the mould. When plastisols are to be casted, the conveyor belt is rotated and passed through another oven kept at higher temperature for final curing. This process is widely used for PVC, epoxy, and phenolic resins and unsaturated polyesters.

Film Casting

Films of thermoplastic polymers can be cast on highly polished drums, belts or bands that are driven by drums. A thermoplast solution in a suitable solvent and with an appropriate concentration is allowed to be poured on a moving drum or belt. The movement of the drum or belt is controlled by motors. The continuous movement of casting device allows the solvent to be evaporated leaving behind a polymeric film. The polishness on the surface, drying time and speed of the moving drums all contribute to the quality of films formed. Longer drying time and slow speed of moving device produce thinner films. The belt method yields thicker films and high production rates can be achieved. The films formed are simply stripped, if needed dried further to remove residual solvent. The different varieties of photographic films and cellophane tapes or sheets are produced by this technique at commercial levels.

Besides above mentioned casting techniques, there are several other methods through which casting of polymeric materials for specialised to very specialised applications can be carried out. These are embedding, encapsulation, potting and impregnation etc. Embedding usually means complete encasement of a device or an assembly in a polymeric sheet matrix. For example, various components and electronic circuits that are used in devices such as television sets, computer monitor or any other display systems or even the instrument circuits, need special mechanical support for their optical functioning. Polymer materials offer not only best mechanical support but also possess other requirements such as better insulation, better resistance from attacks of oxygen, moisture, temperature, current flash over and leakages, salt spray, radiation, solvent, chemicals and microorganisms etc. The polymeric materials used as embedding matrices are either liquids, granular solids or powder solids made of epoxies, silicones, polyurethanes, polyesters and polysulfides etc. The selection of a particular polymer and its form depend highly on its resistance to various factors as mentioned earlier.

Encapsulation is just a modification or an extension of embedding in which the part or item to be encapsulated is dipped into a highly viscous or thixotropic material. Then a thin coating of the polymer material is achieved around the part by keeping it at required temperature for scheduled time. Caution is always taken to minimise the internal stress in the die during the process to avoid any crack in the coated film.

Potting is another form of encapsulation, in which firstly, the part is embedded in a plastic matrix. Then the whole embedded item along with the plastic is encased in the mould. Thus, potting is a special embedding where a double coating takes place on a piece of mould. The part of the mould in which the encasing is done can be used directly for specialised devices.

Impregnation is another specialised method of embedding in which a liquid plastic is forced into the bulk of a component, which is usually porous in nature. In the extreme case of impregnation, a liquid like prepolymer with a catalyst is poured into a small hold made in the device and then allowed to cure in an oven.

5.7.2   Thermoforming

Thermoforming is the process in which the thermoplastic material, which is usually in a sheet form, is first heated till it softens. So obtained flexible plastic material is pressed into moulds having desired shape and dimensions. The die used in the process in fact consists of two parts besides clamps. A male part and a female part combine together to form the die. The male part is used to press the warm and flexible plastic material against the female part, which has a dug in cavity and acts as a store for the hardened material. The hardening or solidification of softened plastic is usually achieved by cooling process. The way the warm and softened plastic is pressed against the female part of the die classifies the thermoforming into four main types: (i) matched-mould forming, (ii) slip forming, (iii) air blown forming and (iv) vacuum forming. In matched-mould forming, as the name suggests, the male and the female parts of the die match with each other but face one opposite another, only a male fold is used in slip forming, the air is blown into the sheet while it sits in the female part of the mould in air blown forming and finally vacuum is applied to the sheet before it is pressed into the cavities of a female part of the mould.

Two important factors are to be considered in processing operation in general and in thermoforming in particular. These are heat transfer and the changes caused in the heated material under stress (i.e. rheological responses). The transfer of heat to the specimen can be achieved by several means, for example by conduction, convection, radiation, or some combination of these. However, when already formed thermoplastics or pre-polymer stage thermosetts are to be processed, the method of heat transfer has to be carefully selected. Since these materials are made up of chemical linkage between small monomer units and also contain several additives that may vary from simple molecules such as catalysts, plasticizers, antioxidants and colorants to complex molecules such as fillers. The means of heat transfer plays a crucial role in achieving the targetted curing or hardening or solidifying processes.

Exposure of plastic mix to radiation for a scheduled and programmed time period in the mould has been proved more suitable among many other means. Polymeric materials depending upon their optical properties i.e. transparent, semi-transparent and opaque etc. interact with radiation differently, hence one can obtain different heat rates by using the same amount of radiation dose. The stretching undergone by a flexible polymer during the thermoforming, can affect the mechanical properties of the article produced. So one must be cautious in choosing the press rates during the thermoforming.

5.7.3 Foaming

Foaming is a process, in which the density of a plastic material is decreased by entrapping air or gas into specially created cell type structures in the material. Thus foaming process results into expandable or spongy materials. Sponginess is due to creation of numerous cells through out the mass of the material. Thus foamed plastics can be considered as two phase systems consisting of a gas phase entrapped in a continuos solid polymer phase. Foamable polymers are of wide variety types. Virtually any thermoplastic or thermosetting polymers can be foamed. The usual and most common foamable polymers are polyurethanes, polystyrene, polyethylene, polypropylene, silicones, epoxy, PVC and styrene and acrylonitrile (SAN) copolymers etc. The foam structure consists of either interconnecting or non interconnecting individual cells. The non interconnecting cells are known as closed cells and the other type are known as open cells. The closed cells are not accessible to the surrounding air or fluids, while the open cells can be filled in by surrounding air or fluids. Foaming process essentially uses the above mentioned cell structures for expanding the polymer materials.

There is a wide variety of methods for obtaining the foamed articles: (i) whipping of air into a molten thermoplastic or its solution or a prepolymer stage thermosetting resin, and then heat cure the expanded product, (ii) dissolution of a gas in a molten polymer kept under pressure and expand the gas into solid melt matrix by releasing the pressure, (iii) heating the compounded polymer mixture which contains a liquid plasticizer, till the latter is volatilised. The volatilised liquid vapor molecules take up the positions in the open cells and (iv) another technique similar to (iii), makes use of the volatilization of water produced curing reaction, within the polymer mass.

However, foaming is usually enhanced by adding foaming or blowing agents. These agents are of two types viz. physical blowing agents and chemical blowing agents. Physical blowing agents consist of compressed gases (N_2 and CO_2 etc) and volatile liquids (n-pentane, n-hexane, toluene and isopropyl ether etc.). The chemical blowing agents are generally solid materials and may come from both inorganic and organic origin. The inorganic agents include carbonates of sodium and zinc. Organic blowing agents are usually hydroquinones, surface active agents, acids, bases and peroxides. The physical blowing agents create the cell structures, which act as reservoirs for storage of gas molecules by a physical process. This process involves the injection of a physical blowing agent into a molten flexible plastic material kept under pressure and then slowly the melt is released to atmospheric pressure or even to lower pressures. Then the surrounding gas expands into flexible polymer matrix and creates foamed structures. The solvents that are used as physical foam agents have low boiling points and thus under above conditions liberate large volumes of gases.

Organic blowing agents are typical in the sense that they decompose at a defined temperature range and produce voluminous gases. The decomposition of these agents can be achieved either by changing the temperature or pressure or both. The decomposition temperatures are usually between 100°C to 400°C. When the decomposition temperature is high, several chemical substances called activators are added to pre-empt the decomposition at faster rate and at low temperatures. Typical activators are zinc oxide, acids, bases and peroxides. Besides the above, the selection of chemical blowing agents is based on following consideration: (i) release of gases over a controlled temperature and pressure range, (ii) the gases produced should not have any harmful effects on the quality and processibility of plastics, (iii) the gases evolved should be as less toxic as possible, should not have bad odor and decolorise the plastic, and (iv) the density of the gases produced should be such that the structure of the cells created in the plastic material should be uniform and the foams formed must be stable and their entrapment in the plastic structure must be stable.

The usage of physical or chemical blowing agents can sometimes create not only problems but is also not feasible especially when the curing process of a thermosett such as unsaturated polyester using styrene as curator. The curing does not yield the required temperature for the chemical blowing agents to split and crack into vapor. In such cases chemical reactions such as oxidation and reduction of hydrazine derivatives and/or a peroxide can be carried out simultaneously. Because the reactions produce gases, that can expand the flexible polymer, which is to be foamed. One of the well known side reactions that facilitates the formation of polyurethane foams is the liberation of carbon dioxide by the reaction between isocynate and water. Since diisocynate is one of the monomers, along with a polyol taken for producing polyurethanes, an excess amount of it and some quantity of water are sufficient enough to produce a large amount of carbon dioxide, which can foam up the polyurethane as it is formed.

Foamed polymers are known by many names such as cellular plastic, plastic foams, expanded plastic foams, and structural plastic foams. Virtually any thermoplastic or thermosett can form a foamed structure under proper conditions. Foamed polymers can behave as rigid, semi-rigid and flexible materials. They can be coloured or obtained plainly. The foamed polymers can be processed, of course during their formation into blocks, sheets, slaps, sprayed coatings and extruded films or sheets. Thus conventional processing aids such as extruders, injection moulding machine, and reservoir moulding machines can be used for processing of foams.

Foamed polymer articles have been put to use in wide forms. For example, foamed polystyrene sheets are useful in building, packaging and insulation applications, expandable polystyrene, polyethylenes (LDPE and HDPE) are used for decorative and insulative wall panes. Expandable polyethylene copolymer which consist of 50% polyethylene and 50% polystyrene resin have been used in many reusable applications such as handling trays and packaging applications which require solvent resistance, impact absorbing and superior toughness. Flexible PVC and polyurethane foams are consumed in large quantities in automobile and furniture industry and as upholstery (i.e. covers, drapiers, curtains, cushions and other interior domestic fittings). Rubber and polyurethane foams are used for mattresses and other cushionable fittings. The rigid polyurethane foams are used as supports for artifical limbs. Foamed polyvinyl chloride is widely used in water resistant clothing, flooring, footware, furniture and packagings etc..

5.7.4 Lamination

The word laminate or lamination is in very common usage and usually refers to heterogeneous materials stick together mostly in a flat sheet. The layers in a laminate can be of same material or of different materials. The most commonly used laminates are plywood, formica and micarta. In plywood the layers of wood (known as veers) are oriented along the linear dimension and hence have good longitudinal strength, but the strength in the lateral sides is weak. The strength of the veers in both the longitudinal and radial directions can be made equal by holding several veers together by bonding them together. Phenol-formaldehyde resins are used for this purpose. Phenol formaldehyde resins taken out from a reaction vessel at pre polymer stage can be transferred into the gaps of stalks of veers and further polymerisation and curing of resins are done under high pressure. Thus these stalked veers bonded by the phenol-formaldehyde resins, have good strength. The laminated stalks are known as plywood sheets, which can be made into different thickness depending upon the number of veers bonded. Besides processing strength, laminates with other special properties such as less moisture pick up, less swollenness and other improved decorative qualities can be prepared by stalking the sheets out of inexpensive wood veers, except the top layer which can made of a rare wood. The modification of inner layers can be done such that the sheets formed have high resistance to moisture, chemicals and fire proofing etc.

Other classes of industrial laminates, which vary in reinforcement and in the binder are produced by using paper, woven or knitted cloth, cotton, glass, asbestos, nylon or rayon layers with binders of phenolic, melamine, polyester, epoxy or other thermosetting resins. Laminates can be produced during extrusion. The process is known as coextrusion. the main advantage of laminates is that multilayered stalks can be fabricated. Laminate systems with upto seven layers are put in operation. These multilayered sheets can be chosen such that each of it or some of them are selected such that it meets special requirements like chemical resistance, water repellence, high strength, toughness and coloured sheets etc. A four layered sheet or laminate made of ABS, polyethylene, polystyrene and rubber modified polystyrene is one of the best commercial examples, which is used as package material for wrapping up of food products. The combination of ABS and high impact polystyrene sheets achieved by thermoforming is used to make the inside door and food compartment parts of refrigerator.

5.7.5   Reinforcing

The term reinforcement implies an improvement of some ultimate property. These properties include all mechanical properties, tear strength and abrasion resistance or fatigue. The reinforcement consists of the polymer material whose properties are to be improved, the material which needs to be combined with polymer i.e. reinforcing materials and an appropriate process which can combine these two and produce high strength material. Almost all the thermoplastic as well as thermosett polymers can be reinforced. The main reinforcing materials are fibers which include natural fibers such as sisal, asbestos, jute, coir and remp etc., glass, graphite, alumina, carbon, boron or beryllia. Synthetic fibers from aromatic polyamides and polyesters are also good reinforcing material. The physical form of the chosen fiber can be anything ranging from chopped fiber pieces, porous mats, woven fabrics, continuous fabrics or filaments etc. The polymeric materials, whose properties need to be strengthened by reinforcing process are polyester, epoxy, polyurethane, phenolics, silicones, melamine, vinyl resins, acetal resins, polycarbonate, polypropylene, acrylonitrile butadiene-styrene terpolymer (ABS) and styrene-acrylonitrile copolymer (SAN) etc. Out of all the fibers, glass fiber is the most commonly used reinforcing material. The glass fiber has following advantages. It has (i) low coefficient of thermal expansion, (ii) good tensile strength, (iii) low dielectric constant, (iv) nonflammability, (v) high chemical resistance, (vi) high directional stability and (vii) low cost of production. The other materials are used when special properties are to be imparted.

Among all thermosett resins, polyester and polyurethane are used in large proportion in the form of reinforced plastics. The properties that get improved after reinforcement are, percentage of elongation, flexural modulus, compression strength, impact strength, hardness, water absorption and mould shrinkage percentage etc. Reinforcing process consists of two steps. In the first step, the fiber is bonded to the polymeric resin matrix and the combined material is cured under pressure and temperature. The second step consists of any of the general processing methods. The following processing techniques of reinforced polymers are widely employed: (i) hand lay up, (ii) spray up, (iii) vacuum bag moulding, (iv) tooling, (v) cold press moulding, (vi) casting, (vii) architectural panelling, (viii) centrifugal casting and (ix) pultrusion. Beside these, filament winding, injection moulding, rotational moulding and cold forming are also used.

A variety of moulds is available for reinforcing process. The type of mould chosen depends upon the kind, form and amount of reinforcement. The shape of reinforced polymer can be obtained from very small to extremely large and from simple to complex. Fiber reinforced plastics (FRP) have high performance utility because of their remarkable high strength to weight ratio, excellent corrosive resistance and nonbiodegradability, light weight and easy processibility etc.. The usage of FRPs range from the body of commercial and military air crafts, space craft, huge surface and underground acid storage tanks, the boat hulls, fiber boats and the protection sheets on the road side which can take up high impact during an accident.

5.7.6   Processing of Fibers

Fibers, both natural and man-made (or synthetic) are produced from polymeric material by a process similar to extrusion. However, the extrusion of polymers through fine holes of a spinneret is called 'spinning'. The spinneret is a special kind of plate made of a metal, gold or platinum and has small round or oval shape holes. There are in principle two ways to obtain fibers. i.e. either from the molten polymer or from polymer solutions. Since most of the natural polymers decompose before melting, they are drawn into small fibers or yarn by dissolving them in suitable solvents. After dissolution, fibers are drawn either by dry or wet spinning processes. Another method of recent origin is however known as melt spinnning, in which especially thermally stable polymers such as polyamides, polyesters and polyolefins are brought to molten state before being extruded through spinnerets to get fibers.

Dry Spinning

This process involves pumping a polymer solution of high concentration through a spinneret, which is hot. The fine filaments that emerge from the orifices of the holes of the spinneret are then passed through a zone in which the solvent is repidly evaporated. The solvent evaporation can be made faster and effective by pumping nitrogen through the zone in a direction opposite to the emerging end of the filament. The filaments or fine fibers thus formed are hardened by solvent evaporation and are directed to a spindle for wounding them up. This process of spinning is very common and fibers in large quantities can be produced. The fibers can be wound up at speeds up to about 100 m/min. Two precautions are to be taken for an efficient dry spinning. The viscosity of polymer solution has to be optimum for the fibers to be formed. The control of viscosity is achieved by changing the temperature of hot solution before it is pumped. The whole process has to be carried out in inert atmosphere to avoid the exposure of hot solution to the air, which can cause oxidative degradation. Fibers of cellulose acetate are drawn from its concentrated solutions maintained at 35–40°C, by this method. Similarly, PAN and PVC fibers are obtained by dry spinning method.

Wet Spinning

Wet spinning also employs a concentrated polymer solution at a higher elevated temperature. The process for wet spinning is identical with dry spinning for initial stages. But the filaments emerging from a spinneret containing large number of holes are directed to a bath filled with a non-solvent. Thus, when the continuously drawn filaments are passed through the non-solvent, the polymer gets precipitated from solution. This process is sometimes called coagulation. In some cases especially, when viscose rayon fibers are drawn, the precipitation stage can involve a chemical reaction leading to the coagulation and solidification or hardening of the fiber. The filaments are then collected from the coagulation bath, washed, dried and finally wound up around the spindles. The coagulation is a slow process and as a consequence, the conventional wet spinning is by far the slowest process. The rate of wet spinning (in fact wounding up) is as low as 50 m/min. But since spinneret has large holes for a given area of the plate, the productivity of fibers is high.

Melt Spinning

Though melt spinning was developed in the late 1930s for producing nylon 6 and nylon 6, 6 it has become popular recently. Polymer either in the form of powder, flakes, chips or granule is heated to a molten state on a heating grid. The polymer melt is then forced through a spinneret by using a metering pump (which is operated at a constant pumping rate). The fine filaments that emerge from the spinneret holes are still hot and are further cooled and solidified by blowing cool air. The blow rate of cool air is such that, it almost blasts into the container and by this way the hot filament solidifies at faster rate. The hardened fibers are then wound up through spindles. One of the utmost cares, that needs to be taken is that, polymers in hot stage can be cross-linked and form lumps, which can block the holes of spinneret. Similarly, any exposure to the air in hot condition can degrade the polymer by oxidation. Hence precautions are taken to guide the polymer melt through appropriate filters and spinning is carried out in an inert atmosphere.

Filaments formed from the polymer melt or dope by cooling or coagulation have poor strength. Their storage stability is also poor and can be stored upto 30–40 hrs. before further processing. Then they are stretched. In this process of stretching the so called drawn filaments are extended by 3–4 times or more depending upon various factors whereby the orientation of molecule takes place in direction parallel to the axis of the filament. The drawing is carried out after spinning. Simultaneous melt spinning and drawing the filament give moderately oriented yarn (MOY) or partially oriented yarn (POY). Such yarns are fully drawn during the texturising process. The POY and MOY have a long storage life and can be produced in single operation of melt spinning combined with stretching. The process of simultaneous production of POY and MOY is called high speed spinning. In another process, as-spun yarns are kept under controlled temperature and humidity and are usually subjected to the process of stretching or drawn-twisting (D/T). Since this process is a two step process and is carried out at the yarn manufacturer's site. However, it has become presently an obsolete technology mainly because of the short shelve life of the as-spun yarn and high cost of production. Most companies manufacture either MOY or POY filaments independently.

In the process of drawing or stretching, molecular orientation takes place along the axis of fibre and can be measured either by x-ray diffraction or birefringence. Continuous filament or yarn does not have same hairiness, bulk and warmth of handle as those found in natural fibres or synthetic staples. Some of such desirable properties are imparted by texturing, wherein the filament is permanently deformed to add bulkiness by the formation of air spaces. The properties such as crimps, coils, loops, curls and/or crinkles into filament are developed by various texturing methods. All texturing processes are mechanical in nature and do not involve any chemical treatment. The process of texturing is carried out at certain temperature depending upon the glass transition temperature of the polymer. Some of the important techniques used for producing textured yarns are: (i) false-twist process (a) spindle technique, (b) friction-texturing process, (ii) air texturing process, (iii) knit-de-knit process, (iv) gear crimping process, (v) knife edge process and (vi) stuffer-box crimping process.

Heat setting, twisting, doubling and cone winding are other processes required before yarn is subjected to the process of weaving or knitting.

5.7.7 Moulding Processes

Moulding processes are those in which a finely divided plastic is forced by the application of heat and pressure to flow into the fill and conform to the shape of cavity (mould). There are five different types of moulding techniques: (a) compression moulding, (b) injection moulding (c) transfer moulding (d) blow moulding (e) extrusion moulding

(a) Compression Moulding

It is one of the oldest methods. The polymeric material is put between stationary and movable parts of a mould. There are two different halves. The lower half usually contains a cavity and the upper half has a projection which fits into the cavity when the mould is closed as shown in Fig. 5.5.

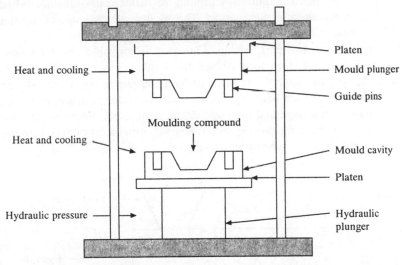

Fig. 5.5   Diagram of compression-moulding press and mould

The gap between the projected upper half and the cavity in the lower one gives the shape of the moulded articles. The compression moulding is very widely used to produce articles from thermosetting materials as well as from thermoplastics. The mould is closed and heat and pressure are applied so that the material softens, flows to fill the mould, and becomes a homogeneous mass. The necessary pressure and temperature vary considerably depending upon the thermal and rheological properties of the polymer material. Moulding themperature can be as high as 150°C and pressure applied is usually in the range of 1000–3000 psi. Slightly excess material is always placed to fill the cavity. As the mould closes down under pressure, the material is squeezed or compressed between the two halves and compacted to a shape inside the cavity. The excess meterial that flows out is known as flash. On cooling or after curing, the moulded article is taken out by opening the mould parts, which can be separated while the mould is still hot to release the finished moulded product.

(b) Injection Moulding

The injection moulding process is best suited for producing articles made of thermoplastic materials. Here the polymer is preheated in a cylindrical chamber to a temperature at which it flows. The molten material is then forced into a relatively cold mould under high pressure applied hydraulically where it solidifies to the shape of the mould. The injection moulding machine has high cost but the production rate of the articles is high and thus economically advantageous. The process consists of feeding the compounded plastic materials as granules, pellets or powder through a hopper at definite time intervals into the hot horizontal cylinder where it gets softened. Pressure moves the piston to push the molten material through a cylinder into a mould fitted at the end of cylinder. While moving through the hot zone of the cylinder, a device called 'torpedo' helps spread the plastic material uniformly around the inside wall of the hot cylinder and thus ensures uniform distribution. The molten plastic material from the cylinder is then injected through a nozzle into the mould cavity.

The mould is a two-part system. One part is movable and the other is stationary. By using a mechanical device, the mould is properly held in position as the molten plastic material is injected under a pressure as high as 1500 kg/cm². After the mould is filled with the molten material under pressure, it is cooled by cold water circulation and then opened so as to eject the moulded article. The whole cycle can be repeated several times either manually or in an automated mode. Figure 5.6 shows schematic drawing of an injection moulding machine.

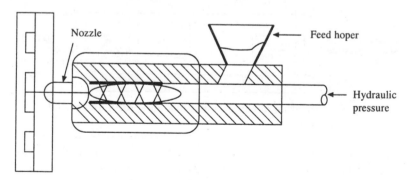

Fig. 5.6   Diagram of a conventional plunger injection-moulding machine

(c) Transfer Moulding

This method uses the principle of injection moulding for thermosett materials. The moulding powder is placed in a heated chamber, maintained at the minimum temperature at which the moulding powder just begins to become plastic. The plastic material is then injected through an orifice into the mould by a plunger, working at high pressure. This moulding technique has the following advantages: (i) since the mix flows into the mould cavity in a highly plasticized condition, so very delicate articles may be handled without distortion or displacement. Fine wires and glass fibers may be inserted into the mould, (ii) intricate shapes which are not attainable by compression moulding, can readily be produced, (iii) article produced is free from flow marks, (iv) even thick pieces can be cured almost completely and uniformly, (v) owing to uniform and thorough cure, shrinkage and distortion rates are at a minimum and hence, the mechanical strength and density of fabricated piece are higher, (vi) finishing cost of fabricated article is almost entirely eliminated, (vii) blistering is almost eliminated since air and excluded gases are expelled in the plasticizing chamber itself and (viii) mould cost is less since it involves very low abrasive action.

(d) Blow Moulding
Most hollow plastic articles are produced by the blow moulding technique which is principally similar to very old glass blowing method used in glass industries. A typical blow moulding process is schematically presented in Fig. 5.7.

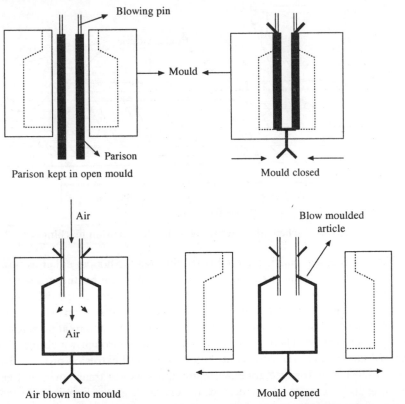

Fig. 5.7 Schematic diagram explaining the steps involved in blow moulding process

In blow moulding, a hot softened thermoplastic tube, usually called 'parison' is properly placed inside a two piece hollow mould. When the two halves of the mould are closed, it pinches and closes one end of the parison and encloses a blowing pin at the other end. The parison is now blown by pressuring from within by blowing compressed air through the blowing pin. The hot parison is inflated like a balloon and goes on expanding until it comes in intimate contact with the relatively cold interior surface of the hollow mould. Under pressure, the parison ultimately assumes the shape of the hollow cavity of the mould. The mould is allowed to cool and the rigid thermoplastic article formed is removed by opening the mould.

Containers, soft drink bottles and numerous other hollow articles are produced by this process. Thermoplastic materials such as polyethylene, polycarbonate, PVC, polystyrene, nylon, polypropylene, acrylics, and ABS plastics can be blow moulded.

(e) Extrusion Moulding
Extrusion technique produces many common plastic products such as films, filaments, tubes, sheets, pipes, rods, etc. quite inexpensively. A simple schematic picture of an extrusion machine is shown below (see Fig. 5.8).

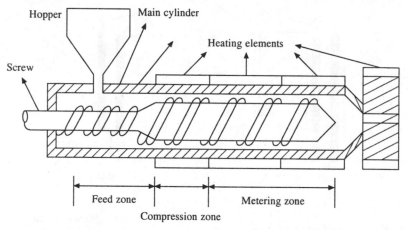

Fig. 5.8   Schematic diagram of a simple extrusion machine

The compounded plastic material is fed through the hopper either as powder or granules into a cylinder having provision for electrical heating for softening the material. The hot plastic charge is further worked through the cylinder by a revolving screw. The temperature of the plastic material rises owing to the frictional heat produced by the compression of the charge between the rotating screw and the cylinder surface. There are three different zones through which the plastic material passes and each zone contributes the overall extrusion process. These are transport zone (or feed zone), compression zone (or transition zone) and a metering zone. The feed zone receives the charge from the hopper and sends it over to the compression zone. In compression zone, the powdered charge melts due to application of heat. The pasty molten plastic material is then sent to metering zone, where it acquires a constant flow rate, imparted by the screw. The pressure built up in this section enables the polymer melt to enter the die and emerge out of it with the desired profile. The extruded material of the required profile emerging from the die is quite hot (usually 125°C to 350°C) and has to be cooled rapidly to avoid de-shaping. Cold air blast and cold water spray are used for cooling. The product formed is cut to the desired length or wound onto rolls.

Calendering

Calendering is a processing technique employed for the production of films and sheets. It, in a simplest form contains a set of highly polished metal rollers that rotate in opposite directions with provision of precise adjustment of gaps between them. Compounded polymeric material is fed between the rollers which are hot. The material here softens and passes through several rollers to give a smooth film or sheet which solidifies on passing through cold rollers. The film can be continuously wrapped on wind up rolls. The schematic drawing of a four-roll calendering machine is shown in Fig. 5.9.

Plastic mass

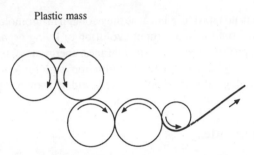

Fig. 5.9   Schematic representation of a four-roll calender machine

Suggested Further Readings

Agassant, J.F., P. Avenas, J. Sergent, and P.J. Carreall, *Polymer Processing: Principles and Moulding*, Cincinnati, Ohio: Hanser-Gardner, 1991.

Astarita, G., and L. Nicolais, (eds.), *Polymer Processing and Properties*, New York: Plenum, 1984.

Biard, D.G., and D.M. Collias, *Polymer Processing: Principles and Design*, Newton, Mass.: Heinemann, 1995.

Birley, A.W., B. Haworth, and J. Batchelor, *Physics of Plastics: Processing, Properties and Materials Engineering*, Cincinnati, Ohio: Hanser-Gardner, 1992.

Chanda, M., and S.K. Roy, *Plastics Technology Handbook*, 2d ed., New York: Dekker, 1992.

Charrier, J.M., *Polymer Materials and Processing: Plastics, Elastomers*, and *Composites*, Cincinnati, Ohio: Hanser-Gardner, 1990.

Florian, J., *Practical Thermoforming*, New York: Dekker, 1987.

Griskey, R.G., *Polymer Process Engineering*, London: Chapman and Hall, 1995.

Klempner, D., and K.C. Frisch (eds.), *Handbook of Polymeric Foams and Foam Technology*, Cincinnati, Ohio: Hanser-Gardner, 1992.

Kresta, J. (ed.), *Polymer Additives*, New York: Dekker, 1988.

Mallick, P.K., *Fiber-Reinforced Composites*, 2d ed., New York: Dekker, 1993.

Michaeli, W., *Plastics Processing: An Introduction*, Cincinnati, Ohio: Hanser - Gardner, 1995.

Middleman, S., *Fundamentals of Polymer Processing*, New York: McGraw-Hill, 1977.

Rosato, D., *Plastics Processing Data Handbook*, London: Chapman and Hall, 1997.

Walczak, Z.K., *Formation of Synthetic Fibers*, New York: Gordon and Breach, 1976.

Ziabicki, A., *Fundamentals of Fibers Formation*, New York: Wiley-Interscience, 1976.

6

Natural Polymers

Living things exist and reproduce because they contain macromolecules (polymeric materials). Throughout the subsequent evolution of living organisms, polymers have been used as protective surfaces, membranes, energy storage systems, skeletal system, and pathways for electrical conduction and for countless of other purposes. These biological/natural polymers are widely known as polysaccharides, proteins, nucleic acids and polyisoprenes etc.

6.1 Polysaccharides

They have macromolecular structure built up of many monosaccharide units joined together with an elimination of a water molecule. Hence, polysaccharides can be broken down on hydrolysis to form monosaccharides. Polysaccharides are of almost universal occurrence in living organisms where they perform a variety of functions. In plant cells, they serve as excellent materials for construction work e.g. cell wall formation. They also serve as food material for the nutrition of cell. Cellulose and starch are amongst the best known polysaccharides.

Cellulose $(C_6H_{10}O_5)_n$
This is a natural polymer of β-d(+) glucose, most widely distributed in plants where it is found in wood, cereal, straw up to about 50% in association with lignin, hemicellulose (polysaccharide of other sugars) and non-cellulose substances. Cotton which contains almost 95% of cellulose, is the oldest natural fibre and is in common use till today. The fiber is resistant to alkali and can be easily dyed using direct, vat, sulfur, reactive or azoic dyes. Filter paper is nearly pure cellulose. Other fibres like jute, flex, hemp and ramie are very rich in cellulose, which can be freed from other substances by fermentation process, called retting. Cellulose can also be produced by bacterial action of certain bacteria in nutrient media containing glucose.

Plants synthesise cellulose out of one of the most basic of all organic substances-glucose, a simple sugar (monosaccharide) which is found throughout the vegetable kingdom and even in the blood of animals. Nature converts glucose into cellulose (in plants) when thousands of glucose molecules are linked together through hydroxyl groups at C_1 and C_4 in a long chain forming a polyanhydride (giant cellulose molecule) and several water molecules (by a process similar to polycondensation). The following is the structure of the repeat unit in cellulose (see Fig. 6.1).

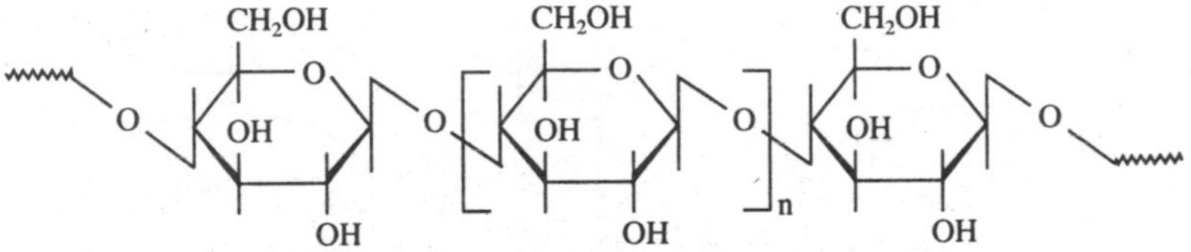

Fig. 6.1 Cellulose molecule

Despite containing several hydroxyl groups (three in each monomeric unit), cellulose is unexpectedly water insoluble. The reason for this is the stiffness of the chains and hydrogen bonding between two -OH groups on adjacent chains. Cellulose differs from starch (polymer of α-glucose) only in arrangement of hydroxyl group at carbon number one in the monomer glucose unit. Both are the polymers of glucose. The two polysaccharides in spite of having only slight difference, possess enormously different properties. Starch being the chief constituent of diet, gets easily hydrolysed by body enzymes to glucose which finally yields CO_2 and H_2O along with the energy, the later being used for sustaining body functions. Cellulose nevertheless is of considerable human dietive value in that it adds "bulk" to the intestinal contents, thereby stimulating peristalsis and eliminates food residues. It thus provides indigestible but necessary roughage in the human diet, and the nourishment for such ruminant animals as sheeps and cows. Indirectly, therefore it also provides nourishment for humans. Sheeps and cows digest the cellulose and absorb it into their body tissues and thus when humans consume meat, cellulose is used up by them.

It has the property of forming long threads and for this reason is largely incorporated as raw material in textile industries Cellulose is highly stereo specific macromolecule. Considering the geometry of the polymer molecule, cellulose can be treated as non-planar molecule, but with a screw-axis and with each glucose unit being at right angles to the previous one. The free rotation about the C-O-C link does not occur. This and the close packing of the atoms result to a fully extended form of conformation. Moreover large intra- and inter-molecular forces (as result of formation of hydrogen bonds) also favour the close packing and thus cellulose possesses an unusually high degree of crystallinity. The melting temperature of crystalline cellulose is far above its decomposition temperature. The fiber forming properties of cellulose can be accounted by its unique conformation. The fully extended cellulose chains are further bonded together to give rise to a sheet structure (see Fig. 6.1). A huge amount of cellulose is used in the paper manufacture. Besides its use in textile and paper industries, cellulose finds its use as celluloid (a hard plastic), cellophane (film) for various purposes. These materials are called semi-synthetic, because they are obtained from the raw cellulose. The synthesis of various cellulose derivatives by chemical reactions is known as regeneration. The various regenerated forms of cellulose are shown in Fig. 6.2.

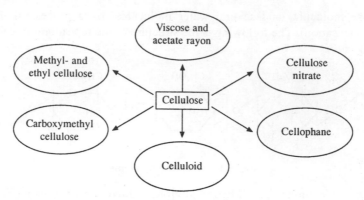

Fig. 6.2   Cellulose derivatives

Starch

It differs from cellulose in the configurational arrangement in glucose units. Starch is a polymer with an α-glucose as the repeat unit and is a widely distributed polysaccharide. It is a major component of human diet in the form of grains e.g. wheat, corn, barley, rice and potatoes etc., This dissimilarity in the structure of strach makes it available as a granular material, with no trace of organised crystalline structure. Starch is in fact, a mixture of two polysacchrides, the one a linear macromolecule called amylose (soluble starch) and the other branched chain called amylopectin in which linear 1–4 linked chains are joined together by 1–6 linked branching points. Besides, its use as foodstuff, starch is extensively used in paper sizing and textiles, and for fermentation to alcohol. Dextrins are partial hydrolytic products of starch obtained by the action of enzyme amylase in acidic media. These have free sugar group and thus reduce alkaline copper sulfate solution. The structures of amylose and amylopectin are shown in Fig. 6.3.

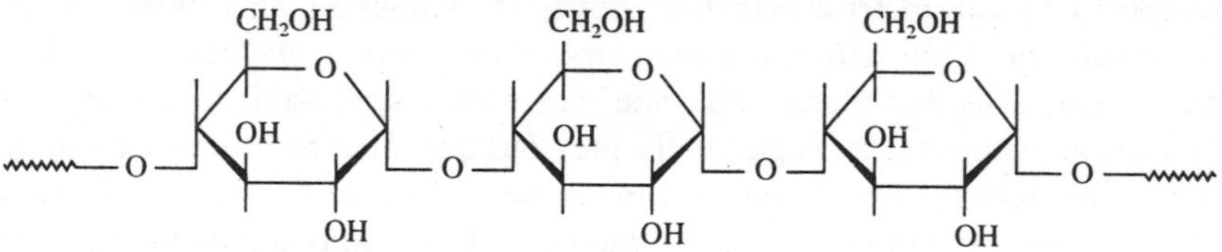

Amylose (soluble starch)

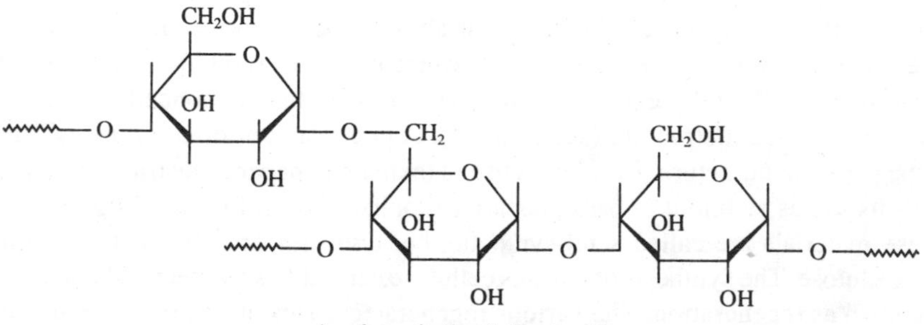

Amylopectin (insoluble starch)

Fig. 6.3   Structures of amylose and amylopectin

Some other natural polymers are briefly mentioned as below.

Glycogen

Glycogen is a polysaccharide, found in animals only. It is an important energy storing material in the body and is found mainly in muscle and liver. This is a glucose polymer similar to amylopectin, except that it is more highly branched.

Dextran

It is another polysaccharide prepared by cultured propagation of various microorganisms on a cane sugar phosphate substrate. It is a useful substitute for blood plasma. During the formation of dextran by the process of fermentation, glucose part of sucrose molecule is polymerised to give dextran.

Table 6(a) shows a comparison between various physical characteristics of some common polysaccharides.

Table 6(a)   Comparison of physical characteristics of some polysaccharides

Characteristics	Cellulose	Starch		Glycogen
		Amylose	Amylopectin	
Monomeric units	β-glucose	α-glucose	α-glucose	α-glucose
Polymer linear structure	linear (sheet like)	branched	spheriod	
Crystallinity	high	high	low	nil
Solubility in water	nil	limited	limited (after boiling)	good
Color in water with iodine	no color	blue	violet-red	brownish-red
Film formation	very strong	strong	brittle films	nil
Fiber formation	very good	poor	nil	nil

The structures of three other commonly occurring polysaccharides namely inulin, levan and mannan are depicted in Fig. 6.4.

Heparin

It is a polysaccharide of animal origin and is widely spread in animal tissues. The extra-cellular matrix mainly consists of heparin and chondroitin sulfates. Heparin is extensively used as anticoagulant for blood and antithrombotic agent. It enhances the natural thromboresistance of endothelium. The biologically reactive nature of heparin is exploited for inhibition of growth of smooth muscle cells and human immunodeficiency virus (HIV), and angiogenesis. The structure of heparin is shown in Fig. 6.5.

Hyaluronic acid

It is a linear high polymer built up of a disaccharide repeat unit, with molecular weight in the range of 2×10^5 to 2×10^6. Gels obtained from this polymer possess good resistance to compression, and appear as lubricants and shock absorbing components for the nose and joints. A similar polymer is chondroitin.

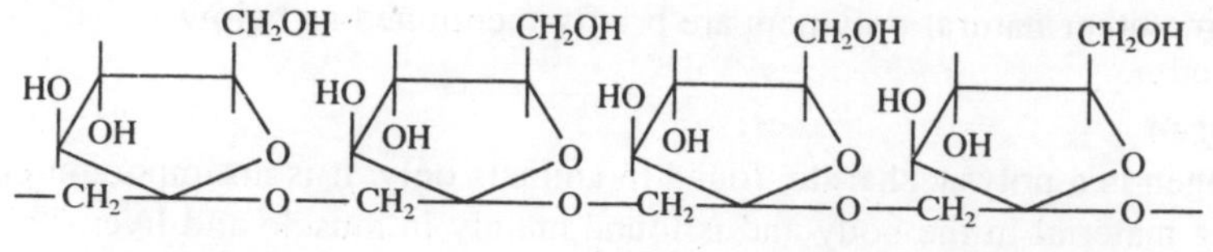

(a) Inulin

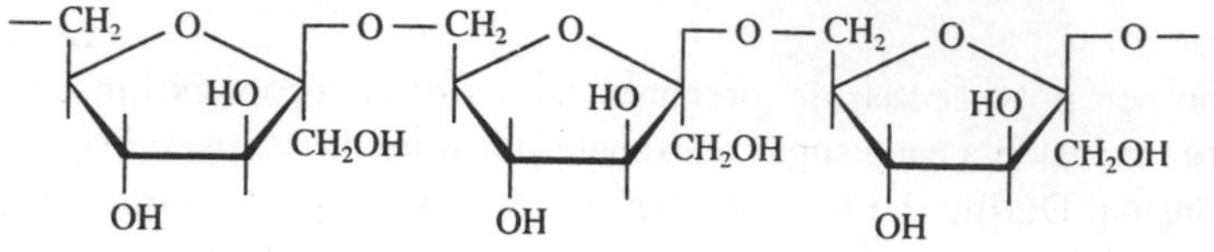

(b) Levan

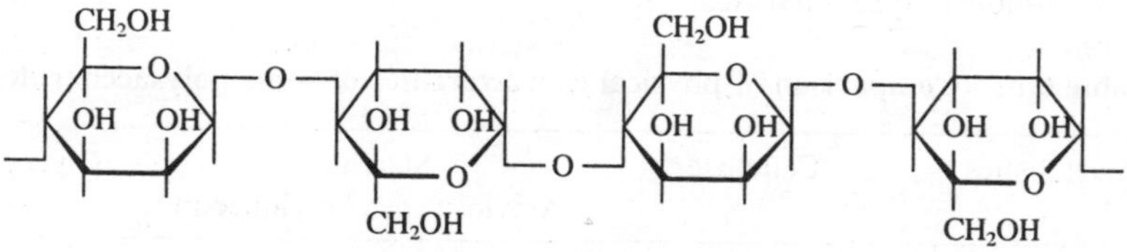

(c) Mannan

Fig. 6.4 Structures of (a) Inulin, (b) Levan and (c) Mannan

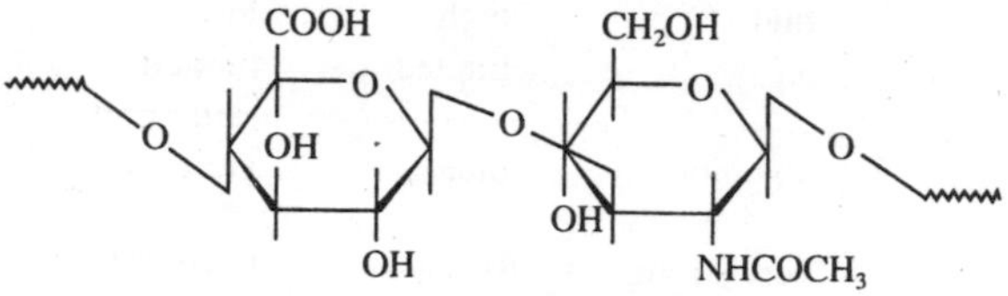

Fig. 6.5 Structures of heparin

Both hyaluronic acid and chondroitin (as sulfate) are often found in combination with protein or lipid molecules. The repeat unit structure of hyaluronic acid is shown in Fig. 6.6.

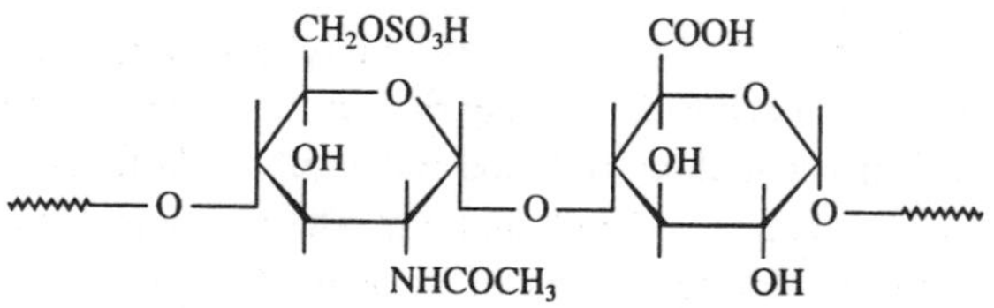

Fig. 6.6 Repeat unit structure of hyaluronic acid

Chitin

It is a tough and brittle polysaccharide, serves as structural component in insects and anthropods (for example in crabs) where it forms so called exoskeleton, playing a role analogous to the protein collagen in higher animals.

Several other polysaccharides help to maintain plant structure. Arabans and xylans, which provide wood many of its typical properties are, made from the pentose sugars viz. arabinose and xylose. Pectins found in fruits are highly branched polymers of glucose, arabinose and galacturonic acid.

Various gums and mucilages from land plants and sea weeds are mainly polysaccharides with applications as emulsifying, gelling and thickening agents.

6.2   Proteins

Proteins are the natural polymers made of α-amino acid units through peptide (-CO-NH-) linkage. In total about twenty of these amino acids occur in nature (of which eight are essential to sustain life). Except the simplest amino acid glycine, all other amino acids are optically active and are of l-series. Aqueous solutions of amino acid can be neutral, acidic or basic. They also form dipolar structure and thus are zwitterionic in nature.

Each protein is a monodispersed polymer with a definite number of amino acids linked with each other in distinct sequence. A simple protein is polyamide (or polypeptide) whereas a conjugated protein consists of a prosthetic group linked with simple protein. Glycoprotein lipoprotein, nucleoprotein and haemoglobin are the examples of conjugated proteins of which a carbohydrate, a lipid, a nucleic acid or haem are the prosthetic groups.

Proteins catalyse a series of biochemical reactions including hydrolysis, group transfer, oxidation and biosynthesis etc. In fact all enzymes are proteins. Proteins are structural elements and function as protective agents in the living systems.

Proteins may be inter-molecularly hydrogen bonded (fibrillar proteins) e.g. keratin of the hair and nails, collagen of the connective tissues, elastin of the sinusos and arteries, fibroin of silk thread; myosin, the muscle protein or intra-molecularly hydrogen bonded (globular proteins) e.g. hormones like insulin and glucagon, enzymes like trypsin, invertase and papain etc., respiratory proteins such as haemoglobin, myoglobin, albumins e.g. egg albumin, bovine serum albumin and antibodies. Fibrillar proteins are water insoluble whereas globular proteins are usually water soluble.

All proteins are hydrolysed finally to a mixture of α-amino acid from which these were built. The reagent ninhydrin gives characteristic coloration to amino acids and thus may be determined chromatographically (a method developed by Nobel laureates Martin and Synge).

The structure of proteins is very complex. Primary structure of proteins shows the chemical formulae of the amino acids arranged in a definite sequence through peptide linkage. The sequence of amino acids in the protein molecule was first successfully examined by Sanger who was awarded two nobel prizes in chemistry. Secondary structure of proteins describes the molecular shape (or conformation) of a protein molecule. Nobel laureate Linus Pauling, has demonstrated that a right handed intra-molecularly hydrogen bonded helical arrangement (α-helix) is an important secondary structure of proteins when many bulky pendant groups are present on the main chain. A β-arrangement or pleated sheet conformation predominates when small pendant groups are present on the chain. The α- and β-forms of proteins and enzyme are shown in Fig. 6.7. The sequence of amino acids in enzymatic protein lysozyme is shown in Fig. 6.8.

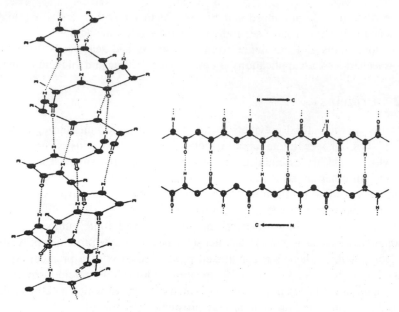

Fig. 6.7 The α-and β-form of proteins

Tertiary structure designates the shape of folding which results due to the sulfur-sulfur cross-links between macromolecular chains. The specificity of enzyme action in fact depends on tertiary structure. Proteins lose their physiological activity because of the disruption of folded structure (denaturation of proteins). Individual protein chain agglomerates in to supramolecular system and this mode of agglomeration is called "quaternary structure".

Although polypeptides can be synthesised by simply heating α-amino acids but the products obtained are random mixtures unless a single amino acid is used. A solid phase technique developed by Merrifield (Nobel laureate, 1984) is a novel method to synthesise polypeptides with desired amino acid sequence.

In Merrifield synthesis, all reactions take place on the surface of cross linked polystyrene beads and the entire reaction for synthesising polypeptides with many programmed sequences of amino acids can be carried out automatically in a simple vessel with a possibility of neat isolation of any intermediate products.

Fig. 6.8 The lysozyme molecule

6.3 Nucleic Acids

These are perhaps the macromolecules of utmost biological importance. Two kinds of nucleic acids exist in the living cell. They are deoxyribonucleic acid (DNA) and ribonucleic acid (RNA). Nucleic acids are usually built up of bases (purine and pyrimidine bases), pentose-a sugar, and phosphoric acid. The purine bases are adenine (A) and guanine (G) whereas the pyrimidine bases are uracil (U), thymine (T) and cytosine (C). The bases (A), (G) and (C) are common in both DNA and RNA but the base (U) is present only in RNA and the base (T) is present only in DNA. The pentose sugar is ribose in RNA whereas it is 2-deoxyribose in DNA. The various constituent chemical species of nucleic acids are shown in Fig. 6.9.

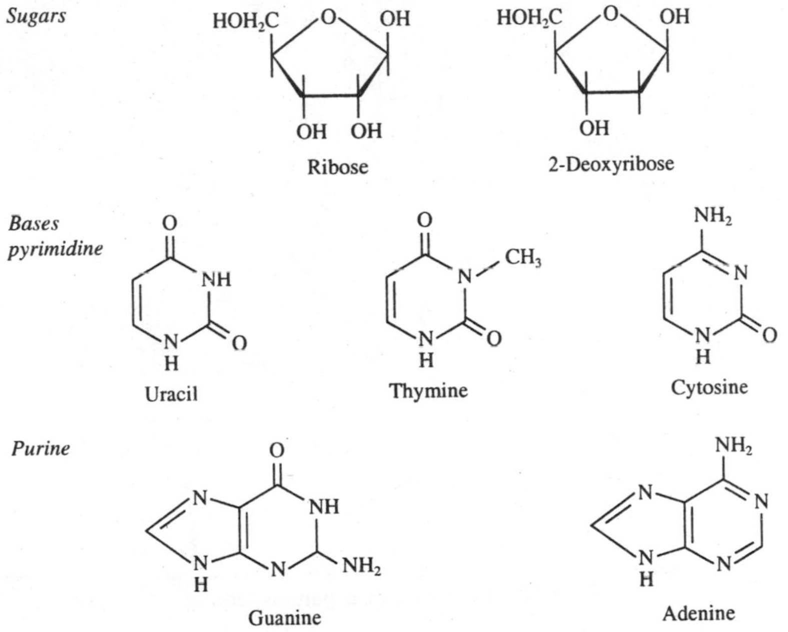

Fig. 6.9a   The constituents of nucleic acids

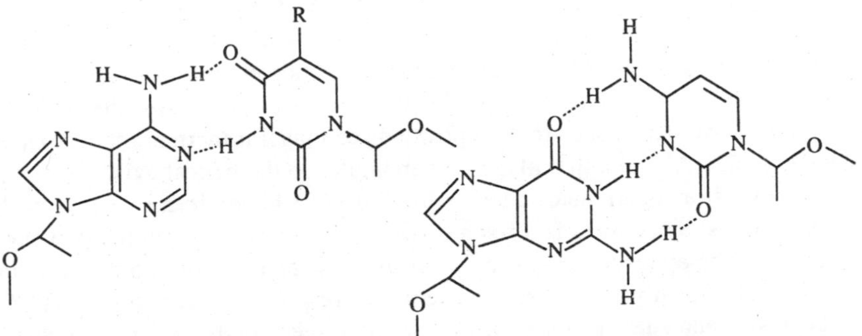

(a)   (b)

Fig. 6.9b   Specific base pairing in (a) adenine–thymine pair and (b) guanine–cytosine pair

The sugar-base combination is called nucleoside. For example the different nucleosides with bases A, G, U, C, T are called adenosine, guanosine, uridine, cytidine and thymidine respectively. Linking phosphoric acid to these nucleosides through the sugar moiety gives nucleotide (Fig. 6.10). The different nucleotides are adenylic acid and guanylic acid. Both DNA and RNA are polynucleotides and consist of chains of repeating units of nucleotides.

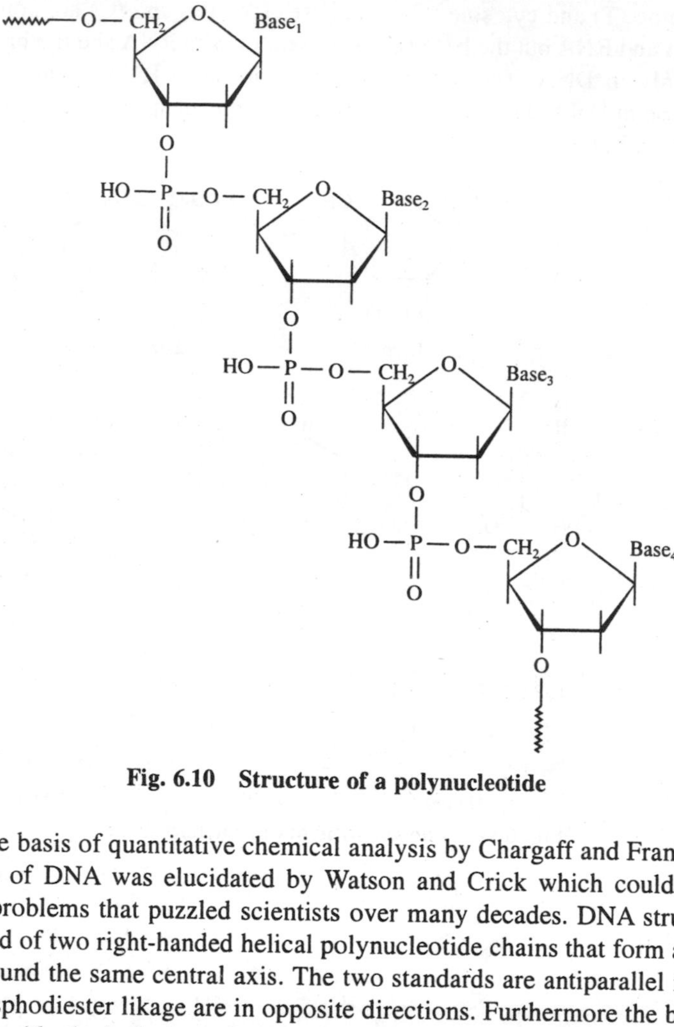

Fig. 6.10   Structure of a polynucleotide

On the basis of quantitative chemical analysis by Chargaff and Franklin, the structure of DNA was elucidated by Watson and Crick which could explain several problems that puzzled scientists over many decades. DNA structure is composed of two right-handed helical polynucleotide chains that form a double helix around the same central axis. The two standards are antiparallel i.e. their 3, 5 phosphodiester likage are in opposite directions. Furthermore the bases are stacked inside the helix in a plane perpendicular to the helical axis.

The two strands are held together by hydrogen bonds formed between the pairs of bases. Since there is a fixed distance (1.08 nm) between the two sugar moieties in the opposite strands, only certain base pair can fit into the structure and as such the only two possible base pairs are adenine and thymine (A)–(T) and cytosine and guanine (C)–(G). The two hydrogen bonds are formed between (A) and (T) and three are formed between (C) and (G). In addition to hydrogen bonds, the hydrophobic interactions established between the stacked bases are also important in maintaining the double helical structure.

The Watson and Crick model has been confirmed with slight correction. X-ray diffraction studies indicate that the base pairs are planar and that the hydrogen bonds are almost collinear, their length lying between 2.8 and 2.9 Å. Each turn of the helix contains 10 nucleotide pairs and the diameter of the helix is bout 20 Å. The spacing between the adjacent pairs is 3.4 Å. The two DNA chains are complementary to each other i.e. a chain with a given sequence of bases can pair only with another chain which has the complementary sequence of bases.

Since (A) = (T) and (G) = (C), (A) + (G) = (C) + (T), while on the other hand there is a considerable variation between species regarding the AT/GC ratio. For example, in higher plants and animals AT is in excess of GC, whereas in viruses, bacteria and lower plants it may be the contrary. For example in man AT/GC = 1.52; in E. coli, it is 0.93.

DNA is one of the largest known molecules. For example, the DNA of the E. coli contains 3×10^6 bases, which corresponds to a sterically specific arrangement of about 10^8 atoms. The propagation, i.e. the self doubling of the DNA takes place by the action of an enzyeme, DNA-polymerase as shown in Fig. 6.11.

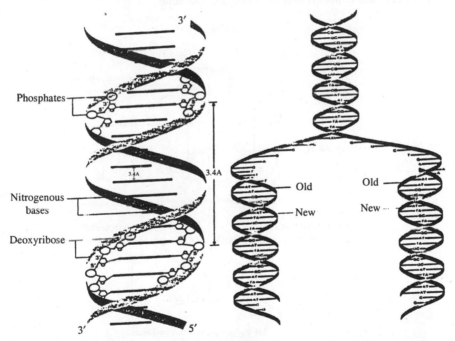

Fig. 6.11 Schematic representation of double helical form of DNA and its reduplication

The parent strands are separated and new complementary strands are formed from monomers by the building up principle. Deoxyribonucleic acid (DNA) is the molecule of heredity; the purine and pyrimidine bases carry genetic information, of whereas the sugar (deoxyribose) and phosphate groups perform a structural role. The function of a DNA molecule is to carry genetic information of a cell in such a way that this information can be passed from one generation to the next. For this to occur the DNA must be a stable molecule, which can be exactly duplicated so that the two daughter cells arising at mitosis may each, receive identical copies. The genetic information is transferred from the DNA to the complementary or messenger ribonucleic acid, m RNA (transcription). This is further translated during protein synthesis from the four letter language of the m RNA into the twenty letter language of the proteins by dictating which transfer RNA with its specific amino acid shall operate next on the growing protein chain.

6.4　Natural Rubber

The polymerisation of isoprene in the rubber plants to natural rubber is highly organised and controlled process which always results 1,4 *cis* product (where the hydrogen and –CH_3 groups are on the same side). In certain types of plants, isoprene gets polymerised to give gutta percha—a hard horny substance with plastic like properties. The possible scheme of the formation of polyisoprene from isoprene in rubber trees is depicted in Fig. 6.12. Also shown in the figure are the *cis*- and *trans*-arrangements of polyisoprene.

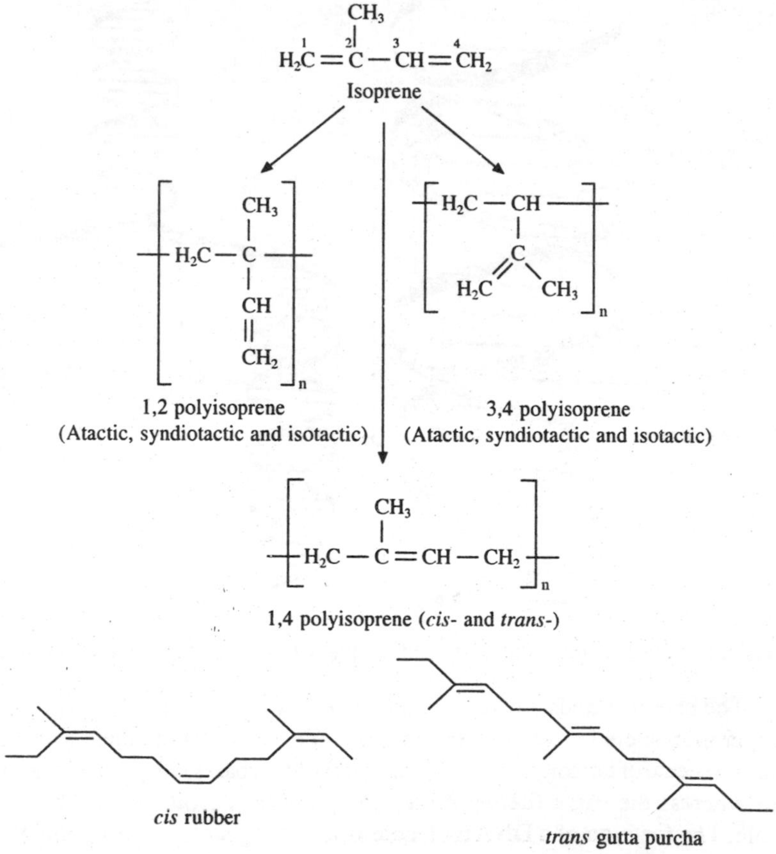

Fig. 6.12　**Probable scheme for the polymerisation of isoprene to polyisoprene in rubber trees**

Rubber occurs in nature in about 200 different plant species distributed over various countries of the world. Most of the commercial natural rubber is obtained from the *Hevea braziliensis* tree which is indigenous to Brazil and is now cultivated in the tropical rain forest regions of all continents.

Rubber is found in tree in the form of milky white fluid called latex, which occurs in the bark of the rubber tree. It is not the sap of the trees, as so many people wrongly believed. It occurs in ting ducts or tubes which spiral upward round the tree from left to right in the bark.

Natural rubber is made from this milky white fluid (latex) obtained by cutting narrow strips from the bark of the rubber tree *Havea Brasiliensis*. The liquid, which exudes from the exposed cut, is allowed to flow into plastic or aluminium cups or coconut shells containing a little preservative and later collected for processing. The process of obtaining the latex is called tapping.

Latex contains about 30–40 percent of solids and 55–60% of water. On coagulation or evaporation to dryness, it gives a solid substance containing approximately 95% polyisoprene, the rest being water and nonrubbery substances such as proteins, sterols, sugar and minerals. The latex is composed of ting rubber particles suspended in the liquid called "serum" which is chiefly water. It is collected from the tree and in coagulated form and de-watered on a roller mill.

The properties of rubber can be improved by heating it with sulphur. This process is called vulcanisation discovered by Charles Goodyear in 1839. The possible mechanism for the vulcanisation and structure of vulcanised rubber has already been discussed in Chapter 4. The vulcanised rubber has greater elasticity and tensile strength than crude rubber. Though the mechanism of vulcanisation is not fully understood, it results in a loss of one double bond with one sulphur atom introduced between two repeat units of neighbouring chains. It is assumed that one sulphur bridge is formed from about 50 initially present sulphur atoms. The vulcanised rubber still retains high elastic properties when the sulphur bridges are spaced sufficiently at long intervals along the chain.

There are two types of vulcanisation processes namely hot (warm) and cold vulcanisations. The hot process occurs at temperatures ranging 120–160°C. It is performed in autoclaves and in the presence of chemical agents called accelerators. For example tetraalkylthiouramedisulphide, zinc dialkyldithiocarbamate and 2, 2-dithiobisbenzothiozole. Various substances promote the vulcanisation. Some such compounds are called accelerators viz. combination of zinc oxide and salts of fatty acids.

The vulcanisation of natural rubber with an excess of sulphur (around 30%) yields a kind of hard and cured rubber known as ebonite. The hardness and the rigidity is due to maximum cross-linking and minimum residual unsaturation left in the chains. This process can be carried out both in cold and hot conditions. Normally sulfur is used as cross-linking agent to vulcanise the rubber at 120–160°C (the mechanism for such hot stage vulcanisation has been discussed in Chapter 4 (4.5)). The vulcanisation can even be carried out at room temperature. Disulfur dichloride (S_2Cl_2) is used as cross-linking agents and produces monosulfide cross-links as shown in the following Fig. 6.13.

The different applications of rubber for different use purposes demand varying

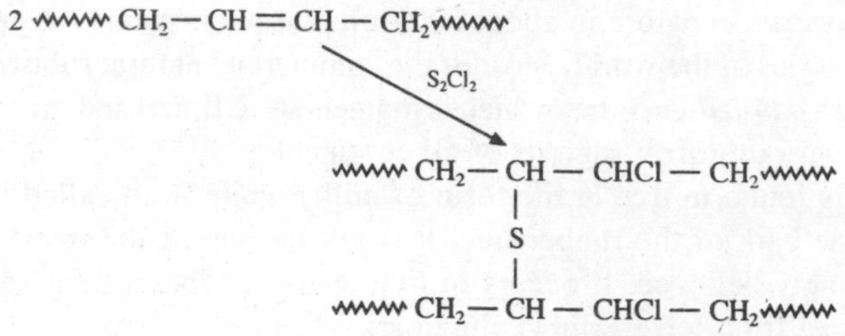

2 ∿∿∿ CH_2— CH = CH — CH_2∿∿∿

S_2Cl_2

∿∿∿ CH_2— CH — CHCl — CH_2∿∿∿

S

∿∿∿ CH_2— CH — CHCl — CH_2∿∿∿

Fig. 6.13   Cold vulcanisation scheme for natural rubber

degrees of cross-linking among the chains. Thus one can select any one of the above mentioned curing processes for obtaining different grades of rubbers. About 50% of all rubbers go into tyre production, 25% for other vehicular applications and the rest goes into other items such as surgical gloves and thin walled articles.

Suggested Further Readings

Dumitriu, S., (ed), *Polymeric Biomaterials*, New York; Marcel Dekker, 1994.

Dumitriu, S., (ed), *Polysaccharides–Structural Diversity and Functional Versatality*, New York: Marcel Dekker, 1998.

Gilbert, R. (ed), *Cellulosic Polymers*, Munich: Hanser 1994.

Kremer, O. (ed), *Biological and Synthetic Polymer Networks*, New York: Elsevier Appl. Sci., 1988.

Radhakrishna, S., A.K. Arof, *Polymeric Materials*, chap. 14, New Delhi: Narosa, 1998.

Saenger, W., *Principles of Nucleic Acid Structure*, Berlin: Springer, 1988.

Vincent, I.V. F., *Structural Biomaterials*, New York: Wiley, 1982.

Young, R.A. and R.M. Rowell, *Cellulose, Structure, Modification and Hydrolysis*, New York, Wiley, 1986.

7

Inorganic Polymers

Inorganic polymers are high molecular weight substances whose molecules are made up of covalent linkages between several small molecules with the absence or near absence of hydrocarbon units in the main backbone. Glass, silicones and polyphosphazenes are some common examples of inorganic polymers of industrial value. Broadly, inorganic polymers consists of several groups to be listed as phosphorous based macromolecules, sulfur polymers, boron polymers, silicon polymers, organo-polymers of silicon, germanium, tin and lead, polymeric metal-alkoxides, organometalloxanes and organometalloxono siloxanes, coordination polymers and electron deficient polymers etc. Some representative examples of inorganic polymers are already discussed in Chapter 1. Inorganic polymers can be classified in a number of ways. Some are based on the composition of the backbone such as silicones (Si-O), polyphosphazenes (P-N) and polymeric sulfur (S-S). Other classification is based on the connectivity, i.e. the number of network bonds linking the repeating units into the network. Thus silicones are based on R–Si–O–, polyphosphazenes are based on N-PX$_2$ and polymeric sulfur has the connectivity of two, while boric oxide (B$_2$O$_3$) has the connectivity of three and amorphous silica has four. This is illustrated in Fig. 7.1.

Sulfur, selenium and tellurium all form high polymers. Polymeric sulfur is elastomeric while polymers of Se and Te are generally crystalline. Inorganic polymers have some unique features, which make them advantageous over organic polymers in some specialized applications. Some of the special characteristics of linear inorganic polymers are an higher Young modulus and a lower failure strain compared to organic polymers. Higher degrees of crystallinity and high glass transition temperatures are often seen in inorganic polymers. Inorganic polymeric materials are growing in importance as a result of combination of the following two major factors, (i) the depletion of the world's fossils fuel resources (which forms the basis of petrochemical industries and thus polymer industry) and (ii) ever increasing demand of modern technology coupled with environmental and waste disposal problems associated with organic polymers. Inorganic polymers are widely employed in the construction and building materials, as abrasives and cutting tools, as fibers, as elastomers, as coatings, as lubricants and as catalysts. The unique properties of inorganic polymers and ever decreasing fossil fuel resource which

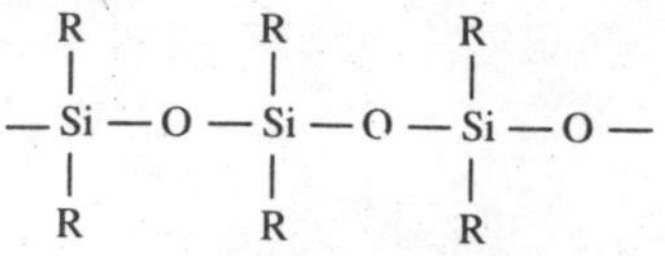

(a) Polysiloxanes (2)

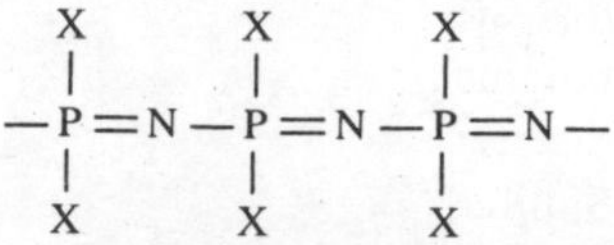

(b) Polyphosphazenes (2)

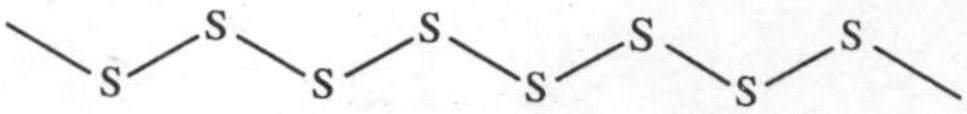

(c) Polymeric sulfur (2)

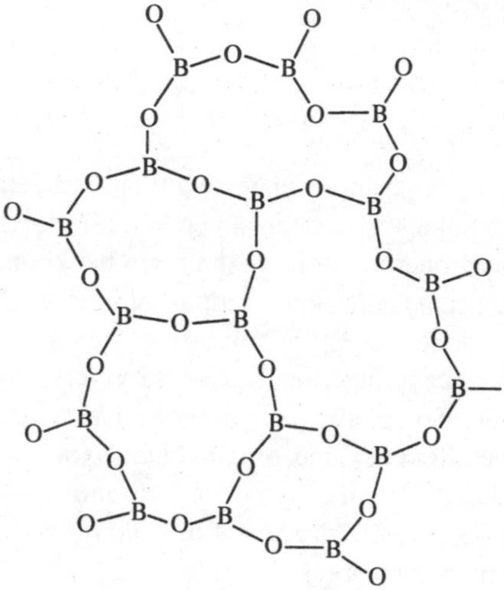

(d) Polyboric oxide (3)

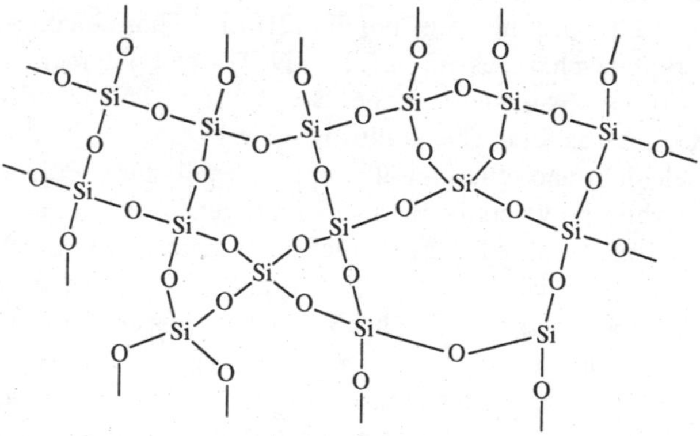

(e) Silica (4)

Fig. 7.1   Inorganic polymers with varying connectivities (given in the parenthesis)

forms the basic raw material for the synthesis of modern organic polymers, has
led to a new branch of chemistry 'Organometallics'. The application of chemistry
of polymerisation to organometallic compounds led to the development of
organometallic polymers. When direct organometallic polymers can not be
synthesised, attempts are made to produce novel and new combination of properties

that are not present in classical organic polymers and conventional inorganic polymers. These attempts resulted into metallation of organic polymers and also metal filled polymers. The metallation of natural rubber is a case in point. Similarly, synthesis of exotic mono and multi metal-sandwiched organic polymers such as poly vinyl ferrocene and its derivatives can also be cited.

Some salient features of most common inorganic polymers are briefly discussed in the following manner.

7.1   Silicones

Silicone is a popular term used to describe a whole family of organo silicon polymers based on a backbone or a molecular chain of alternate silicon and oxygen atoms. Polydimethylsiloxane (PDMS) is a most common example.

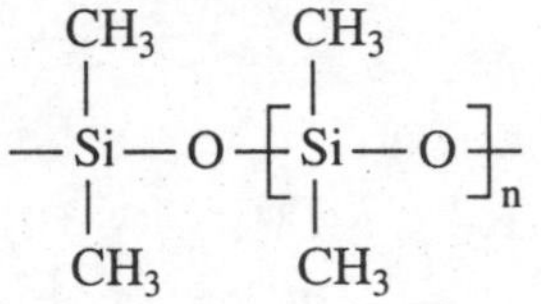

$$-\overset{\displaystyle \underset{|}{CH_3}}{\underset{|}{Si}}-O\left[\overset{\displaystyle \underset{|}{CH_3}}{\underset{|}{Si}}-O\right]_n$$

Depending upon the length of the chain and the organic groups attached to the silicon atoms, the silicone polymers may be like water thin liquids, heavy oil like fluids, greases, gels, rubbers and solid resins etc. The basic reactions in the synthesis of silicones are usually the hydrolysis of chloromethylsilanes which can be produced by the reaction of Grignard reagent with silicon tetrachloride;

$$R\ Mg\ X + Si\ Cl_4 \rightarrow R\ Si\ Cl_3 + XMg\ Cl$$

$$R\ Si\ Cl_3 + R\ Mg\ X \rightarrow R_2\ Si\ Cl_2 + X\ Mg\ Cl$$

The hydrolysis of dialkyldichlorosilanes gives silicones which subsequently polymerise to polydimethylsiloxane as shown below (see scheme 1).

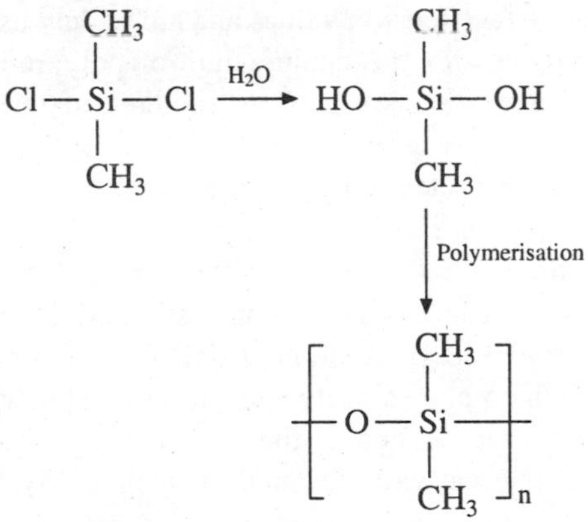

$$Cl-\overset{\displaystyle \underset{|}{CH_3}}{\underset{|}{Si}}-Cl \xrightarrow{H_2O} HO-\overset{\displaystyle \underset{|}{CH_3}}{\underset{|}{Si}}-OH$$

Polymerisation

$$\left[-O-\overset{\displaystyle \underset{|}{CH_3}}{\underset{|}{Si}}-\right]_n$$

Scheme I

When the chain becomes sufficiently long, the reaction can be stopped by terminating end functional group (–OH) by trimethylchlorosilane (see scheme

II). Such end capped low molecular weight linear silicones are oily materials. Adding some trichloromethylsilane (a trifunctional compound) during the polycondensation of dialkylsilanediol results in a crosslinked silicon, which can be end capped by trimethylchlorosilane to give a silicone elastomer or resin.

Silicones are exceptionally stable and highly bio-compatible polymers. The characteristic resistance of silicones to elevated temperatures is related to the siloxane backbone, which has high heat resistance capacity.

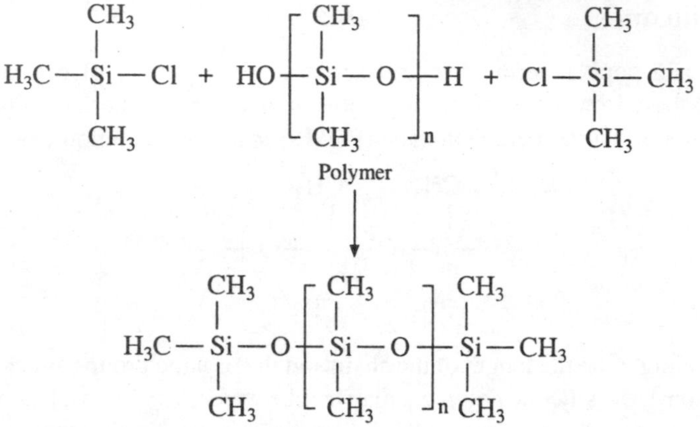

$$
\underset{CH_3}{\overset{CH_3}{H_3C-Si-Cl}} \;+\; HO\!\left[\underset{CH_3}{\overset{CH_3}{Si-O}}\right]_n\!\!H \;+\; Cl\!-\!\underset{CH_3}{\overset{CH_3}{Si-CH_3}}
$$

Polymer

$$
\downarrow
$$

$$
\underset{CH_3}{\overset{CH_3}{H_3C-Si-O}}\!\left[\underset{CH_3}{\overset{CH_3}{Si-O}}\right]_n\!\underset{CH_3}{\overset{CH_3}{Si-CH_3}}
$$

Scheme II

The lubricant action and water repellency are related to the pendant lipophilic alkyl groups, which encase the abrasive backbone. Surface related applications of silicones include their use as water repellents. Siloxanes modified with polyethers are now commercially available and possess solution properties similar to conventional low molecular weight nonionic surfactants. These modified siloxanes are known as silicone glycols or silicone surfactants. They also undergo aggregation in aqueous and non-aqueous solutions to yield micelles. Silicone surfactants are characterized by low surface tension values and are widely used as anti-foaming agents in one shot process for the manufacture of polyurethane foams. Being biocompatible, silicones are also used as implant materials in biomedical applications.

Silicones are often characterized by combination of chemical, mechanical and electrical properties which taken together are not common to any other polymers available commercially. Silicones show unique rheological properties and are inert to most ionic reagents. They behave as liquids at fairly high molecular weight and possess low freezing point, pour point and boiling point. They have low T_g and possess high compressibility and gas permeability. Silicone rubber is a thermosetting elastomer produced by the cross-linking high molecular weight silicone gums and is of great value in medical technology. It can be obtained over a broad range.

7.2   Polyphosphazenes

These are very important phosphorus containing inorganic polymers.

Polyphosphazenes with linear, cyclolinear and cross-linked and cyclo-matrix structures have been produced. Polyphosphazenes have low glass transition temperature and thus have potential use as elastomers. These polymers show fire resistant properties (the P—N chain in these polymers has resistance to combustion). Polyphosphazenes are synthesized by ring opening reaction of hexachlorocyclotriphosphazene and subsequent substitution of the labile phosphorus-chlorine bond with alkoxy or alkylamino groups (see Fig. 7.2).

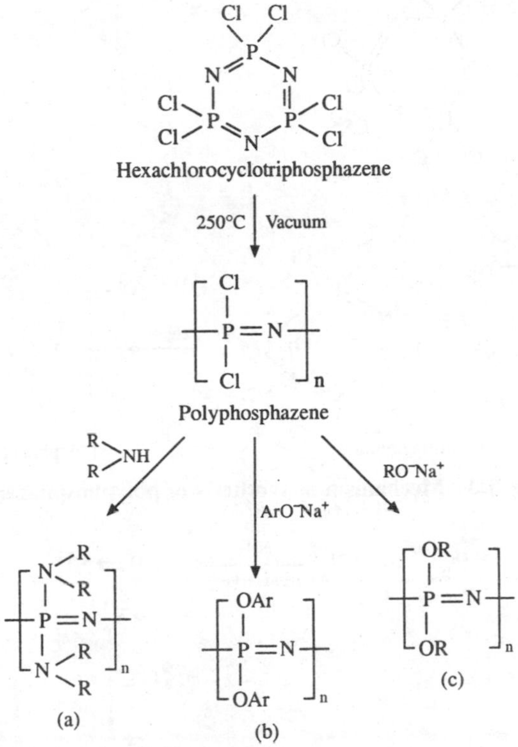

Hexachlorocyclotriphosphazene

250°C | Vacuum

Polyphosphazene

Substituted polyphosphazenes

Fig. 7.2 Synthetic route for polyphosphazenes

The ring opening polymerization mechanism for the formation of polyphosphazenes is shown in the following steps (see Fig. 7.3).

Polyphosphazenes can also be synthesized through polycondensation. The reaction between PCl_5 and NH_3/NH_4Cl proceeds in a step wise fashion by elimination of HCl to first form a monomer [A], then a dimer [B], trimer [C], tetramer . . . . The cyclization could occur to give cyclic chlorophosphazenes at any stage beyond dimer as shown in the scheme below.

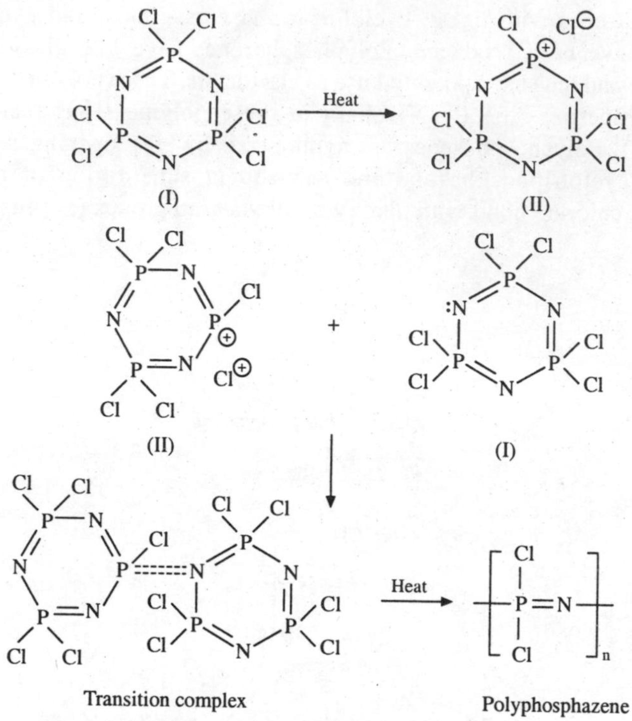

Fig. 7.3   Mechanism of synthesis of polyphosphazene

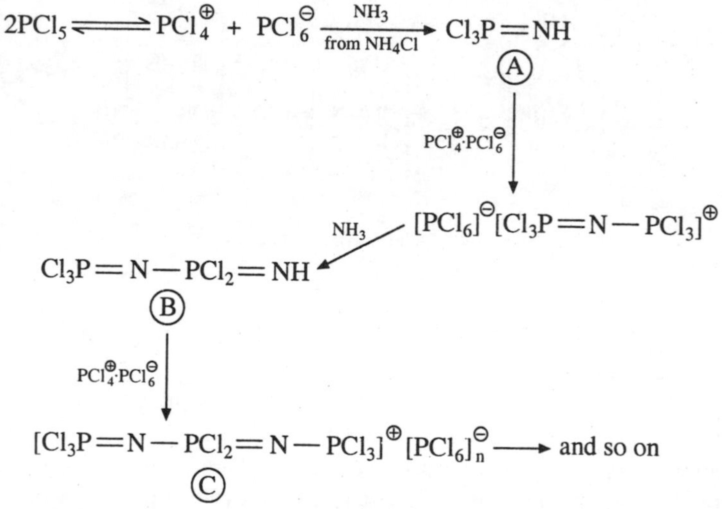

$$2PCl_5 \rightleftharpoons PCl_4^{\oplus} + PCl_6^{\ominus} \xrightarrow[\text{from NH}_4\text{Cl}]{\text{NH}_3} Cl_3P=NH$$

(A)

$$\downarrow PCl_4^{\oplus} \cdot PCl_6^{\ominus}$$

$$[PCl_6]^{\ominus}[Cl_3P=N-PCl_3]^{\oplus}$$

$$Cl_3P=N-PCl_2=NH \xleftarrow{NH_3}$$

(B)

$$\downarrow PCl_4^{\oplus} \cdot PCl_6^{\ominus}$$

$$[Cl_3P=N-PCl_2=N-PCl_3]^{\oplus}[PCl_6]_n^{\ominus} \longrightarrow \text{and so on}$$

(C)

Scheme III

Polyphosphazenes when refluxed with nucleophiles like sodium trifluoroethoxide or sodium cresylate produce amorphous elastomers as per the below shown reactions; Several substitutions with possibilities of two or more different organic groups either simultaneously or subsequently give rise to polyphosphazenes with useful properties.

Bifunctional reactants such as hydroquinone (or dihydroxybenzene) produce cross-linking in polyphosphazenes. Polyphosphazenes are usually soluble in common organic solvents such as $CHCl_3$, THF and can be formed into fibers by extrusion or into films by solution casting. Polyphosphazenes with microcrystallinity and high strength can be made by single substitution. These polymers can be fabricated more easily by solution casting or solution spinning. The biodegradability of these polymers leads to their biomedical use as specialty blood compatible implants and as surgical sutures etc. A significant advantage of polyphosphazenes over other biomedical polymers is the ease with which different organic side groups can be linked to polyphosphazene chain. Thus by changing the type of side groups present, it is possible to generate virtually any combination of properties needed for a particular biomedical use including the ability to hydrolyze inside the body to nontoxic small molecules that can be metabolized or excreted. Surface modification of polyphosphazenes can be achieved by the pendant organic groups and thus polyphosphazenes can be made hydorphilic or hydrophobic. These polymers are quite useful in the immobilization of biomolecules such as enzymes, cells and drugs etc. Other useful inorganic polymers are briefly mentioned below:

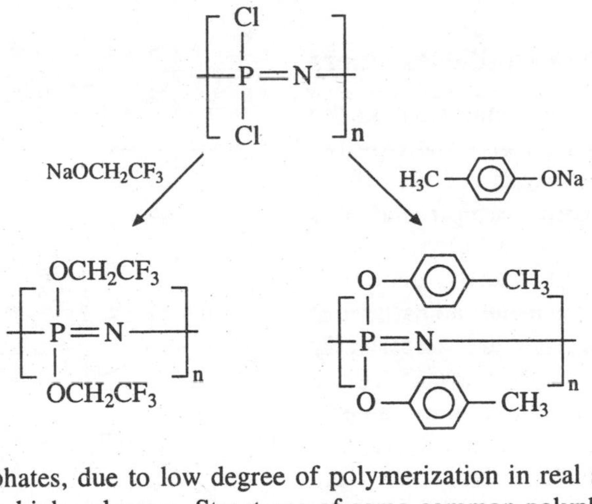

Polyphosphates, due to low degree of polymerization in real sense are not considered as high polymers. Structures of some common polyphosphates are shown in Fig. 7.4.

Boron polymers are structurally related to graphite and consist of hexagonal boron nitride (BN) (see Fig. 7.5). These BN polymers possess outstanding chemical, thermal and dielectric properties. Borate glasses and boron phosphate can be considered as other examples of boron polymers.

Silicate polymers mainly consist of naturally occurring fiber like asbestos and sheet like mica. Silicate glass is an inorganic polymer that is formed by the *trans* polycondensation of SiO_2 (quartz sand) with the aid of washing soda and lime. The presence of sodium in glass disturbs the crystalline structure of quartz and thus they are amorphous under ordinary conditions.

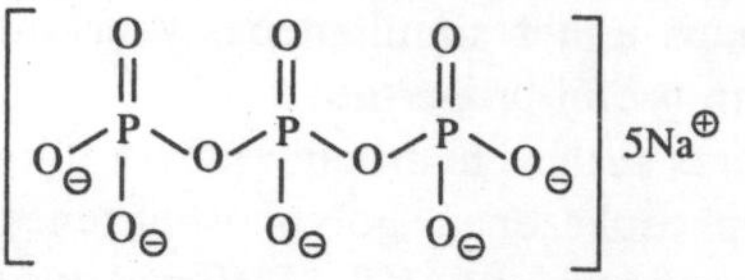

$$\left[\begin{array}{c} O_\ominus^{\substack{O\\||\\P}} \diagdown O \diagup \underset{O_\ominus}{\overset{O\\||\\P}{P}} \diagdown O \diagup \underset{O_\ominus}{\overset{O\\||\\P}{P}} \diagdown O_\ominus \end{array} \right] 5Na^\oplus$$

Sodium tripolyphosphate ($Na_5P_3O_{10}$)

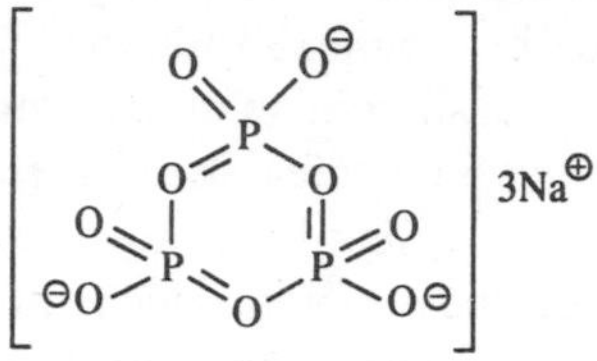

$$\left[\begin{array}{c} \text{(ring structure)} \end{array} \right] 3Na^\oplus \qquad \left[\begin{array}{c} \text{(chain structure)} \end{array} \right]_n 3Na^\oplus$$

Soluble form ($Na_3P_3O_9$) 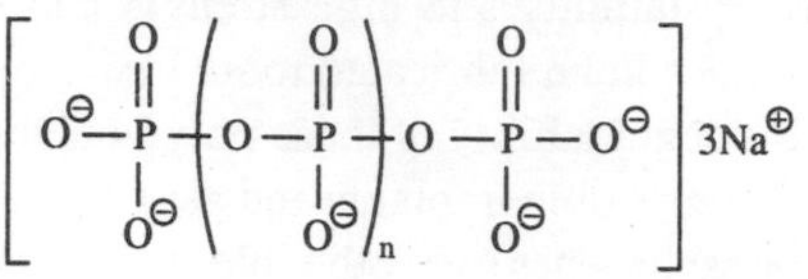   Insoluble ($Na_3P_3O_9$)

Fig. 7.4   Some common forms of polyphosphates

Sulfur nitride polymers $[SN]_n$ can be prepared by solid phase polymerization of dimer S_2N_2 at room temperature. These are crystalline polymers, which are superconducting at 0.25 K.

7.3   Organometallic Polymers

The presence of metal atom in the polymer molecule may increase the stability or rigidity of polymer. If the metal atom carries easily replaceable ligands such as carbonyl groups, catalytic activity can be achieved. Polymers where metal atoms or ions lie close together are electrically conducting. Methyl methacrylate copolymerized with tributyltinmethacrylate shows piezoelectric and pyroelectric properties due to the orientation of dipoles in the influence of high d.c. electrical field applied above T_g. The copolymer is a chief constituent of anti-fouling paint for ship bottoms.

Ferrocene polymers have structures as shown in Fig. 7.6. These are polymers made from vinyl ferrocene (ferrocene is a class of compound such as di-π-cyclopentadienyl metal complex, $[(C_5H_5)_2\,Fe]$).

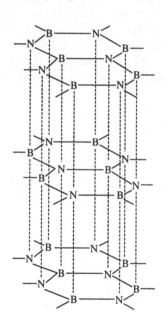

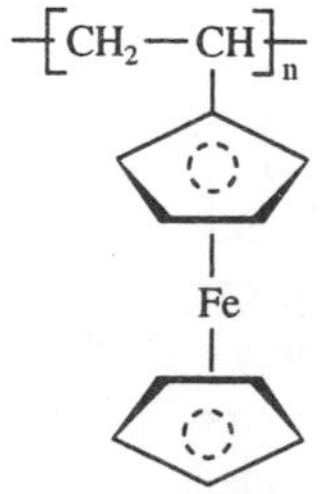

Fig. 7.5   Hexagonal crystal structure of boron nitride polymer

Fig. 7.6   Structure of polyvinyl ferrocene

7.4 Coordination Polymers

Coordination (or chelate) polymers are high molecular weight substances whose main chain is built from units which represent chelate rings. Metals can be coordinated with organic ligands to form a polymer or to form a coordination complex with an already formed polymer. These polymers show little flexibility due to the presence of metal ions. The metal-ligand bonds have enough ionic character. The coordination number and the stereochemistry of the metal ion determine the polymer structure. These polymers can be made by the reaction of preformed metal complexes polymerized through functional groups or by coordination of metal ion by a polymer containing chelating groups. These inorganic-organic polymers have exceptionally good thermal stability. A best known example of coordination polymers is that of phthalocyanine polymers produced when copper metal is heated with tetracyanothylene. The reaction scheme is depictod as follows;

Polyvinylmacrocyclic ethers form complex with metal ions viz. Zn, Cd, Fe and Ni etc. These complexes are commonly known as crown ether complexes and have cage structure as shown in Fig. 7.7.

Coordination polymers do form during the tanning of leather where the coordination of metal ions with the protein (which make the hide) takes place. These protein-metal ion complexes make leather resistant to bacterial attack, wear and weathering. The blood protein, haemoglobin, is another classical example of a coordination polymer.

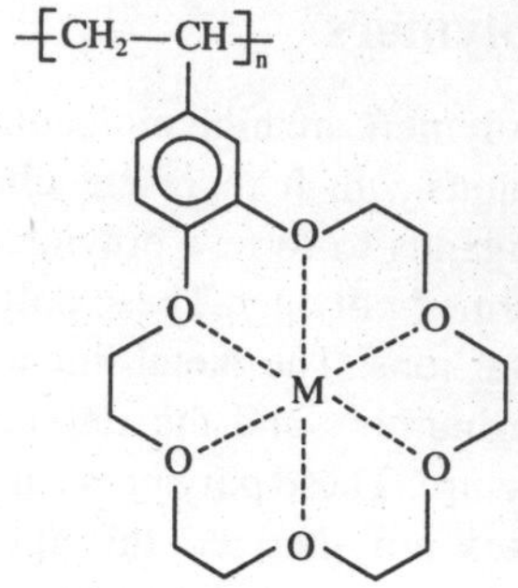

Fig. 7.7   Cage structure of polyvinyl crown ether

Suggested Further Readings

Barry, A.J., and Beck, H.N., in F.G.A. Stone and W.A.G. Graham (eds.), *Inorganic Polymers*, New York: Academic Press, 1962.

Bhattacharya, S.K. (ed.), Metal Filled Polymers. *Properties and Applications*, New York: Marcel Dekker, 1986.

Carraher, Jr. C.E., J.E. Sheats, and Pittman, Jr. C.U., (ed.), *Organometallic Polymers*, New York: Academic Press, 1978.

Mark, J.E., H.R. Allcock, *Inorganic Polymers, An Introduction*, Englewood Cliffs, NJ: Prentice-Hall, 1992.

Ray, N.H., *Inorganic Polymers*, New York, Academic Press, 1978.

Stone, F.G. A., and W.A.G. Graham, *Inorganic Polymers*, New York: Academic Press, 1962.

Wisian, D., H.D. Stenzenberger, and P.M. Hergenrother (eds.), *Inorganic and Organometallic Polymers II: Advanced Materials and Intermediates*, Washington, DC: ACS, 1995.

Zeldin, M., K.J. Wynne, and H.R. Allcock (eds.), *Inorganic and Organometallic Polymers*, Washington, DC: ACS, 1988.

8

Specialty Polymers

8.1 Polyelectrolytes

The water soluble charged polymers
Polyelectrolytes are polymers containing functional groups capable of dissociating into ions in solution e.g. poly (sodium acrylate), poly(styrene sulphonate) (see Fig. 8.1). Polyacids such as polyacrylic acid, hydrolysed copolymers of maleic anhydride, hydrolysed polyacrylamide and polybases such as polyvinyl amine, polyethylimine, polyvinylpyridinium halides and polyphosphates are also some common polyelectrolytes. Proteins and nucleic acids are the biopolymers that behave like polyelectrolytes in aqueous solutions. It is in fact the solvent that facilitates the polyelectrolyte like behavior of polymers. For example, polyacrylic acid acts as a polyelectrolyte in aqueous solution but the same polymer is non-ionizable in solvent such as dioxane and thus behaves like a normal uncharged polymer.

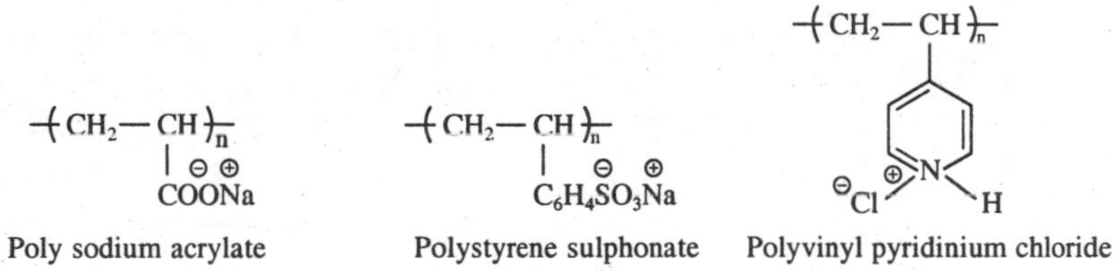

Poly sodium acrylate   Polystyrene sulphonate   Polyvinyl pyridinium chloride

Fig. 8.1 Structures of some polyelectrolytes

A polyelectrolyte may be anionic or cationic depending on the charge on the macroion. Due to the compact arrangement of ionogenic groups, polyelectolytes undergo considerably great electostatic attraction in solution which highly deform the chains of flexible molecules. This deformation depends upon the extent of group ionization, which is caused by low molecular weight electrolytes and pH of the aqueous media. Almost all polyelectrolytes dissolve in polar solvents. The repeat units in many natural and synthetic macromolecules are acidic or basic and may therefore be charged or uncharged depending upon the pH of the solution. It is possible to obtain pure polyacids such as polyacrylic acid which is negatively charged at high pH and uncharged at low pH or, pure polybase such as polyvinylimine which is positively charged at low pH and uncharged at high pH. All proteins and some synthetic polymers contain some acidic and some basic side groups, thus making the molecule amphoteric. Such amphoteric molecules have a net positive charge at low pH, net zero charge at intermediate pH (this pH is called the isoelectric point) and a net negative charge at high pH. A single molecule may have hundreds or even thousands of charged groups on it at the same time. Most rigid macromolecules and some flexible macromolecules are incidentally polyelectrolytes in nature. Owing to the attractive forces acting between charged groups of different sign, the density of coil will be greater than for a conformation, which corresponds to the most statistically probable shape of a macromolecule or to its maximum entropy.

Polyelectrolytes are increasingly used as coagulants and flocculants for colloidal dispersions e.g. in water treatment and for complex formation etc.. Ion-exchange resins are cross-linked polyelectrolytes and are employed extensively.

The shape of a macromolecule is affected not only by a change in the pH of the medium but also by the introduction of an indifferent electrolyte in the solution. When the amount of electrolyte is not too large, the ionisation of ionogenic groups is suppressed and then the macromolecule acquires the statistically most probable conformation. For high salt concentration, their salting out effect reduces the solubility of polymer and causes the formation of compact macromolecular coil.

The viscosity behaviour of dilute polyelectrolyte solutions is usually monitored by plotting the reduced viscosity as a function of concentration of polymer solutions. Such dependence of reduced viscosity, η_{sp}/C on concentration is depicted in Fig. 8.2 for dilute solutions of polyelectrolytes, in different pH and in the presence of added salts

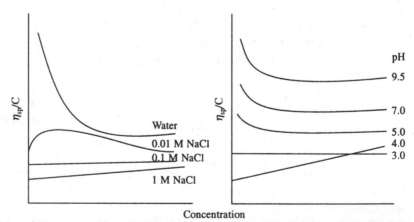

Fig. 8.2   Reduced viscosity versus concentration curves for (a) dilute polyelectrolyte solution in water and at different salt concentrations and (b) dilute aqueous polyelectrolyte (polyacrylic acid) solution at different pH

The following salient features of the curves are observed: (i) the reduced viscosity, η_{sp}/C increases with decreasing polymer concentration when only water is used as solvent, (ii) the curves, however, take an upturn by addition of small amount of salt and become linear with increasing salt concentration and (iii) the η_{sp}/C versus concentration curves for a polyelectrolyte (e.g. polyacrylic acid) aqueous solutions are highly sensitive to pH of the medium. The –COOH functional group responsible for the generation of charge, dissociates to different degree with increasing pH. At lower pH values, most of the –COOH groups are protonated with no net charge. But at high pH deprotonation followed by ionisation occurs and polymer chain has fully polyelectrolyte nature. The large increase in viscosity in lower concentraton is normally attributed to chain expansion due to electrostatic repulsion among the ionised groups.

An important application of polyelectrolytes is in destabilisation of dispersions. Besides making stable dispersions for various applications, the knowledge of colloid science enables to destroy unwanted colloidal dispersions. The problem of water, associated with drilled oil as water/oil emulsion is quite common. Some industries need to eliminate particulate dispersions, for example, in the final stage of water purification which is achieved by the neutralisation of charge (alum is used for this purpose) or by using high molecular weight polymers (polyacrylamide etc.) which cause flocculation through bridging mechanism. The long chains of a polyelectrolyte adsorb onto several oppositely charged colloidal particles making the entire mass sedimenting fast, which causes coagulation. Polyelectrolytes are also used in coal and nuclear industries too. The coal dust is eliminated from wash water using polymeric flocculants. The stable concentrated coal suspensions with desired rheological properties can be made using polyelectrolytes.

8.2 Ionomers (Ion Containing Polymers)

The generic term ionomers has been used for thermoplastics containing ionizable carboxyl groups, which can form ionic cross-links between the chains. Ionomers, the ion-containing polymers, are synthetic organic polymers with ionic content of up to 10–15 mol% and are insoluble in water. These polymers differ from polyelectrolytes, which have higher ion content and are water soluble where upon the molecule dissociates into macro-ion and counter-ions. More specifically, ionomers are ionized copolymers whose major component is a non-ionic backbone and the minor component is of ionic monomers with associated counter ions. The ionic groups in the polymer chain can be introduced by (a) polymerizing a monomer with small quantities of ionic co-monomer or (b) modifying a non-ionic polymer through appropriate chemical process. Several ionomers are generated from polymers containing polyethylene and polystyrene back-bones. The presence of ionic groups in the polymer chain has profound effect on the properties of the non-ionic polymers. The glass transition temperature, mechanical and rheological properties are greatly modified making polymers useful for some specific applications These changes in polymer properties arise due to the aggregation of ions or microphase separation, which takes place within the matrix consisting of non-ionic part of the polymer. Small as well as large aggregates do form depending upon the nature of the ionomer.

The T_g of a polymer is considerably increased in the presence of ionic groups. For example, 9 mol % of sodium methacrylate introduced into polystyrene chain raises the T_g of later from 100 to 130°C. Usually, T_g increases linearly with increase in ionic content. Also, the incorporation of ions in non-ionic polymers significantly increases Young's modulus. For example, a polystyrene sample with mol. wt. ranging from 10,000–1,00,000 is liquid at 180°C but with presence of ~9 mol% sodium methacrylate gives a modulus of ~10^8 N/m^2 at that temperature. The melt viscosity of polymers also shows an increase with the introduction of ionic content. The extent to which the properties of a polymer are altered depends on dielectric constant of the backbone, position and type of ionic group and its concentration, type of counter-ion and the degree of neutralisation.

At sufficiently high ion content, most ionomers show resemblance to non-ionic block copolymers and form microdomains due to strong incompatibility of ions and organic backbone material.

8.3 Conducting Polymers

Most polymers are electrical insulators. Great strides have been made in preparing polymer films, fibres and single crystals that are electrically conductive or that can be doped to make them so. Hence, a few polymeric materials made during past two decades, showed high enough electrical conductivity (> 0.1 S cm^{-1}). Some polymers become conducting after incorporation of some 'dopant' (usually an inorganic electron acceptor like AsF$_5$ or I$_2$). Such polymeric material which show electron conductivity are called conducting polymers (sometimes also called as synthetic metals). The electrical conducting behaviour exhibited by such polymers and that of copper are shown in Figure 8.3

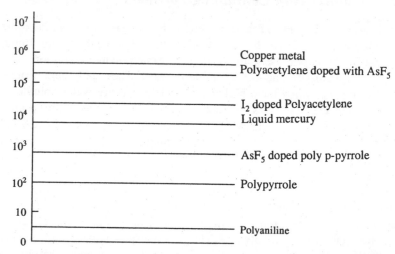

Fig. 8.3   Conductivity (S m^{-1}) range for different polymers and copper metal

The examples of some conducting polymers are given in Table 8(a).

trans-Polyacetylene is a weak semiconductor as such in bulk form but behaves like conducting when doped with AsF$_5$. The incorporation of AsF$_5$ is believed to

Table 8(a)   **Structures of CRUs of some potential polymers and dopants used to make them conducting**

Polymer	Dopant	Conductivity range, S cm^{-1}
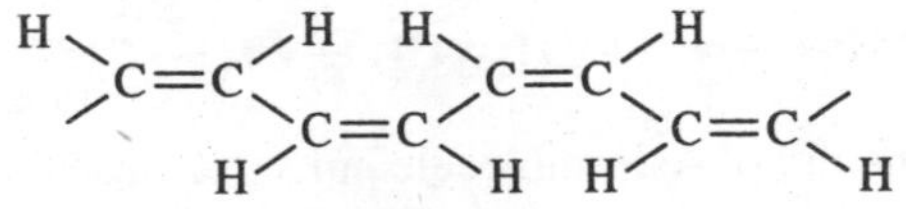 Polyacetylene	Br$_2$, I$_2$, AsF$_5$, HClO$_4$ and H$_2$SO$_4$ (dopant)	10^{-12}
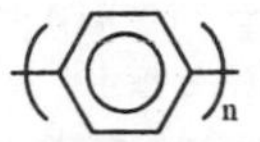 Poly p-phenylene	AsF$_5$ (dopant)	10^{-12}–10
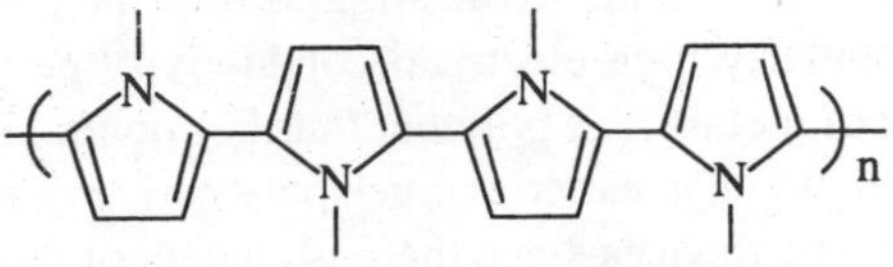 Polypyrrole		10^{-4}–10^2
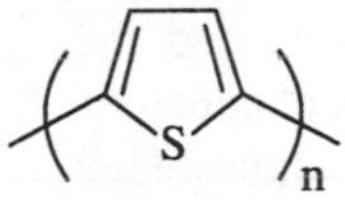 Polythiophene		10^{-4}–10^3
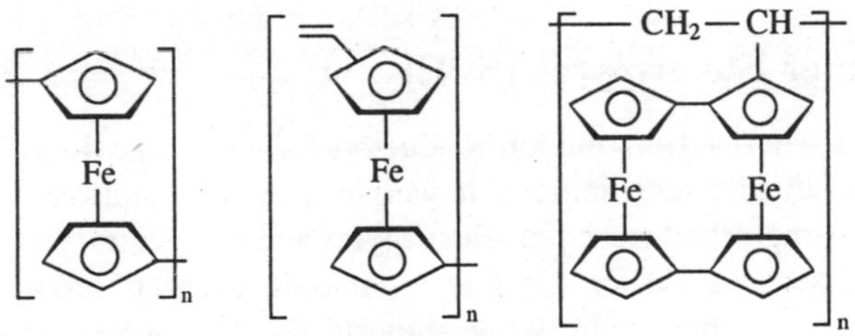 Ferrocenylene polymers		10^{-10}–10^{-3}
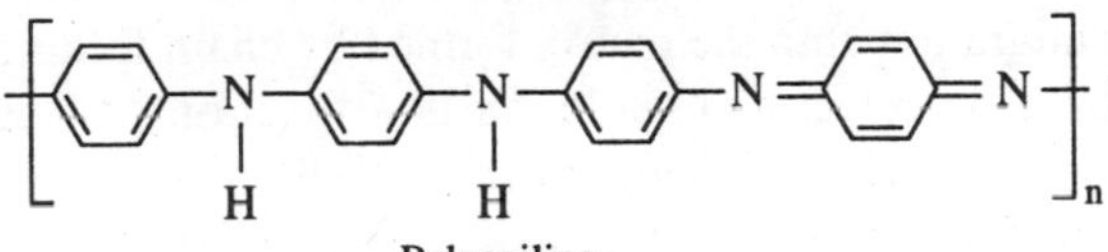 Polyaniline	HCl (dopant)	10^{-9}–10^2

Photo conducting polymers

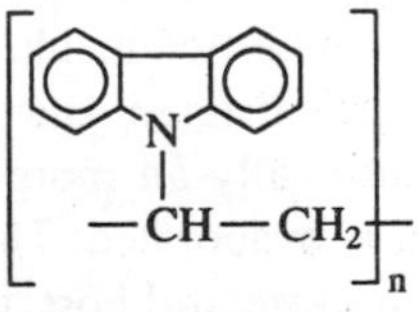

Poly N-vinyl carbazole

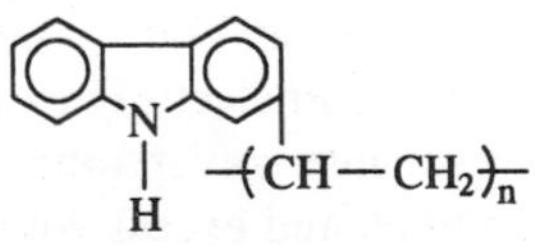

Poly 2-vinyl carbazole

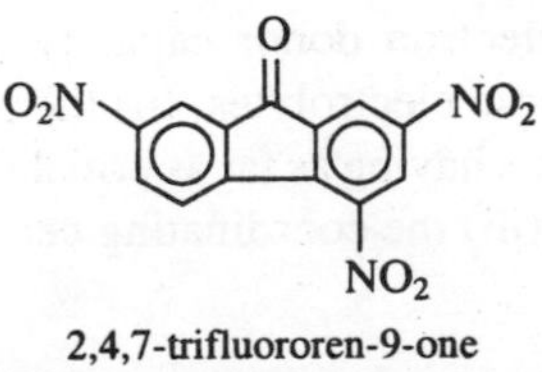

2,4,7-trifluororen-9-one

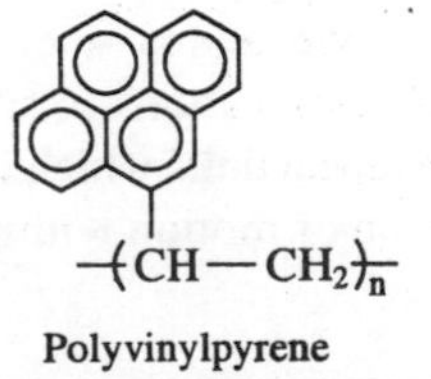

Polyvinylpyrene

give rise to charge transfer complexes at various points along the chains as shown below;

$$-\!\!\left(CH\!\!=\!\!CH\right)_{\overline{n}} CH\!\!=\!\!CH\!- \;+ AsF_5 \;\longrightarrow\; -\!\!\left(CH\!\!=\!\!CH\right)^{+}_{\overline{n}} \overset{\displaystyle AsF_5^{-}}{\overset{\displaystyle \uparrow}{CH}}\!-\!CH\!-$$

The strong electron withdrawing nature of AsF_5 molecule pulls out electronic charge from the conjugated chain. The resultant positive charge is stabilised by its delocalisation over a section of $-\!\!\left(CH\!\!=\!\!CH\right)_{\overline{n}}$ with n = 14. Thus, the positive chage can be considered as the mobile 'positive hole' which moves and transfers electric chage. Interestingly, the conducting behaviour of doped polyacetylenes increases when it is stretched due to the alignment of the chains. Doped polyacetylenes offer a particularly high electrical conductivity per unit density of material (even thrice that of metal). The polymer thus has potential for use as electric wiring or as electrode material in batteries. Such polymeric materials may help the scaling down of electronic devices since there is change of current leakage in the wrong direction. Organo-metallic polymers with novel electrical and optical properties are being prepared and their mechanical and thermal characteristics improved for their use as conductors. The disadvantage of using the conducting polymers is their high cost, problems in processing (due to their insolubility, in-fusibility and brittleness) and long term instability.

8.4   Solid Polymer Electrolytes (SPE)

Polymer electrolytes are defined as solid ion conductors formed by the dissolution of inorganic salts in suitable high molecular weight polymer solutions. Such materials shoud be distinguished from polyelectrolytes and gel electrolytes. In a polyelectrolyte, as discussed earlier cationic or anionic charged groups are chemically linked to a polymer chain, while importantly the counter ions are solvated by a high dielectric constant solvent media (usually water). These charged groups are free to move or migrate within the matrix formed by chain foldings. The conductivity shown by a polyelectrolyte solution is a function of water content and arises due to ion transport through aqueous media in which the polymer is dissolved. Gel electrolytes are mixed polymer–solvent –salt systems, where polymer plays a role of a mechanical support in the solvent media in which cations and anions are in motion resulting into electrolytic conductivities.

Polymer electrolytes are thus different and consist of a host polymer dissolved in suitable solvent and additionally an inorganic salt is added to the polymer solution and excess solvent is evaporated. Thus, solid polymer electrolytes are complex systems. To act as a successful host, the polymeric material or an active part of it must meet the following essential characteristics; (i) the polymer chain must have atoms with sufficient electron donor capacity to form coordinate bonds with cations of simple inorganic electrolytes, (ii) the polymer chain must have repeat units which form segments having as far as possible free and sufficient rotation or motion within them and (iii) the coordinating centers on the polymer chain should be suitably placed so that formation of multiple intrapolymer ionic bonds can take place. A number of polymers satisfy this criterion, but to date polyethylene oxide (PEO) has been most widely studied. The range of salts that may dissolve in PEO is large but better and higher conductivities are obtained when salts of lithium and sodium are employed. For development of practical devices, it has been found that high molecular weight PEOs have limited applicability because these polymers are highly crystalline and thus have restricted segmental motion. This leads to poor conductivities ($\sigma \sim 10^{-8}$ S cm^{-1}) at ambient temperature. However, it is found that reasonable conductivities in the range of $\sim 10^{-5}$ S cm^{-1} can be obtained at temperatures close to 100°C, but at this temperature, the polymer loses mechanical strength and can not provide any support what so ever to the device, for example a film used in rechargeable battery. So the technologists face with a tough and typical problem i.e. when improvement in conductivities is achieved the system loses the necessary mechanical properties and similarly with good mechanical strength, conductivity is lost. One has to compromise between these two to optimize the electrolytic and electrochromic properties of SPEs.

A number of potential polymer hosts for SPEs have been tried. These include both polyether as well as nonpolyether classes of polymers. Among the polyethers besides PEO, polypropylene oxide (PPO), block copolymers of type, PEO-PPO, PEO-PPO-PEO, siloxane based polymers such as copolymers of dimethyl-siloxane -ethylene oxide and comb polymers of poly {(ω-methoxy oligo(oxyethylene) ethoxy)methyl siloxane}s have also been successfully tried for reasonable conductitivites. The nonpolyether polymers mainly include polyesters such as polyethylene sebacate, polyethylene succinate and poly β-propiolactones and have shown good promise. Polyurethane linked polyethers networks have also shown good conductivities when dimethylsiloxane units are incorporated into network structures. Polyphosphazene based polymers exhibit good ionic conductivites and they suffer from very low T_g values and hence can not be applied in practical devices. Polyphosphazene polymers with substituted ethylene oxide units having following structure (Fig. 8.4) have been reported to exhibit ionic conductivites of two or three orders of magnitude higher than those of pure PEO based polymers.

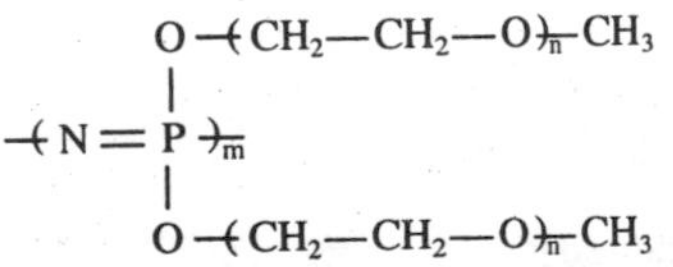

$$O \!-\!\!\left(CH_2\!-\!CH_2\!-\!O\right)_{\!\overline{n}}\!-\!CH_3$$
$$|$$
$$-\!\!\left(N = P\right)_{\!\overline{m}}$$
$$|$$
$$O \!-\!\!\left(CH_2\!-\!CH_2\!-\!O\right)_{\!\overline{n}}\!-\!CH_3$$

Fig. 8.4 Structure of polyether modified polyphosphazenes

Table 8(b) lists structures of repeat units of some homopolymer architectures that have been used as hosts in polymer solid electrolytes. Not only homopolymers but also polymers of varied architecture have been tried as hosts in solid polymer electrolytes. Care that must be taken in designing the polymers is that the branches to the backbone should not add rigidity to the over all chain. Similarly, the added branches should not have chemical constitutions that will decrease the T_g. These conditions ultimately leave not many alternatives than combinations of polyethers based on PEO and dimethylsiloxanes. The advances in synthetic polymer chemistry have enabled the scientists to tailor make several variations in the molecular architectures of potential polymers that can be used SPEs. Based on these variations, one can try for example, comb or branched or network copolymers in place of homopolymers. The constituent components of such heterogeneous polymers can be chosen in such a way that both the desired electrolytic and mechanical properties can still be retained. Given below structures of some of the copolymers which have shown potentiality in the construction of SPEs.

Table 8(b)   Molecular architecture of repeat units of some polymers used in solid polymer electrolytes

Type	Name	Structure of repeat unit
Homopolymer Polyethers	Polyethylene oxide	$H{-}(O{-}CH_2{-}CH_2)_n{-}OH$ $n \leq 150$
	Polypropylene oxide	$H{-}(O{-}CH_2{-}CH)_n{-}OH$; CH_3
Polyesters	Polyethylene succinate	$-(O{-}CH_2{-}CH_2{-}O{-}\overset{O}{\overset{\|}{C}}{-}CH_2{-}CH_2{-}\overset{O}{\overset{\|}{C}})_n$
	Polyethylene sebacate	$-[O{-}CH_2{-}CH_2{-}O{-}\overset{O}{\overset{\|}{C}}{-}(CH_2)_8\overset{O}{\overset{\|}{C}}]_n$
	Poly β-propiolactone	$-(CH_2{-}CH_2{-}\overset{O}{\overset{\|}{C}}{-}O)_n$
Nonpolyethers	Polyethylimine	$-(CH_2{-}\underset{CH_3}{CH}{-}\underset{CH_3}{N}{-}CH_2{-}CH_2{-}NH)_n$ $CH_3{-}CH_3$
	Polyphosphazene	$-(N{=}\underset{Cl}{\overset{Cl}{P}})_n$

Copolymers (Comb-branched):

Poly(bis-(methoxyethoxy ethoxide)) phosphazene

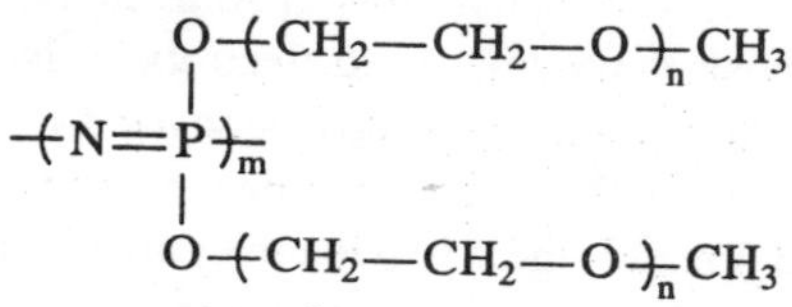

$-(N{=}\underset{O-(CH_2-CH_2-O)_n CH_3}{\overset{O-(CH_2-CH_2-O)_n CH_3}{P}})_m$

Poly (dimethylsiloxane-co-ethylene oxide)

$$\left[\begin{array}{c} CH_3 \\ | \\ Si-O-(CH_2-CH_2-O)_n \\ | \\ CH_3 \end{array}\right]_m$$

Poly {(ω-methoxy oligo(oxyethylen)ethoxy)methyl siloxane}

$$\begin{array}{c} CH_3 \\ | \\ (Si-O)_{\overline{n}} \\ | \\ O-(CH_2-CH_2-O)_{\overline{n}} CH_3 \end{array}$$

Poly {(ω-methoxyoligo(oxyethylene)ethoxy)propyl siloxane}

$$\begin{array}{c} CH_3 \\ | \\ (Si-O)_{\overline{x}} \\ | \\ (CH_2)_3 \\ | \\ O-(CH_2-CH_2-O)_n CH_3 \end{array}$$

Poly γ-methyl L-glutamate

$$\begin{array}{c} O \\ \| \\ (C-CH-NH)_{\overline{n}} \\ | \\ (CH_2)_2 \\ | \\ O=C-O-(CH_2-CH_2-O)_{\overline{3}} CH_3 \end{array}$$

Ethylene oxide substituted methacrylate polymers

$$\begin{array}{c} CH_3 \\ | \\ (CH_2-C)_{\overline{y}} \\ | \\ O=C-O-(CH_2-CH_2-O)_{\overline{n}} CH_3 \end{array}$$

Ethylene oxide substituted polyitaconate backbone

$$\begin{array}{c} O=C-O-(CH_2-CH_2-O)_{\overline{n}} CH_3 \\ | \\ CH_2 \\ | \\ (CH_2-C)_{\overline{y}} \\ | \\ O=C-O-(CH_2-CH_2-O)_{\overline{n}} H \end{array}$$

In addition to above mentioned polymer molecular architectures, following cross-linked polymers based on siloxane structures have also been reported to show ionic conductivity.

$$(H_3C)_3\text{-Si—O}\underset{x}{\overset{\overset{\displaystyle CH_3}{|}}{+\text{Si—O}}}\underset{y}{\overset{\overset{\displaystyle CH_3}{|}}{+\text{Si—O}}}\text{Si+(CH}_3)_3$$

with pendant $(CH_2)_3$ — $(CH_2—CH_2—O)_z\text{H}$

$$\text{PEO-(CH}_2)_3\overset{\overset{\displaystyle CH_3}{|}}{\underset{|}{\text{Si}}}\text{—O—}\overset{\overset{\displaystyle CH_3}{|}}{\underset{|}{\text{Si}}}\text{-(CH}_2)_3\text{-PEO}$$

$$\text{PEO-(CH}_2)_3\overset{\overset{\displaystyle O}{|}}{\underset{\underset{\displaystyle CH_3}{|}}{\text{Si}}}\text{—O—}\overset{\overset{\displaystyle O}{|}}{\underset{\underset{\displaystyle CH_3}{|}}{\text{Si}}}\text{-(CH}_2)_3\text{-PEO}$$

$$\text{+(Si—O)}_n\overset{CH_2CH_2CN}{\underset{}{\text{Si}}}\text{—O—(Si—O)}_m$$

$$\text{+(Si—O)}_n\text{Si—O—(Si—O)}_m$$

Formation of polymer—salt complexes

The simplest way of introducing ionic conductivity into a polymer is to solvate the inorganic ion to the polar groups of polymers. The polar groups in the polymer are thus known as solvating groups. The most usual solvating group is an ether oxygen which is invariably found in polyethylene oxide (PEO). The ether group solvates most of the metal cations by establishing a coordination link between them. Though several metals can solvate with the ether groups through such coordination links, the requirement is that the metal coming from a salt with minimum lattice energy is preferred. Thus most of the reported studies involve the use of lithium and to a lesser extent sodium salts. Majority of experiments conducted so far have employed mostly salts based on lithium perchlorate ($LiClO_4$), lithium trifluromethanesulfonate (triflate) ($LiCF_3SO_3$), lithium thiocyanate (LiSCN), sodium iodide (NaI), sodium thiocyanate (NaSCN), sodium perchlorate($NaClO_4$) and sodium tetraphenylborate, ($NaB(C_6H_5)_4$). Other wide variety of salts can be used to form soluble complexes depending upon the molecular architecture of polymer. The solvating power of host polymer and the extent of dissociation of salt into its constituent ions are very important factors before an effective solvation complex is formed. The salts with large anions such as perchlorate and trifluromethanesulfonate have been found to form effective complexes because these anions have maximum solubility and large quantities of dissociated ions. The structures of these important ions are shown below;

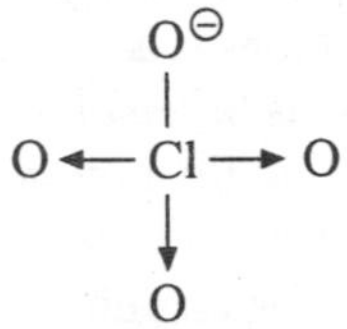

Perchlorate 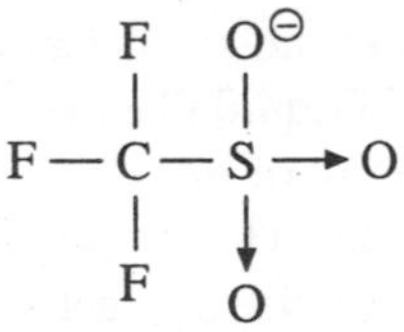   Triflate 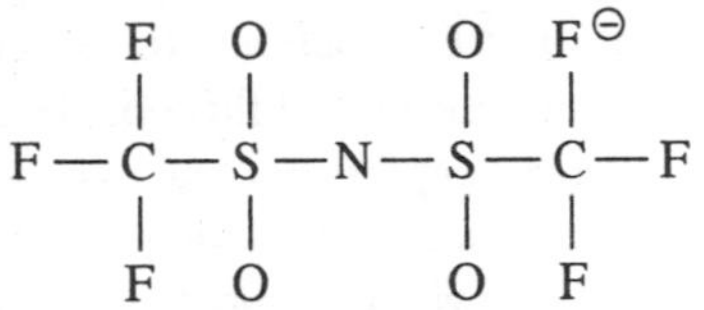   TFSI

Synthesis of polymer-salt complexes is achieved by three ways;

(a) Addition of salt to solvating polymer
The two components are dissolved in a common solvent usually acetonitile and the mixture is homogenised and the solvent is evaporated off. This results in to a fine film of SPE.

(b) Synthesis of salts of polymeric ions
The ionomers as discussed earlier are ion containing polymers. They can be soaked in a solution of the appropriate inorganic salt containing a cation other than sodium (e.g. lithium) for an ion exchange reaction. A modern polymer material known as nafion (a polymer membrane based on polytetrafluoroethylene) has following structure;

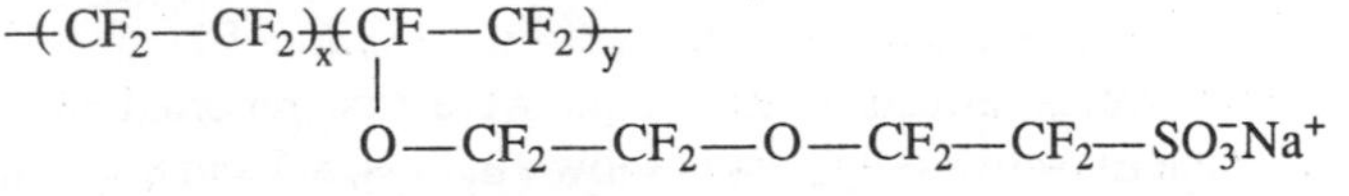

$$-(CF_2-CF_2)_x(CF-CF_2)_y$$
$$O-CF_2-CF_2-O-CF_2-CF_2-SO_3^-Na^+$$

The Na^+ can be easily exchanged with other cations by an ion exchange process in aqueous medium. Water not only acts as solvent but also as a plasticizer and plays a dual role of enhancing the solvating power of polymer and dissociation capacity of added salt.

(c) Usage of co-solvents or plastcizers and salts together
Polymers with little or no ion-solvating properties are first dissolved in a co-solvent which acts as a plasticizer. The co-solvent is absorbed entirely in to polymer and produces a soft single phase material. This material is then added to an electrolyte solution for solvation of cations. Polyvinylidene difluoride (PVDF) can be plastcized with propylene carbonate containing dissolved lithium perchlorate to give a lithium ion conducting polymer.

Mechanism of ion conductivity
The ion transport in polymer electrolytes may occur by a combination of several

events namely ion motion, local motion in the polymer segments and inter- and intra-polymer transitions between ion coordinating sites. Thus, ion conductivity in solid polymer electrolytes depends not only upon ion solvation but also on the morphology of physical structure, nature of phase states of polymers, transport properties of ionic species and their mobility. Several macroscopic models are proposed to explain and understand the ionic conductivity mechanism in polymer electrolytes. These models are based on liquid like and solid like mechanisms for the ionic conductivity. In solid like mechanism, the motion of ions is activated by thermal means, while consideration of liquid like mechanism is based on free volume concept. From macroscoping point of view, the particle/ionic motion is due to continuous movements in the matter. This motion consists of discrete and random jumps. The jump distance is the distance between suitable ion coordination sites.

Application of SPEs

The SPEs are thought to be ideal medium for a wide range of electronic processes. They include primary and secondary rechargeable batteries, ambient temperature fuel cells, electrochromic devices, modified electrodes/sensors, solid state reference electrode system, super capacitors, thermoelectric generators, high vacuum electrochemical devices and electrochemical switching.

8.5   Electroluminiscent Polymers

Electroluminiscence (EL) is a phenomenon involving generation of light by electrical excitation. This is seen in a wide range of semiconductors of low molecular weight inorganic and to a lesser extent in organic materials. The process of electroluminiscence occur through recombination of electrons from one electrode and their capture by the holes from the other electrode and subsequent radiative decay of the excited electron hole state (exciton).

Electroluminiscence effects in polymers were first reported in 1990. The structure of some EL active polymers is shown in Figs. 8.5 and 8.6, respectively.

These polymers are basically conjugated polymers and derive their semi conducting properties by having delocalised π electron bonding and π^* antibonding orbitals from delocalised valence and conduction bands which support the mobile charge carriers. Poly p-phenylene vinylene (PPV) is synthesised by the so called precursor route, in which a soluble precursor polyelectrolyte is thermolysed *in situ* into PPV. The polyelectrolyte is a bis-sulphonium salt obtained by Wessling condensation polymerisation of p-xylene halides and base. However, this process inevitably introduces chemical and morphological effects in the chains with the result that there is a distribution of effective conjugation lengths and these are far shorter than the normal degree of polymerisation. To solve the porblem copolymers in which short phenylene vinylene sequences of determined length are alternated with inert methylene sequcnce in a multi block architecture (see Fig. 8.7). These alternating block copolymers have good solubility, are homogeneous in terms of conjugation length and can be designed to emit any portion of visible spectrum.

One of the major applications of EL active polymers is as light emitting diode (LED) devices. The LED fabrication is shown in Fig. 8.8. The performance of

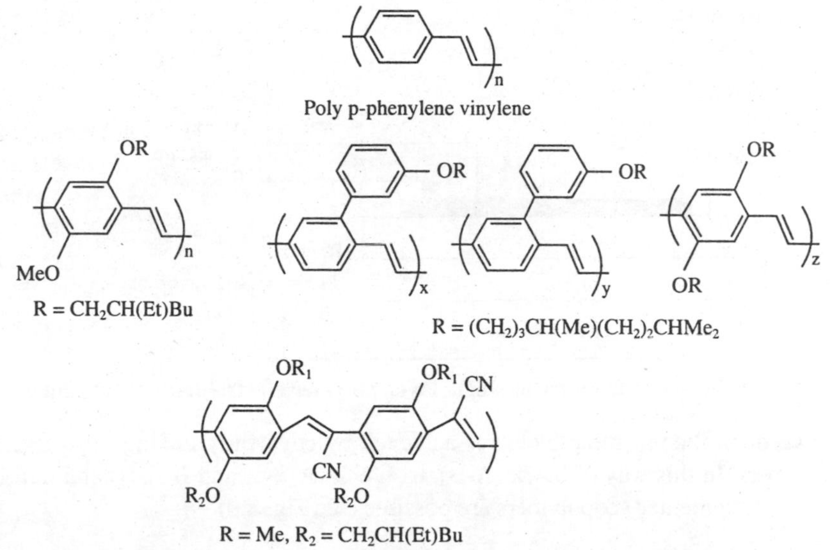
Poly p-phenylene vinylene
OR
MeO
n
R = CH₂CH(Et)Bu
OR
OR
OR
x
y
z
OR
R = (CH₂)₃CH(Me)(CH₂)₂CHMe₂
OR₁
OR₁ CN
CN
R₂O
R₂O
R = Me, R₂ = CH₂CH(Et)Bu

Fig. 8.5 EL active poly p-phenylene vinylene based polymers

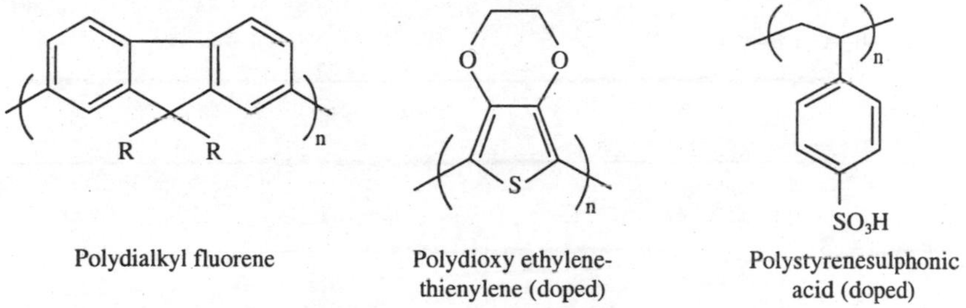
R
R
n
O
O
S
n
n
SO₃H
Polydialkyl fluorene
Polydioxy ethylene-thienylene (doped)
Polystyrenesulphonic acid (doped)

Fig. 8.6 EL active non poly p-phenylene vinylene (PPV) polymers

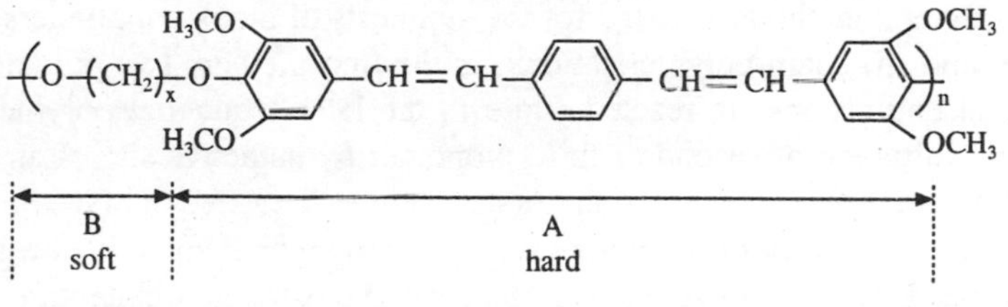
H₃CO
H₃CO
O
(CH₂)
x
O
CH=CH
CH=CH
OCH₃
OCH₃
n
B
soft
A
hard

Fig. 8.7 Typical alternating block copolymer structures based on PPV

LEDs meets many of the targeted functions necessary for display applications. Attempts are made to obtain full color display using EL polymers. The range of potential applications of polymer LED, would be in backlighting segment displays, alphanumeric displays, full color graphic display (for computer monitors and video display) and color pixels in ink-jet printing etc.

8.6 Block Copolymers (Important multiphase polymers)

A block copolymer has a molecular architecture in which a linear arrangement

Poly p-phenylene vinylene

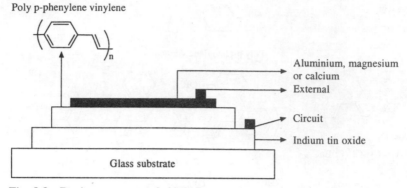

Fig. 8.8   Device structure of single layer polymer electroluminescent diode

of two often incompatible blocks is achieved by covalently linking of different monomers. In this way di-block (A-B), tri-block (A-B-A and B-A-B) and multi-block (or segmented) copolymers are possible (see Fig. 8.9).

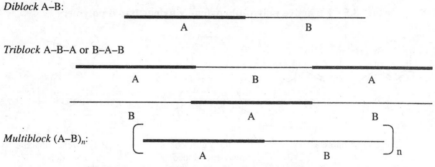

Fig. 8.9   The different types of block copolymers

Two general methods are used for the synthesis of block copolymers namely (i) step- and (ii) chain-polymerisations. In the first method, two polymers with functional end groups are reacted while in the later sequential polymerization involving initiation of second or third monomer by an active site of an already formed macromolecular chain of the first monomer is carried out (see Fig. 8.10). Schwarz developed living anionic polymerization method and it is now possible to tailor make block copolymers with the desired molecular architecture and low polydispersity by this method.

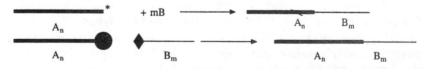

Fig. 8.10   Synthetic routes for block copolymers

Block copolymers possess unique structural features resembling to that of conventional low molecular weight surface active agents which have two distinct moieties in their molecule that behave differently. Thus, they do adsorb onto interfaces and self-assemble to form micelles in solutions and in addition show microdomain formation in solid state.

This makes these multiphase systems useful for a variety of applications. It is therefore not surprising that block copolymers have aroused world wide interest at academia (by physicists, chemists and chemical engineers) and also industrial level. Several products based on block copolymers are commercialised. These products are available in the market for the users. They are; thermoplastic elastomers e.g. styrene- diene block polymers (**solprene**, *Philips*; kraton, *Shell*), segmented polyester-polyurethane (**estane**, *Goodrich*; **spandex**, *Dupont*) and surfactants, polyalkylene block polymers e.g. ethylene oxide-propylene oxide (pluronics *BASF, ICI*) and ethylene oxide-butylene oxide and ethylene oxide–butylene oxide– ethylene oxide, **Polyols**, *Dow*).

Unique features of block copolymer in solid state

Block copolymers possess incompatible sequences within their molecule and exhibit characteristic morphological behaviour and interesting properties. Incompatible and noncrystallizable A and B blocks form microdomain structures due to their tendency of microphase separation. Block copolymer morphological features are reviewed extensively in the literature. The conditions of microphase separation, equilibrium domain size and their various morphologies viz. spherical, cylindrical and lamellar are predictable in terms of interaction parameters and molecular characteristics of the copolymer. The various common structures are schematically shown in Fig. 8.11.

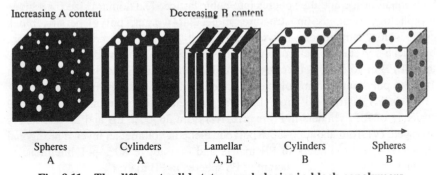

Increasing A content		Decreasing B content		
Spheres	Cylinders	Lamellar	Cylinders	Spheres
A	A	A, B	B	B

Fig. 8.11   The different solid state morphologies in block copolymers

The repulsive interactions between A and B blocks cause the domain to grow in thickness (thickness of the lamellar domains and radius of the spherical and cylindrical domains) so as to reduce surface to volume ratio. This is counter balanced by two forces of entropic origin, i.e. loss of conformational entropy in maintaining uniform segment density and the loss of placement entropy in confining A–B junction in the interfacial region. Partial miscibility in the interfacial zone depends on the nature of the blocks, type and molecular characteristics of block copolymer and temperature (see Fig. 8.12).

Morphologies of copolymers with amorphous blocks in general, and styrene-diene block copolymers in particular as studied by different methods have shown

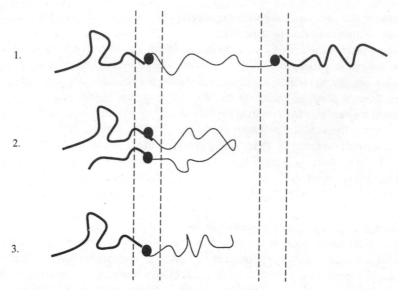

Fig. 8.12 **Schematic representation of microphase separation of diblock (3) and triblock copolymers (1 and 2)**

that segregated microphases can be sphere, cylinder or lamella depending upon the relative composition of the two blcoks. While lamellae form a regularly repeating lamellar order, the cylinders arrange themselves in a two dimensional hexagonal lattice and the spheres into cubic lattice. The domain size for a given morphology depends on structure, molecular weight, polydispersity, block composition, temperature, polymer-polymer interaction parameter. For styrene-butadiene-styrene triblock copolymer (PS-PB-PS), it has been observed that (i) up to 16% PB in volume, a cubic structure of PB spheres in PS matrix, (ii) from about 16–18% polybutadiene in volume, an orthorhombic structure formed by short rods of PB in PS matrix, (iii) from 18–36% polybutadiene in volume, a hexagonal structure of PB cylinders in PS matrix, (iv) from 36–60% PB and PS alternately arranged, (v) from about 60–80% polybutadiene in volume, an inverse hexagonal structure of PS cylinders in PB matrix and (vi) above 80% polybutadiene in volume, in inverse cubic structure of PS spheres in PB matrix.

Morphological studies on copolymers with amorphous and crystalline blocks reveal crystalline lamellae formed by one or two layers of folded crystallizable block. Such structures depend on crystallization temperature, molecular weight and the nature of amorphous and crystalline blocks. Here, for di- and triblock copolymers e.g., styrene-ethylene oxide block copolymer, well organised periodic structures have been observed while for multiblock copolymers e.g. segmented polyurethanes or polyesters, microphase separation without any periodic structure is noticed. For block copolymers with both the crystallizable blocks, lamellar structures are obtained but the morphology is more complicated.

At certain temperature when the free energy of mixing of both the blocks is zero, complete mixing of the phases would occur. Thus order-disorder transition from microphase separated domains into disordered would occur which can be clearly observed from mechanical behaviour. The phase transition is evident for copolymers when the chains are sufficiently short. With increase in temperature, thermal agitation predominates and the system gets disordered. At this stage, the molecules diffuse into the entire volume and the copolymer now accommodates stresses, by restructuring itself much more rapidly. It behaves like a conventional viscoelastic polymer liquid. The force required to maintain it in a deformed shape decreases rapidly in a few milliseconds. The deformation is then permanent.

Block copolymers in solution

Due to structural resemblance to the surface active substances, block copolymers also exhibit adsorption characteristics and micelle formation in solution. These two fundamental properties which can be suitably controlled provide unique application possibilities for block copolymers.

(i) Adsorption from solution

The adsorption of polymers onto surfaces plays an important role in many industrial processes. When a surface is exposed to a polymer solution, one of two diametrically different processes generally occurs. In the first, the polymer is enriched in the surface zone (adsorption), while in the second, the surface zone is depleted with respect to the polymer. Since the change in entropy on adsorption is negative, due to the restriction of the translational freedom and of the number of possible conformations of the polymer, adsorption would occur only if the gain in energy is larger than the loss in entropy. Thus there is a critical energy of adsorption which determines whether adsorption or depletion occurs (this energy is generally quite small, typically a few tenth of a kT due to a large number of adsorption sites). Polymers are known to adsorb onto surfaces as trains, tails and loops. Block copolymers, due to their amphiphilic nature are best in that they adsorb strongly and extensively onto a large variety of interfaces (see Fig. 8.13).

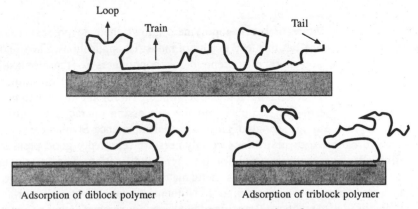

Adsorption of diblock polymer   Adsorption of triblock polymer

Fig. 8.13 Polymer adsorption at interfaces

While soluble block forms a mantle, the insoluble block acts as an anchor; the efficient adsorption of block copolymers from solution onto the surface of colloidal particles provides a long-term stability, *Napper* has pointed out that the mechanism of polymeric stabilisation of colloidal solutions is governed by steric stabilisation or depletion stabilisation. In the former case the particles are stabilised by the attached or adsorbed polymer. The depletion stabilisation is provided by unanchored, unattached polymer molecules in the dispersion phase.

(ii) Micellization in selective solvents

Block copolymers in a selective solvent (a good solvent for one block but precipitant for the other) do aggregate in solution. This aggregation process is analogous to the micelle formation of conventional non-ionic surfactants (see Fig. 8.14).

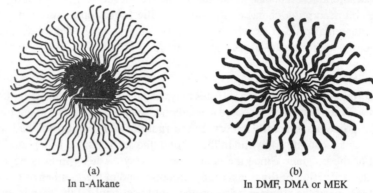

| (a) | (b) |
| In n-Alkane | In DMF, DMA or MEK |

Fig. 8.14   **Polystyrene-polybutadiene micellar structures (a) micelle core is of polystyrene block and outer shell is of polybutadiene block and (b) micellar core is of polybutadiene blcok and outer shell is of polystyrene block**

Block copolymer micelles do solubilize substances, which are otherwise insoluble in the selective solvents. Solubilization of various substances (both of low and high molecular weight) in copolymer micelles has been investigated.

Applications of block copolymers

The unique architecture of block copolymers allows them to possess some interesting novel properties that lead to a lot of industrial applications. The use of styrene-butadiene and styrene-isoprene block copolymers as thermoplastic elastomers in footwear is well known. The other thermoplastic elastomers are polyether-polyester and polyether-polyamide. Thermoplastic elastomers combine the properties of a rubber and the processing characteristics of a thermoplastic make these materials of great industrial importance. Styrene-diene and styrene-hydrogenated diene triblock copolymers are reportedly good pressure sensitive adhesives and are used for bitumen modification. The interfacial characteristics make block copolymers useful materials as polyblends, as adhesives, sealants, and binders for laminates, as constituent of coating materials, surface modifiers for fillers and fibers. Blending of block copolymers with other polymers provides new polymeric materials with desired properties. Some grades of high impact polystyrene are such systems. Block copolymer blends with homo- or copolymers of different kinds (elastomers, thermoplasts and thermsets) lead to interesting mechanical properties. Block copolymers are used in surfce coating for metals, glass and in surface modification of fillers like calcium carbonate, carbon black, silica and different fibers. Surfactant based applications of block copolymers are as colloid stabilizers, in dispersion polymerization in aqueous and non-aqueous media (particularly in the later case where the conventional surfactants fail to perform satisfactorily), in emulsion stabilization, microemulsion polymerization and polymeric emulsions. EO/PO block copolymers find numerous applications in medical, pharmaceutical, textile, cosmetic, detergents and pesticidal formulation. Their reversible thermorheological feature makes them potential candidates as vehicles for controlled release of pharmaceutical formulations.

8.7 Polymer Colloids

Lyophobic colloidal dispersions are often desirable in stable form but some times they are useful in their coagulated state. This makes the understanding of colloidal stability and instability an important topic. Almost all industrial concerns often face the problem of stability/instability of colloidal dispersions and suspensions: A colloidal dispersion (where the particles of dispersed phase are in the size range 10^{-5}–10^{-7} cm) remains stable due to the surface charge onto the dispersed particles and offers potential applications. The need of breaking (or coagulating) a colloidal form arises in several instances e.g. in the removal of suspended impurities of water from the drilled oil (where it emulsifies with oil to from water-in-oil emulsions) before sending the oil to the refineries etc. Stable colloidal dispersions, on the other hand, are required for various cosmetic, pesticidal, paints, pharmaceutical preparations. It is important here to consider what contributes to the stability of colloidal dispersions. In fact, the most widely accepted theory proposed independently in 1940s by the Russians (Derjaguin and Landau) and Dutch (Verwey and Overbeek) commonly known as DLVO theory considers the attraction and repulsion terms between two isolated colloidal particles and the net interaction between them determines whether the system would be stable or not. Usually, the repulsive interaction is dominated by the charge on the colloid surface (measured as zeta potential) and thus the colloidal systems with high values of zeta potential would tend to be stable as the repulsion would dominate over attraction. The typical energy profile showing the attractive, repulsive and net energy is schematically shown in Fig. 8.15 as a function of inter-particle distance.

In the foregoing discussion we will see how a polymer influences the stability

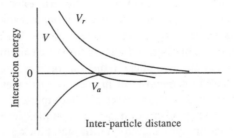

Fig. 8.15 **Energy profile as a function of inter-particle distance**

of a lyophobic colloid. Polymers can be conveniently used for both the purpose, for example, to stabilize a dispersion and to flocculate a dispersion. The mixed system of a polymer and colloidal dispersion together is usually called as polymer colloid, i.e. colloidal dispersion of particles containing polymers as sensitizing/de-sensitizing agents. When the surface of dispersed particles is exposed to a polymer solution, one of the two diametrically different porcesses generally occur. Either the polymer is enriched in the surface zone (adsorption) or the surface zone is depleted with respect to polymer. As the adsorption involves negative entropy change (due to the restrictions of the translational freedom and of the number of polymer conformations) polymer adsorption would occur only when the gain in energy is larger than the loss in entropy. Thus there is a critical energy of adsorption that would determine whether the adsorption or depletion would occur (the value is however, very small, typically a few tenths of kT, due to high number of adsorption sites).

Polymer adsorption leads to a drastic change in conformations of its molecules. A polymer molecule adsorbs on a surface as trains, loops and tails as shown below (see Fig. 8.16). Whether or not the polymer will adsorb on the surface of a colloidal particle would depend upon the nature of the polymer molecule (the lyophobic and lyophilic sites in the polymer), nature of the colloidal particle and of course more importantly in the quality of the solvent. For polymers dissolved in good solvents (e.g. xanthan in water) adsorption is very unlikely, in the so called theta solvents, it is likely and in poor solvents; it is all but certain.

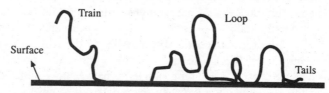

Fig. 8.16   Conformation outline of an adsorbed polymer chain on a surface

The polymer adsorption onto the surface of colloid particles may impart stability (or instability) and modifies (hydrophilize or functionalize) surface and is thus of interest from basic (theoretical and experimental) and applied points of view. Often basic studies provide information on the amount of the polymer adsorbed and surface coverage, conformation of adsorbed polymer molecule, thickness of the adsorbed polymer film, kinetics of polymer adsorption and most importantly the segment density distribution. Several experimental techniques have been employed to investigate the polymer adsorption characteristics. However, modern methods viz. ellipsometry, neutron reflectometry, dynamic light scattering and small angle neutron scattering, and surface force measurements have provided valuable information. Among theoretical studies on polymer adsorption, Scheutjens-Fleer (SF) theory based on mean-field model, in particular has proven to be successful in predicting adsorption features of polymers.

The situation of polymer adsorption is more complex when the polymer is a block or graft polymer or a polyelectrolyte. In fact, block and graft copolymers are most efficiently adsorbed onto surfaces due to their structural analogy to conventional surface active agents in that each has two dissimilar moieties in their molecules. Polyelectrolyte adsorption is influenced by the ionic charges present on the back-bone of the chain and the extent of their dissociation.

When a colloidal solution and a polymer solution are mixed, the dispersion may be stabilized or destabilized (flocculated). Those polymers which adsorb, usually at low polymer to particle concentration ratio, instability (also called bridging flocculation) occurs. For high polymer to particle ratio leading to complete surface coverage, solvency determines the stability or instability. For nonadsorbing polymers, flocculation can be seen but a very large amount of the polymer may be necessary to bring this effect.

Bridging flocculation

As the name suggests the flocculation takes place as a result of bridging of colloidal particles by polymer molecules. As discussed before, the concentration of polymer causing bridging flocculation should be very low such that the surface coverage is below its saturation value. Here the polymer molecule adsorbs onto several colloidal surfaces as shown in Fig. 8.17 and flocculates the dispersion.

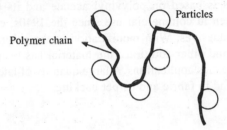

Fig. 8.17   Schematic depiction of bridging flocculation of colloidal particles by an adsorbing polymer chain at low surface coverage

There exists a class of polymers that is widely exploited industrially as flocculants for colloidal dispersions. These polymeric flocculants may be anionic, cationic or non-ionic. Use of high molecular weight polyacrylamides in the domestic and industrial waste water treatment is the widely known example of bridging flocculation.

Depletion flocculation

The origin of depletion flocculation is quite different from that of bridging flocculation and does not involve the adsorption of polymer. However, a colloidal system undergoing depletion flocculation by addition of non-adsorbing polymers can be re-dispersed on dilution with dispersion medium. The origin of force that is responsible for this type of flocculation was first recognised by Asakura and Oosawa in 1954 who accounted for the fact that the interaction between two collidal particles would result if the particles approach so closely that the polymer chains present in the inter-particle region are excluded. This depletion of polymer chain would be possible if the inter-particle distance is less than the diameter of the non-adsorbing polymer causing the flocculation (the size of the polymer coil can roughly be treated as the root mean square end-to-end distance or radius of gyration). At such separations, the polymer chains can only fit between the particles by undergoing a significant decrease in spatial extension. This compression is accompanied by a loss in configurational entropy which is thermodynamically unfavorable. The depletion of the polymer chains from the inter-particle zone, means that the micro-reservoirs of polymer free dispersion medium are generated in the bulk. For colloidal systems with dispersion medium working as good solvent for the polymer, closer approach of the colloidal particles would be possible due to the migration of dispersion medium from the inter-particle zone towards the polymer chains outside that leads to the reduction in free energy.

The adsorption and ability of polymers to stabilize or de-stabilize the colloidal dispersions have been exploited in many commercial and industrial formulations. Following some of the examples are worth considering.

Adhesives

In order to eliminate pollution problems associated with oil based products, the trend now is towards water based formulations. It is thus not surprising that adhesive industries are also shifting to water based adhesives. The most important of these adhesives is based on polyvinyl acetate and its copolymers. These adhesives have been in commercial use since the 1940s, especially as wood, paper, and glass adhesives. With modification in the polymerization stage by copolymerisation and other additions, the material has been made usable in a large number of plastics applications. A large quantity of latex is consumed as a textile adhesives and in fabric and carpet backing.

Coatings

Polymer colloids have always been essential materials for the coating industry, whether the coating is oil-or water-based. The polymer or copolymer in the latex, the particle size, size distribution of the colloid, and the nature of the surface functional groups (whether from surfactant or copolymer) are all important variables, as are the chemical nature of the substrate and the mode of application that need to be considered for effective coating.

Paper

Paper making and printing processes are also important application of polymer colloids. When applied to paper they enhance the quality and the printability of the paper surface.

Pharmaceutical and medical applications

Pharmaceutical coatings have always been important in the manufacture of tablets. Tablets coated with polymer (e.g. polysaccharides) have been found superior to sugar coatings. Solvent-based polymer coatings have been important for some 20 years, but the pharmaceutical industry has slowly begun to change to water-based systems to eliminate the necessity of using organic solvents. Now aqueous colloidal polymer dispersions are used almost exclusively. An interesting and important extension of the tablet coating by polymer colloids is in controlled release of tablet composition. When the polymer in the latex particle, for example is hydrolysable and strong acid groups are present at the surface, self catalyzed hydrolysis may also occur, resulting in controlled release of the material. Immuno-microspheres are specially designed colloidal particles that have specific molecules such as antibodies chemically bound at their surfaces. These particle are capable of attaching themselves to tissues or cells that contain the corresponding antigen. These target cells could be viruses or other antigenic agents, and under favorable conditions no non-specific binding to other kind of cells takes place. These immuno-microspheres may be synthesized to incorporate compounds that are highly radioactive, intensely fluorescent, magnetic, opaque, highly colored and pharmacologically active. This type of polymer colloidal particles provide a multitude of applications in medicinal research, diagnosis, and ultimately in therapeutical practices.

8.8 Thermoplastic Elastomers (TPE)

These are polymeric materials which function as a thermoplastic during the processing and as elastomer in the finished product. Several types of the polymers belong to this class but each of it has a common feature in that they are relatively low molecular weight elastomers comprising of hard segments of sufficiently high molecular weight. These hard segments are incompatible with elastomer's part (thus form discrete microdomains) and act as a sort of cross-link at service temperature. The cross-linking increases the effective molecular weight many times. These physical cross-linking dissociate at elevated temperatures and low molecular weight polymer is regenerated and flows under the conditions encountered in machine used by thermoplastic processing industries. This behaviour avoids the usual time consuming processing of rubbers (e.g. polyisoprene) through vulcanization. Thermoplastic elastomers have diverse applications and are in use as adhesives, reinforcing agents for plastics, shoe soles and elastic bands etc.

The first commercial production of TPEs began by Shell Co. in 1965 and thereafter other thermoplastic elastomers have entered the market. They include styrene-diene block copolymers by Shell and Phillips, polyesters by Du pont and several polyurethanes (TPUs) and polyolefins (TPOs). These are briefly described below.The rubbery polymer e.g. polybutadiene is associated with polystyrene end blocks PS-PB-PS. These end blocks associate to form the hard phase and provide the polymers strength. These block copolymers with desired molecular characteristics can be prepared by anionic living polymerisation technique (Shell-Kraton, Phillips–Solprene). Rigid polystyrene end blocks which are large in dimension, aggregate or form domains which are joined by the elastomeric polybutadiene. At room temperature, these block copolymers behave almost like cross-linked rubber. However, at temperature above 100°C (T_g of polystyrene), the domains soften and may be disrupted by applying stress allowing the block copolymer to flow (for more see block copolymers).

8.9 Polyblends (Heterogeneous Plastics)

When two or more polymers are mixed together, polyblends or polymer alloys are obtained. This physical mixing or blending of two polymers produces an alloy with quite different properties, which can be potentially useful. Two polymers are generally noncompatible as they have very low combinatorial entropy of mixing for the components. This is insufficient to overcome the positive heat of mixing of polymers to make the Gibbs free energy of mixing negative. Only in the presence of specific interaction between two polymers (e.g. hydrogen bonding, acid-base type interactions etc.), heat of mixing is negative that makes the free energy of mixing a negative quantity and then the mixing is spontaneous. Unlike the mixing of small molecules, the dictum 'like likenes like' does not hold good for mixing of macomolecules. That is why most polyblends result to incompatible systems. However, both compatible and incompatible blends are industrially important materials.

The two polymers may from a compatible blend, which exists as a single phase. The incompatible blends on the other hand exist as two phase system. Since most blends combine immiscible components and so the material that results contains tiny particles of one polymer in a matrix of the other. Controlled mixing and cooling of the blend makes it possible to form the particles in the optimum concentration and range of sizes. Blending makes it possible to combine the good properties of several polymers.

The most direct method to obtain a polyblend is to mix the component polymers in the molten state (melt mixing). In this case the extent of mixing depends on the rate of diffusion of the molecules. Since such a mixing requires high temperature, the polymer may decompose and undergo chemical transformation. It is, therefore, this mixing process is restricted to thermally stable polymers. Another process involves the mixing of two polymer lattices pre-formed by emulsion polymerisation technique. The polyblend can be obtained by coagulating the mixed latex system. A polyblend can also be obtained by mixing the polymers in a mutual solvent followed by the removal of solvent e.g. by drying.

Usually, blending can be made easier by the incorporation of a block or graft copolymer containing two moieties similar to the component homopolymers. For example, styrene-butadiene block copolymer shows adhesion and as a result act as stabilizer/compatibilizer for a polyblend made by mixing polystyrene and polybutadiene. Usually, such blends are called 'polymer alloys'. In compatible polyblends, the glass transition temperature can be easily altered by varying the composition of the component polymers and thus desired physical and mechanical properties can be achieved. For example, a small amount of PVC when mixed with natural rubber, the product shows improved resistance for heat, light and chemical substances. Similarly, the presence of small amount of nitrile rubber in PVC significantly lowers the T_g of the product making it flexible and easily processible. Polyblends made of polystyrene and polyphenylene oxide commercially known as 'Noryl' are compatible blends, which provide improved toughness, mechanical properties and heat resistance as compared to polystyrene alone. Polyelectrolyte blends (often called polyelectrolyte complexes) have tremendous applications in medical science.

The main use of incompatible polyblends is to improve the impact strength of brittle and glassy polymers. The most known example belonging to this class of polyblends is of high impact polystyrene (HIPS). In the presence of the rubber phase, the impact strength of polystyrene is considerably improved. The inclusion of rubbery material in the matrix of polystyrene forms fine 'cracks'. Nylon-phenolic blends find application as ablative heat shields on space vehicles. A new class of immiscible blends that has been developed is the so called high performance blends. A promising example is of polycarbonate blended with polybutylene terephthalate. Because both the polymers are in almost equal amount, the two phase coexists in an intricate interpenetrating network. It is highly resistant to heat and solvent and possesses long durability. It is used in place of metals in automotive components such as bumpers.

8.10 Polymer Composites

A composite, in the broadest sense, is simply a material containing more than one component. Any such composition that comprise two or more materials as separate phases, at least one of which is a polymer can be called as polymer composite. By combining a polymer with another material, such as glass, carbon, or another polymer, it is often possible to obtain unique combinations or levels of properties. Typical examples of synthetic polymeric composities include glass-, carbon-, or polymer-fiber-reinforced thermoplastic or thermosetting resins, carbon-reinforced rubber, polymer blends, silica- or mica- reinforced resins, and polymer-bonded or -impregnated concrete or wood. Coating formulations (pigment-binder combinations) and crystalline polymers (crystallites in an amorphous polymer matrix) can also be called as composites. Typical naturally occurring composites include wood (cellulosic fibers bonded with lignin) and bone (minerals bonded with collagen). On the other hand, polymeric compositions compounded with a plasticizer of very low proportions of pigments or processing aids are not ordinarily considered as composites.

Usually, composites have properties superior to that (particularly strength-to-weight) of its components. In polymer based composites, the reinforced fibres and fillers add tensile strength and stiffness. The dimensional stability (low shrinkage during parts fabrication) is also improved. The behaviour of composites depends upon the volume fractions of the phases, their shape, and on the nature of the constituents and their interfaces. High-modulus reinforcements can stiff a low-modulus matrix, but fibers and platelets are more effective than spheres. Rubbery inclusions lower the stiffness through they are inherently less strong.

8.11 Inter Penetrating Network (IPN) Polymers

IPNs are exciting polymer materials and can be designed as per the intellectual imagination and practical requirement. IPNs represent a unique topology of entanglement of two polymers out of which at least one is synthesized or cross-linked in presence of the other. The entanglement of polymer chains is such that each of it macroscopically or their phases microscopically cannot be separated physically. Thus IPNs belong to a broad class of polyblends but possess unique and diverse properties than the later. The primary condition for IPN formation is that the constituent polymers should be compatible with each other and should not form any chemical links. The IPN network results only by physical forces. The first synthetic IPN reported is based on two polymers of cross-linked polystyrene and polydivinylbenzene. Over the period of time several synthetic routes have been developed for preparing IPNs. In fact, IPNs are classified mostly by the method of their synthesis. Following types can be conidered; (i) Sequential IPNs—a cross-linked polymer is swollen in a solvent media which it self is a monomer. Then other active agents such as cross-linking agents and initiators are added to solvent media. Then the second monomer is polymerized in situ to get the IPN, (ii) Simultaneous IPNs—the synthesis of this type of IPNs involves two independent non-interfering reactions that can simultaneously be run under the same conditions and in the same reaction vessel. The reactions for example polycondensation and polyaddition which take place with different mechanisms, are carried out in the same container using different monomers, IPNs can be formed as a result of physical links between the polymers and (iii) Semi IPNs—these networks are formed essentially by grafting of polymers to another linear counter part. Then curing or cross-linking of the grafted part yields a semi IPN. Semi IPNs in fact are of two types; first type consists of a semi IPN with polymer I in cross-liked form and polymer II in linear form and the second type has a semi IPN polymer I in linear form and polymer II is cross-linked.

Besides the above IPNs, two more additional classes of networks have been prepared. In the first class an emulsion, (for example) of a elastometric polymer I is mixed with the emulsion of elastomertric polymer II and then cross-linking is done by evaporating the solvent media and allowing a film formation on a substrate. These networks ar called as interpenetrating elastometric networks. The second class of networks are obtained by emulsion polymerisation of a monomer II in a original seed latex of already cross-linked polymer I. It is however essential to add cross-linking agent during the subsequent emulsion polymerisation for the formation of IPN structures. This type of IPNs are called latex IPNs.

Most IPNs synthesised so far are based on alkly acrylates, styrene, divinylbenzene, urethane, urea, acrylic acid or acrylates, dimethylsiloxanes, propylene oxide and acrylonitrile monomers. However, it is to be kept in mind that the actual IPN synthesis involves a preformed cured/uncured polymer out of any of the above monomer and polymerisation of second monomer. The combination of two polymers is chosen such that both of them should have good compatibility and phase homogeneity. The pre- and post cross-linking are achieved in the usual way i.e. either by a cross-linking agent to the unsaturation in the chain or generating macro free radicals within the chain.

IPNs have great advantage over the other polymer combinations. If in the IPN, one of the polymers is elastomeric and the other is plastic at the use temperature, the combination of both of them yields synergistic behaviour in properties. For example, depending upon the phase composition in the final IPN, a reinforced rubber or high impact resistance thermoplastic can be prepared. The factors that play a decisive role in the formation of any class of IPNs are (i) compatibility of polymer chains, (ii) weight fraction of each polymer as well as the mobility of the units within their chains, (iii) relative rate of reaction and network formation within the chain and (iv) average molecular weight of cross-link.

Applications of IPNs

IPNs represent the zenithal ingenuity of synthetic polymer chemists as well as polymer technologists. The major advantage of these systems is that the combination of two dissimilar polymers results into a network with useful properties, which are not found in the respective individual polymers. The IPNs can be easily casted into tough films. Several applications have been forecasted out of IPNs. Table 8 (c) summarises the possible application areas of IPNs.

Table 8(c)   Areas of application of IPNs

Area of application	Types of IPN
Ion exchange resins	(i) Network of polystyrene and polydivinybenzene and polymethacrylic acid and polydivinylbenzenes (ii) Network of polyethylene and cross-linked polystyrene (chloromethylate followed by aminated) and divinylbenzene polymers
Microencapsulation (drug delivery and carbonless carbon paper, etc.)	Hydrogel beads of 2-hydroxymethacrylate and N-vinyl pyrrollidone copolymers cross-linked with poly n-(butylene oxide) swollen in mixture of diisocynate and diol and triols to yield polyurethane based IPNs
Semipermeable membrances for reverse osmosis	Anion exchange resins based on polyethylene and cross-linked poly (N-vinyl pyridine) and cation exchange resins on polyethylene and polystyrene sulfonates
Electrical application as insulators	Polyurethane and polymethyl methacrylate or polyurethane, epoxy resin and polymethyl acrylates

8.12   Thermally Stable Polymers

Unlike metals, polymeric materials are not stable at high temperatures and decompose. However, some polymers can withstand high temperature, and these thermally stable polymers are quite useful in aeronautic and aerospace industries. Some requirements of a thermally stable polymer are (a) it should have high melting point and degree of crystallinity, (b) it should have structural features that do not allow degradation by low energy process, (c) it should possess high bond dissociation energy.

Polymers like poly (para-phenylene), polythiazoles, polyimides, polyphenylene oxide, polyphenylene sulphide etc. are some examples of thermally stable polymers (for structures see appendix). Thermal stability often results from aromatic rings in the chain which impart rigidity, high bond energy and low degree of reactivity.

Ladder and spiro polymers (schematically shown in Fig. 8.18) are non-cross-linked polymers in which two molecular chains are joined to each other.

Such polymers could be soluble and fusible and may possess superior thermal

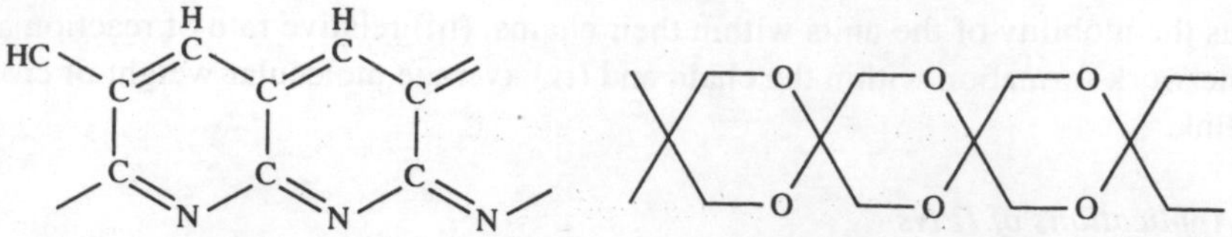

Fig. 8.18 (a) **Ladder polymer obtained from the pyrolysis of polyacrylonitrile** (b) **spiropolymers**

stability. Among ladder polymers are cyclized/dehydrogenated polyacrylonitrile (carbon fiber), polybenzimidazopyrolones and polyquinoxalines etc.

8.13   Liquid Crystalline Polymers

The molecules in a liquid crystalline material have a restricted mobility which maintains a degree of order compared to a true liquid. This results to anisotropy (i.e. the properties depend on the direction in which they are measured in contrast to isotropy shown by liquids). Liquid crystals are often called mesomorphic phases and are apparently between a crystalline solid and a normal liquid. Low molecular weight substances, which exhibit liquid crystalline behaviour tend to be rather rigid with rod- or disc- like shapes. The liquid crystalline behaviour originates in the form of long-range order extending over many millions of molecules. Here, although the individual molecules are in constant motion, as those are in liquids, there is a rapid fall off in probability that the axis (or plane) of any particular molecule will deviate far in terms of the angle from the most probable direction at any instant.

There are two broad classes of liquid crystals, nematic (have only longitudinal orientation) and smectic (possess a pronounced layering tendency). Further the liquid crystals can also be classified by the method of developing the crystalline order in the material. The materials made by heating a substance are called thermotropic. Those which do organise into liquid crystalline form (often surface active agents in concentrated solution form such structures) are called lyotropic liquid crystals. The organisation of molecules in nematic and smectic classes of liquid crystalline order is depicted in Fig. 8.19.

Polymeric materials possessing liquid crystalline (LC) behaviour are called liquid crystalline polymers. The best known example of liquid crystal polymer is that of an aromatic polyamide, 'kevlar'. Apart from the liquid crystal polymers where the main chain is responsible for this behaviour, interest has also been shown to produce liquid crystal polymers where LC property originates rather due to its rigid side chains. Liquid crystal polymers have a varied field of applications.

8.14   Telechelic Polymers (Functional Polymers)

Functional polymers are low molecular weight (~ 500–10,000) substances and contain functional groups like $-OH$, $COOH$, $-SH$, $-HC=CH_2$, $-N=C=O$ etc. Thus they are also called as functional oligomers. These functional groups are present at the ends of the molecule or distributed over the chain. These substances

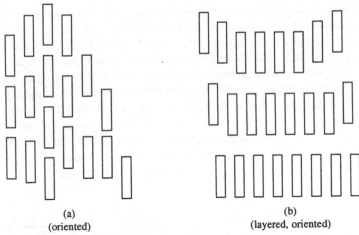

(a)
(oriented)

(b)
(layered, oriented)

Fig. 8.19 **General representation of liquid crystalline order in (a) nematic and (b) smectic forms of liquid crystals**

can be obtained using known methods usually employed in preparing polymers by suitably adjusting the reaction, whereby the introduction of functional groups takes place simultaneously or afterwards. The controlled degradation of high molecular weight polymers also serves as the synthetic route for functional oligomers. The introduction of functional groups takes place in free radical addition polymerisation by using initiators, chain transfer agents and/or by monomers containing functional groups. For example, an initiator of peroxide or diazo type can lead to a functional polymer. The polycondnsation method can be employed for making functional oligomers. The growth of an high molecular weight chain can be prevented by the nonstochiometric quantities of two monomers e.g. a reaction between a diol and a diacid with excess of diol will result in oligomers with functional hydroxyl groups. These functional oligomers are also called as telechelic polymers.

Functional oligomers can possess different chain composition and different end groups. Thus, a variety of chemical reactions are possible. These can be conveniently used to prepare block copolymers and to develop cross-links. Due to their low viscosity, low toxicity (free from solvent) and presence of chemically active functional groups, these oligomers have a large number of applications. These can be used directly in the synthesis of new polymeric materials. The polar functional groups make these telechelic polymers behave like amphiphilic molecules and are used as emulsifiers, dispersing agents, in surface modification (e.g. of fillers pigments, fibres etc.), to increase compatibility with solvents or with polymers. The uses of telechelic polymers as plasticizer (due to their low volatility), in liquid/liquid chromatography for purification of enzymes, in the synthesis of useful block and graft copolymers, cross-linkable elastomers in adhesive, paint and printing technology are well known.

8.15 Polymer Microgels

Polymer microgels do form in solution depending upon the structure of polymer and the interaction of polymer molecules with solvent. The microgel structures are intermediate between branch and microscopically cross-linked systems. Their overall dimensions are still comparable with those of high molecular weight linear polymers. It is difficult to precisely define a microgel but these can be illustrated well by schematic means (see Fig. 8.20). Such microgels are often observed for biopolymers like proteins and polysaccharides. Microgels possess characteristic features such as they dissolve (networks only swell) in certain solvents but their conformation is nearly fixed (unilke linear polymers which show innumerable conformations) and they can be processed like linear polymers but the final product shows characteristic network behaviour. Microgels from synthetic polymers can be obtained by the following two routes.

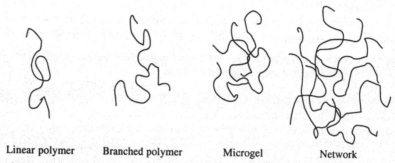

Linear polymer   Branched polymer   Microgel   Network

Fig. 8.20   Schematic representation of gels of polymer networks

Type A. microgels
The addition of multifunctional monomers to a polymerising system usually forms network poymers. However, the advantage of network polymers is that they show gelation when dissolved in solvents even in low concentrations.

Type B. microgels
These can be obtained by cross-linking polymers in restricted volume to avoid network formation. An easy approach is to do cross-linking in microemulsion polymerisation where the microemulsion droplets (swollen surfactant micelle, containing polymer droplets) grow and form microgels. Besides these microgel structures, which are homogeneous, more complex microgels can also be prepared. These differ in molecular chemistry and can have different features, each with potential applications. Such complex structured microgels are schematically shown in Fig. 8.21.

Homogeneous microgels are filtered for optically transparent 'high impact' polymers. These microgels embedded in the matrix of linear polymer stabilise the whole structure by distributing the mechanical tension. A polymer can be achieved in different useful topological forms using microgels. Functionalised microgels are good catalysts in nonaqueous media and act as stabilizers in complex coating formulation.

8.16   Biomedical Polymers

Polymers have become most active materials and cover a multitude of different

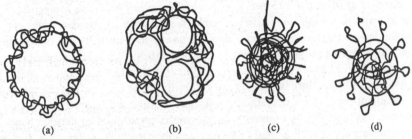

(a)   (b)   (c)   (d)

Fig. 8.21 Structure of (a) hollow micronetwork, (b) macroscopic microgel, (c) core-shell microgel and (d) functionalised spheres

aspects from inorganic macromolecules to natural polymers; the latter are of utmost importance for living systems. One needs only to think of natural macromolecules such as the proteins (enzymes), the nucleic acids (DNA and RNA), the polysaccharides and others. It is not unlikely that many of the diseases have their origins actually in abnormal macromolecules, leading to abnormal reactions in the living systems. Synthetic polymers have gained much medical interest in recent years and are being used for a variety of purposes viz. as implant materials, as carriers for biomolecules, as drugs etc.

As implant materials

Several synthetic polymers have been used as implant materials; a few important ones are polyethylene, polyetrafluoroethylene, polyesters, polysulfones, polyurethanes, acrylics and silicones. Polyethylene is used for articulation surfaces in joints, in implants and in repair of chest wall and diaphragm. Polytetrafluoroethylene, a most inert polymer with lowest frictional coefficient is used as a blood vessel replacement and in otology prosthesis. Polyethyleneterephthalate, the fibrous polymer, is used in knit arterial prosthesis and as non-absorbable sutures. Polysulfones, which are transparent thermoplastic material and exceptionally tough and resistant to high temperatures, are used in heart valve and pace components as well as in neurological implants. Polyurethanes are noted for their high strength, aberration and tear resistance and are employed as implant materials. Acrylic polymers, e.g. PMMA (better known as Plexiglas or Lucite), polyhydroxyethylmethacrylate are transparent plastics and often used as implant in the eye after cataract surgery. These polymers constitute hard and soft contact lenses. Applications of silicones as implant materials are diversified due to their nontoxic nature and biocompatibility. Silicone rubber serves a variety of purposes in eye surgery. Silicone rubber tear ducts can be implanted to serve as a means of controlling glaucoma. Silicone adhesives are used to cover the third degree burns and making artificial skin as well. Use of silicones in prosthesis, e.g. mammary and panile prosthesis is well known. Artifical breast and a totally implantable hydraulically driven total cardiac prosthesis are constructed with silicone-urethane copolymers. Carbon fibres, obtained by heating polyacrylonitrile at high temperatures, find application as implant materials as such or as composites with high molecular weight polyethylene or polytetrafluroethylene.

As carriers of bioactive substances (Immobilisation of bioactive molecules)
It is known that various bioactive materials like enzymes, pesticides, drugs can be attached to an inert polymeric carrier without any loss in their activity. Enzyme attachment onto a polymeric support (immobilisation) is a most fascinating area of research and has several medical, analytical and industrial applications. The immobilization not only imparts greater stability to the enzyme but also presents an excellent method for investigation of enzyme function in metabolic process. Further, repeated use of immobilised enzyme is possible. The various ways of immobilisation of enzyme on to a polymeric support are adsorption, cross-linking, covalent bonding and microencapsulation etc. The polymers which have been used as support materials are many, a few important of them are cellulose and its derivatives, agarose, polystyrene, nylon, ethylene-maleic anhydride copolymers etc. Medical applications of contolled drug release systems also involve polymeric materials. Polymers are ideal candidates for subcutaneous implants since they are not absorbed orally and generally have short *in vivo* half lives. Examples of subcutaneous implants are of insulin, interferon, growth hormones and vaccines. Among several polymers used in controlled drug release are nylon, polyacrylamide, cellulose and its derivatives, biodegradable polymers, like polylactic acid and polyglycollic acid.

Cell labelling
The influence of polymeric microspheres in the studies of biological systems is just in the beginning stage. Polymeric spheres to which biological moieties have been attached are being used as immunoadsorbents for antigen and antibody purification and labelling at the cell surface. Cell labelling can be achieved by electrostatic or immunological interaction between the polymeric microspheres and the cell surface.

Polymeric drugs
With the development of polymer research it has now been possible to synthesize a tailor-made macromolecule, i.e. a polymer with predetermined structure, discrete molecular weight and with specific functional groups with desired physiological activity etc. In spite of many positive results in the investigation on the compatibility or long lasting activity, no decisive breakthrough for a clinical application of polymeric drugs has so far been realised. Polymeric drugs are advantageous in respect that they exhibit delayed action, prolongation of activity, decreased rate of drug metabolism and drug excretion. A number of polymeric drugs have been developed as antibiotics such as those based on the structure of penicillin. Polyvinylpyridine-N-oxide has been put in clinical use to inhibit silicotic fibrosis. Polyvinyl pyrrolidone is known for its use as blood plasma extender. Many physiologically active monomers lose their activity in the polymeric form but a few others exhibit enhanced activity. L-lysine, which has apparently no antibacterial activity as a monomer, is very active against *E. Coli* and *S. Aureus*. Urea-formaldehyde polymer has been used as antibacterial and antifungal agent. Polyanionic drugs are very valuable substances. Various surface active polymers, e.g. nonyl phenyl ether of polyoxyethylene (Tritons) have been tested for antituberculosis. Table 8(d) lists medicinal uses of some polymeric drugs. In

brief, it can be realised that polymeric materials, both natural and synthetic, play crucial role in life processes.

Table 8(d)   Polymers as drug materials

Drug type	examples of polymers
Antibacterial agent	Quaternary ammonium polymers, polyanionic polymers, polypeptides
Antifungal agent	Polyanionic polymers
Antiviral agent	Polyanionic Polymers
Anticoagulatns	Polyanionic polymers, heparin
Antitumor agent	Polyanionic polymers
Plasma extenders	Polyvinylpyrrolidone, dextran, gelatin
Antisilicosis	Poly-N-oxides

8.17   Polymeric Supports for Solid Phase Synthesis

In 1963, Merrifield introduced a new concept for the synthesis of peptides using cross-linked polystyrene beads with chlromethyl groups. The first amino acid of the proposed polypeptide chain was attached by its carboxylic groups to the reactive site. The second amino acid, with all but one of its reactive groups protected was coupled to the first amino acid, leaving the protective dipeptide formerly bonded to the support. This could be filtered and washed thoroughly. In this way the different amino groups were attached in the desired sequence. After the desired sequence of amino acids was assembled, polypeptide was removed from the support materials by selectively breaking the bond that held the polymer and polypeptide together. Now the polypeptide with desired sequence of amino acids was free and dissolved in solution, could be removed from the insoluble solid support and was purified by conventional methods.

The use of polymer supports for organic synthesis can be easily understood because, the functional group X, linked to the polymeric support can be replaced by the desired functional group Y, which can be recovered from the reaction medium for further reaction with other substances.

The solid phase synthesis has the following advantages: (a) excess quantities of unattached reagents may be employed to obtain high yields, (b) the polymeric support can be recycled and (c) the reaction is simplified by the insoluble polymer support. Majority of solid phase synthesis reactions have been carried out using cross-linked polystyrene (easily and economically prepared by copolymerising styrene with small amount of divinylbenzene). The stable polystyrene beads can be conveniently functionalised by halogenation, halomethylation, acylation, metallation etc.

Merrifield's solid phase synthesis has resulted to an automatic polypeptide synthesizer where the synthesis can be achieved at greater ease. This method is likely to provide large scale mnufacture of polypeptides, like insulin, enzymes and nucleic acids which have potential applications. These can contribute to the production of insulin (for treatment of diabetic patients), more efficient synthetic enzymes for biomedical and industrial use and in the understanding of mechanism of enzyme action versus their structures. Merrifield's Nobel prize (1984) winning work on the synthesis of polypeptides using insoluble cross-linked polymeric materials as supports has stimulated the use of this technique for organic synthesis.

8.18 Polymers in Combating Environmental Pollution

Polymers paly an ever-increasing role in the field of pollution. These are becoming more prominent in controlling all types of pollution namely water, air and solid waste. To combat water pollution, polyelectrolytes such as high molecular weight polyacrylamides are frequently used as flocculants and coagulants. Their use as ion-exchange resins, e.g. cross-linked polystyrene sulfonate is well known. Membranes for reverse osmosis (for desalination purpose) are made of polymers and complex formation between metal ions and polyelectrolytes may provide good product. In air pollution, with the advancement of adsorption-desorption process, new polymers are designed with side groups which react very fast with air polluting gases and releasing them under different conditions for recycling into respective industrial processes. Pyrolysis of polymers may be useful for solid waste pollution which is becoming a serious problem.

8.19 Polymers as Chemical Reagents

Polymers are now being used as dehydrogenating agents, as catalysts in dehydration, antioxidants and decomposition reactions. The advantage of using polymers as catalyst is that with polymers, no compound other than the dehydrogenated product is formed. Heat resistant polymers, e.g. pyrolysed polyacrylonitrile (with condensed pyridine rings acting as strong hydrogen acceptor) are used in dehydrogenating olefinic compounds in vapour phase at high temperatures. Pyrolysed polyacrylonitrile causes a double bond shift as well as *cis-trans* isomerisation in olefinic compounds such as 1-butane. Polymers with conjugated double bond or rings are good catalysts for autooxidation reactions as shown in the following scheme.

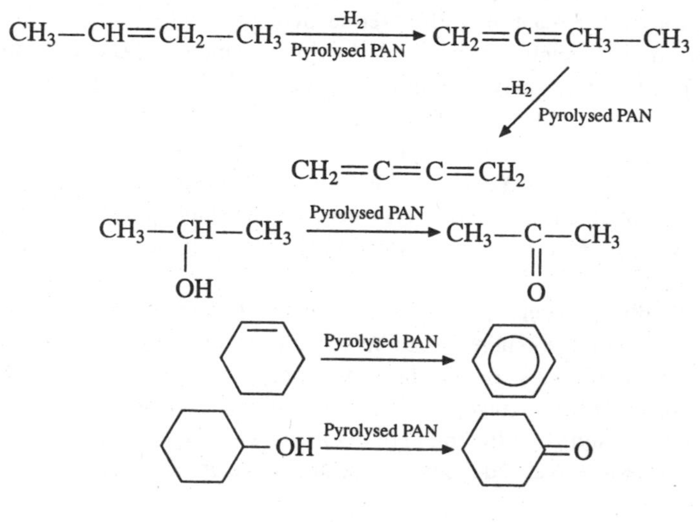

Suggested Further Readings

Polyelectrolytes

Dautzenberg, H., W. Jaeger, J. Kotz, B. Phillip, C. Seidel and D. Stscherbina, *Polyelectrolytes*, Munich: Hanser, 1994.

Glass, J.E. (ed.), *Polymers in Aqueous Media, Advances in Chemistry Series* 223, Washington DC: Am. Chem. Soc., 1989.

Hara, M. (ed.), *Polyelectrolytes Science and Technology*, New York: Marcel-Dekker, 1993.

Mandel, M., *Polyelectrolytes*, in Encyclopedia of Polymer Science and Engineering, Vol. 11, 2nd ed. (Mark, H.F., N. Bikales, C.G. Oberberger and G. Menges eds.), p. 739 New York: Wiley, 1988.

Oosawa, F. *Polyelectrolytes*, New York: Marcel and Dekker, 1971.

Selegny, E. (ed.), *Polyelectrolytes*, Dordrecht, D. Riedel, 1974.

Ionomers

Eisenberg, A., and M. King, *Ion-Containing Polymers*, New York: Academic Press, 1977.

Hara, M., (ed.), *Polyelectrolytes: Science and Technology*, New York: Marcel-Dekker, 1993.

Schmitz, K.S., *Macroions in Solution and Colloidal Suspension*, Weinheim: VCH, 1992.

Conducting Polymers

Blythe, A.R., *Electrical Properties of Polymers*, Cambridge: Cambridge Univ. Press, UK, 1980.

Dyson, R.W. (ed.), *Specialty Polymers*, New York: Chapman and Hall, 1987.

Goosey, M.T. (ed.), *Plastics for Electronics*, London: Elsevier Appl. Sci. Pub., 1985.

Kiess, H. (ed.), *Conjugated Conducting Polymers*, Berlin, Springer, 1992.

Radhakrishna, S., and A.K. Arof, (eds.), *Polymeric Materials*, New Delhi: Narosa, 1998.

Salaneck, W.R., I. Lundstrom, and B. Roanby, (eds.), *Conjugated Polymers and Related Materials*, New York: Oxford Univ. Press, 1993.

Seymour, R.B. (ed.), *Conductive Polymers*, New York: Plenum Press, 1981.

Skotheim, T.A., (ed.), *Handbook of Conducting Polymers*, (2 vols.), New York: Marcel-Dekker, 1986.

Solid Polymer Electrolytes

Gray, F.M., *Solid Polymer Electrolytes*, New York: VCH, 1991.

MacCallum, J.R., and J.R. Vincent, (eds.), *Polymer Electrolyte Reviews*, 1st and 2nd ed., London: Elsevier Applied Science, 1987.

Radhakrishna, S., and A.K. Arof, (ed.), *Polymeric Materials*, New Delhi: Narosa, 1998.

Scrosati, B., (ed.), *Second International Symposium on Polymer Electrolytes*, London: Elsevier Applied Science, 1990.

Skotheim, T.A. (ed.), *Electro Responsive Molecular and Polymer Systems*, Vol. 1, New York: Marcel-Dekker, 1988.

Electroluminiscent Polymers

Friend, R.H., R.W. Gymer, A.B. Holmes, J.H. Burroughes, R.N. Marks, C. Taliani, D.D.C. Bradley, D.A. Dos Santos, J.L. Bredas, M. Logdlund and W.R. Salaneck, *Electroluminiscence in Conjugated Polymers*, a review article, Nature, 397, 121–128, 1999.

Block Copolymers/Polyblends/Polymer Composites

Folkes, M.J. (eds.), *Processing, Structure and Properties of Block Copolymers*, New York: Elsevier, 1985.

Goodman, I. (ed.), *Developments in Block Copolymers*, Barking: Appl. Sci. Publ., 1982.

Manson, J.A., and L.H. Sperling, *Polymer Blends and Composites*, New York: Plenum, 1976.

Meier, D.J. (ed.), *Block Copolymers. Science and Technology*, New York: Harwood Academic Publ., 1983.

Nohay, A., and J.E. McGrath, *Block Copolymers, Overview and Critical Survey*, New York: Academic Press, 1976.

Riess, G.P. Bahadur, and G. Hurtrez, *Block Copolymers* in Encyclopedia of Polymer Science and Engineering, 2nd ed., New York: Wiley, 1985.

Thermoplastic Elastomers

Legge, N.R., G. Holden, H.E. Schroeder, (eds.), *Thermoplastic Elastomers*, Munich: Hanser, 1987.

Liquid Crystalline Polymers

Ciferri, A., W.R. Kringbaum, and R.B. Meyer, (eds.), *Polymer Liquid Crystals*, New York: Academic Press, 1982.

Collyer, A.A., (ed.), *Liquid Crystal Polymers*; *From Structure to Applications*, New York: Chapman and Hall, 1992.

Donald, A.M., and A.H. Windle, (eds.), *Liquid Crystalline Polymers*, Cambridge; Cambridge Univ. Press, 1993.

Keller, A.M., Warner, A.M. Windle, (eds.), *Self Order and Form in Polymeric Molecules*, London: Chapman and Hall, 1995.

For Other Topics

Burchard, W., S.B. Ross-Murphy, (eds.), *Physical Networks*, London: Elsevier Appl. Sci., 1990.

Cowie, J.M.G., (ed.), *Specialty Polymers*, Vol. 1, New York: Plenum, 1981.

Miles, I.S., and S. Rostami, (eds.), *Multicomponent Polymer Plastics*, New York: Wiley, 1993.

Sperling, L.H., *Interpenetrating Polymer Networks and Related Materials*, New York: Plenum, 1981.

Utracki, L.A., and A.P. Plochoko, *Industrial Polymer Blends and Alloys*, Munich: Hanser, 1985.

Utracki, L.A., *Polymer Alloys and Blends*, Munich: Hanser, 1989.

Appendix A

Laboratory Experiments

I Aim: Preparation of polystyrene by a free radical polymerisation process

Requirements

Apparatus   Balance, heavy walled glass tubes, stopper for glass tube, Buchner funnel, filtration flask, separating funnel and filter papers etc..

Reagents and chemicals   Styrene (monomer), 25% aqueous sodium hydroxide solution, molecular sieves/drying agents ($CaCl_2$) or CaO), benzoyl peroxide, toluene and methanol.

Procedure

(1) Wash styrene twice with 25 ml portions of 25% aqueous sodium hydroxide solution in order to remove the inhibitor. Then wash twice with the 25 ml of distilled water to remove any residual sodium hydroxide (2) Dry the inhibitor free monomer using molecular sieves or other drying agents ($CaCl_2$) and store immediately with a nitrogen blanket in a refrigerator (3) Add 50 ml of inhibitor freed dry styrene to a heavily glassed test tube. Flush the tube with nitrogen and add 1.0 g of benzoyl peroxide. Gently shake the contents of the tube, stopper and place the tube in an oil bath at 80°C for 1–2 hrs (4) When liquid becomes syrupy and viscous before it gels, dissolve the contents in 50 ml of toluene and then pour into 500 ml of methanol in order to precipitate polystyrene that is formed (5) Isolate the polymer by filtration and dry the product (6) Weigh the amount of polystyrene and report the yield.

II Aim: Preparation of polystyrene by an emulsion polymerisation process

Requirements

Apparatus Three necked round bottom flask, mechanical stirrer, nitrogen cylinder

Reagents and chemicals   Styrene monomer, triple distilled water, potassium persulfate, sodium lauryl sulfate and ferric alum

Procedure

(1) Add 12 ml of distilled water, 7 ml of styrene monomer (inhibitor freed and dried as in Exp. I), 4 ml of 0.7% potassium persulfate solution, 10 ml of 4% sodium lauryl sulfate solution to a three necked round bottom flask equipped with mechanical stirrer (2) Purge the contents of the flask with nitrogen till all the dissolved air is removed (3) Place the flask in a water/oil bath and raise the temperature to 80°C and allow the process to go on for three hours with continuous stirring and to a maximum conversion of monomer to polymer (4) The contents in the flasks would be quite turbid to milky appearance. Coagulate the emulsion by adding alum solution to isolate the polystyrene formed. If needed, boil the mixture for complete isolation (5) Filter the contents, wash with water and dry the polystyrene powder. Determine the yield percentage based on the weight of the polystyrene isolated.

III Aim: Preparation of polystyrene by an anionic polymerisation method

Requirements

Apparatus Three necked round bottom flask, magnetic stirrer and bar (preferably glass enclosed), water condenser, thermometer, septum, syringe, Buchner funnel, Buchner flask, vacuum pump etc.

Reagents and chemicals Pre-dried inhibitor freed styrene monomer, n-butyl lithium in n-hexane (2.5 mol dm^{-3}), tetrahydrofuran, methanol and 2-butanone, (all the three solvents are to be completely dried and freshly distilled), stoppered separating funnel with standard point tail.

Procedure

(1) Fit the three necked flask with septum, water condenser and thermometer (through a mercury seal). Add a glass encased magnetic stirring bar (2) Add 6 ml of dry and freshly distilled THF using a syringe into the flask (3) With a careful control of temperature around 20°C, add the styrene monomer drop wise (0.2 ml/sec) until 2.0 ml has been added. The reaction mixture turns deep red due to the formation of styrllithium. Continue adding the remaining styrene at the same rate. Once all the styrene is added, wait for 10–15 minutes (4) Add 1.0 ml of the methanol to quench the reaction (5) Precipitate the polystyrene by adding excess of cold methanol (about 5. ml). Further purification of polystyrene can be done by dissolving it in 2-butanone and re-precipitating again by addition of cold and dry methanol. Vacuum dry the purified polystyrene and estimate the yield of polymerisation.

Note It is very important to keep the equipment and reactants dry as well as oxygen free during the course of anionic polymerisation, otherwise chain termination can take place. Solvents and monomer have to be delivered via syringe and no additions can be made by keeping the flask open to air.

IV Aim: To polymerise styrene by cationic polymerisation process

Requirements

Apparatus Two large thick walled test tubes, septum for test tubes, hypodermic syringe, sintered glass filter, filtration flask, ice water bath, Erlenmeyer flask and vacuum pump etc.

Reagents and chemicals Pre dried and freshly distilled styrene monomer, 25% (w/v) aqueous sodium hydroxide solution, methylene chloride, stannic chloride ($SnCl_4$) and methanol.

Procedure

(1) Take two large thick walled test tubes fitted with septum and needle to act as pressure relief valve. Label them as I and II. To each add 40 and 30 ml of dry methylene chloride respectively via a hypodermic syringe (2) Lower both the test tubes into an ice water bath so that they attain a temperature close to 0°C (3) Inject 0.9 ml of stannic chloride to the tube I. Inject 6.5 ml of styrene monomer to the tube II and keep it cold for about 15 to 20 minutes at 0°C. Again through syringe add 4–5 ml of cooled stannic chloride solution from test tube I to test tube II. This marks the start of cationic polymerisation process (4) After 20–25 minutes remove the test tube II from the ice bath. Precipitate the polystyrene by taking out the tube II from ice water bath and pouring the polymerised reaction mixture into methanol (about 10 ml) taken in another flask (5) Dry the product in a vacuum oven for 30 minutes after filtering and giving methanol washes (6) Weigh polystyrene and report its yield

V Aim: Preparation of solid epoxy resin

Requirements

Apparatus Three necked flask, water condenser, mechanical stirrer and thermometer etc.

Reagents and chemicals Bisphenol-A, epichlorohydrin and 10% aqueous solution of sodium hydroxide

Procedure

(1) Charge the three necked flask with 15 g of bisphenol-A and 5 ml of 10% aqueous sodium hydroxide. Equip the flask with stirrer, condenser, dropping funnel and thermometer via a mercury seal (2) Place the flask with its contents on a water bath. Warm the contents of the flask to 45°C and allow the mixing to go on for 10–15 minutes (3) Then add 10 g of epichlorohydrin rapidly maintaining uniform stirring (4) Heat the mixture then to a temperature of about 85°C and allow the polycondensation to go on for about 1 hr (5) The reaction mixture separates into two phases. Siphon off the aqueous layer. Wash the solid product with hot water until the washed water is neutral to the litmus paper. Dry the resin in an air oven at 130°C and report the yield.

VI Aim: Preparation of liquid epoxy resin

Requirements

Apparatus Three necked flask (500 ml), stirrer, thermometer, vacuum distillation assembly, water condenser, evaporating dish, dropping funnel and Buchner funnel etc.

Reagents and chemicals—recrystallised bisphenol-A, epichlorohydrin (good grade), sodium hydroxide and toluene

Procedure

(1) Set up the three necked flask reactor on a water bath. Add bisphenol-A (25 g) and epichlorohydrin (15 ml) with stirring on. Heat the mixture to 70°C by raising the temperature of the water bath (2) Meanwhile prepare 40–45% aqueous sodium hydroxide solution and add the alkali to the reactor drop wise through a dropping funnel over a period of 1 hr (3) Keep the temperature of the reactor around 70°C through the addition of alkali. Continue the reaction for another 30 minutes after the addtion of alkali (4) Place the reactor in a cold water bath to allow the contents to reach room temperature (5) Set up the vacuum pump assembly through the receiver and water condenser. Heat the contents of the reactor to 70°C (6) Vacuum distill water and un-reacted epichlorohydrin from the reaction vessel (7) Disconnect vacuum distillation set up and place the reactor in cold water bath to bring back the contents to room temperature (8) Add 20 ml of toluene to the reactor and filter the mixture on a Buchner funnel to remove precipitated NaCl (9) Place the filtrate back into the reactor and again set up the vacuum distillation assembly. Raise the temperature of the reactor to 70°C under vacuum and distill out the toluene completely (10) Transfer the resin left to an evaporating dish and gently heat it using low Bunsen flame in a fuming cupboard. Continue heating to 140°C till all the volatile matter disappears. Weigh the liquid epoxy resin and report its yield.

VII Aim: Determination of epoxide equivalent of the given epoxy resin by the pyridinium chloride method

Requirements

Apparatus   Round bottom flask, water condenser, burette, sample test tubes (small glass vials)

Reagents and chemicals   Epoxy resin sample, 0.2 N pyridinium chloride— pyridine reagent (by dissolving 17 ml of conc. HCl into 1 liter of pyridine) standard 0.5 N methanolic NaOH, phenolphthalein indicator solution

Procedure

(1) Dissolve about 1 g of epoxy resin in 25 ml pyridinium chloride—pyridine reagent by gentle stirring and if necessary warming (2) Take 25 ml of pyridinium chloride in pyridine solution in another flask (this set is to be considered as

blank) (3) Fit the water condenser for both the flasks and reflux the contents for 1 hr (4) Cool the solution with condenser fitted on and add 15 ml methanol. Add few drops of phenolphthalein indicator and titrate the contents with 0.5 N methanolic NaOH (5) Calculate the epoxide equivalent using the formula

$$\text{epoxide equivalent} = \frac{16 \times \text{wt. of the sample in gram}}{\text{gm of oxirane oxygen in the sample}}$$

gm of oxirane oxygen in sample = [B − S] × N × 0.016
where B = ml of NaOH used for titration of the blank solution till end point and S = ml of NaOH consumed for the titration of the sample solution till the end point, N = normality of NaOH.

VIII Aim: Preparation of polysulphide rubber (thiokol)

Requirements

Apparatus   Round bottom flask, water condenser, water bath, boiling chips, thermometers etc.

Reagents and chemicals   Sodium hydroxide, powdered sulphur, 1, 2-dichloroethane

Procedure
(1) Dissolve about 40 g of NaOH in about 800 ml of distilled water. Place the solution in the round bottom flask and warm it on a water bath with water filled condenser fitted (2) Add 13–15 g of powdered sulphur to the flask (3) The contents are heated to reflux temperatures for 30–35 minutes (4) Cool the red brown solution of Na_2S_x by adding 500 ml of water to a temperature of 65–70°C (5) Add 2 ml of 1, 2-dichloroethane and place 2–3 boiling chips. A two phase mixture is formed. Reflux the mixture for 45 minutes by using mild heating rates. A yellow rubber thiokol is formed (6) Cool the contents to room temperature and wash the product well with water and dry it.

IX Aim: Preparation of polyacrylamide (PAM) by free radical polymerisation

Requirements

Apparatus   Three necked round bottom flask, reflux water condenser, nitrogen inlet bent glass tube, nitrogen cylinder and water bath

Reagents and chemicals   Recrystallised acrylamide monomer, AIBN, methanol, ethyl acetate and distilled water

Procedure
(1) Dissolve about 2 g of pure acrylamide monomer in 500 ml of methanol and place the solution in the round bottom flask. Prepare separately 0.2 mol dm^{-3} AIBN solution in dry methanol. Add 2–5 ml of this solution to acrylamide solution (2) Flush the nitrogen into the reaction flask with the reflux condenser

fitted on. Heat the contents to 65 °C for to $1^1/_2$ hours. Intermittently shake the flask (3) After the polymerisation is completed, pour the mixture into equal amount of dry ethyl acetate with efficient stirring to precipitate the polyacrylamide. The formed polymer is purified further first dissolving it in water and re-precipitating twice with methanol. Dry the polymer and report the yield.

X Aim: Preparation of methacrylic acid-methacrylamide copolymer by free radical polymerisation

Requirements

Apparatus—same as for PAM (Exp. IX)

Reagents and chemicals—pure methacrylic acid, recrystallised acrylamide, potassium persulfate ($K_2S_2O_8$), ethyl acetate and distilled water

Procedure

(1) Take both the methacrylic acid and methacrylamide monomers in the weight ratio of 9:1 (total weight of both the monomers may be taken about 15–25 g) in the three necked round bottom reaction flask. Add about 5 ml of 0.1% aqueous solution of $K_2S_2O_8$ initiator (2) Flush the contents with nitrogen purging and conduct the polymerisation by heating the contents to 70°C on water bath for 1 hr (3) The product obtained is cooled and re-precipitated twice with ethyl acetate for removing the un-reacted monomers. The copolymer formed is then dried and weighed for reporting yield.

XI Aim: Grafting of acrylamide onto polyvinyl alcohol

Requirements

Apparatus   Three necked reaction round bottom flask, glass water condenser, mercury seal, thermometer, water bath, electric motor driven stirrer, nitrogen cylinder

Reagents and chemicals   Recrystallised acrylamide, polyvinyl alcohol (mol. wt. > 30,000), ceric ammonium nitrate, 1 N nitric acid, acetone and methanol

Procedure

(1) Set up the reaction assembly on a water bath with three necked flask fitted with a glass water condenser and electric motor driven stirrer (2) Prepare the polyvinyl alcohol (PVA) solution by adding 10 g of PVA into 10 ml of distilled water and gently heating to 80°C. It may take 1–1.5 hours for clear solution to be formed and hence this solution has to be prepared well ahead of starting the practical (3) Introduce 10 ml of PVA solution into the reaction flask and purge the nitrogen gas for about 10 minutes at 25°C. Then add 8 ml of 0.1 mol dm^{-3} solution of ceric ammonium nitrate prepared in 1 N HNO$_3$. Raise the temperature of the contents to 45–50°C (4) Allow the polymerisation and grafting for another 1 hour with continuous stirring and under nitrogen atmosphere. The solution's

yellow color will then disappear. Stop stirring and allow the contents to cool down to room temperature (5) Pour the solution with rapid stirring into a litre of acetone to precipitate the copolymer. The precipitated copolymer is to be washed with acetone and methanol and finally dried in vacuum at 70°C. Determine the mole percentage of acrylamide in the copolymer by nitrogen estimation method.

XII Aim: Preparation of polymethyl methacrylate beads by pearl polymerisation

Requirements

Apparatus Three necked flask, glass condenser, electric motor driven slow speed stirrer, thermometer (with mercury seal), Buchner filtration apparatus

Reagents and chemicals Methyl methacrylate monomer, sodium hydroxide, benzoyl peroxide, sodium phosphate, liq. ammonia, calcium chloride, polyvinyl alcohol, hydrochloric acid and double distilled water.

Procedure
(1) Free the freshly vacuum distilled methyl methacrylate monomer from the stabiliser by washing it thrice with 20 ml of 2% (w/v) sodium hydroxide aqueous solutions and with 50 ml of distilled water (2) Set up a three necked reaction flask in the water bath. Fit the stirrer and glass condenser (3) Place a mixture of 120 ml of distilled water, 3.5 ml of 10% (w/v) sodium phosphate aqueous solution, 5 ml of ammonia in the reaction flask. Stir the mixture with specially designed thick walled glass stirring rod. The shape of the glass stirring piece is very important. It should be in the form of a thick glass sheet with two portions out of which the bottom portion is in oval shape and the upper part is of square shape. Maintain a speed of 60–100 rpm with the low speed stirrer unit (4) Add 50 ml of distilled water and 7.5 ml of 10% (w/v) calcium chloride solution drop wise through the dropping funnel over a period of 1.5–2 hours. Add 10 ml of 1% (w/v) of PVA aqueous solution. Then fill water bath with water at a temperature of 70°C and raise the temperature to 80°C during the first 30 minutes of polymerisation (5) Add 100 ml of stabiliser freed methyl methacrylate monomer drop wise using a droping funnel over a 15 minutes time period in which 1.0 g of benzoyl peroxide is dissolved. Extreme care should be taken to maintain the slow speed of the glass stirrer piece, so that a stable dispersion of methyl methacrylate droplets is formed (6) Allow the polymerisation to continue at a constant stirrer speed and temperature for a minimum time of 1.5 hours. Care should be taken that the stirrer does not stop during this time and the suspension does not gellify. After this time observe visually that small tiny and transparent beads are formed (7) Dismantle the reaction flask assembly and filter off the beads of polymethyl methacrylate by means of a Buchner vacuum filtration apparatus. Wash the beads with 2% (w/v) hydrochloric acid and then with distilled water and finally dry at 40–45°C in an electric oven. Weigh the dried beads and report the yield.

302 *Principles of Polymer Science*

XIII Aim: Preparation of Polyvinyl Alcohol (PVA)

Requirements

Apparatus Three necked round bottom flask, glass condenser, water bath, Buchner funnel and thermometer

Reagents and chemicals Polyvinyl acetate (good commercial grade), anhydrous freshly distilled methanol, potassium hydroxide, calcium chloride, distilled water and acetone

Procedure

(1) Set up the three necked reaction flask on a water bath. Charge it with 3 g of polyvinyl acetate (PVAc) and 75 ml of anhydrous methanol. Fit the flask with a reflux water condenser attached to a $CaCl_2$ drying gaurd tube. Reflux the contents until PVAc dissolves (25–30 minutes). Add 25 ml of anhydrous methanol containing 2 g of potassium hydroxide when the contents of the flask are still warm. Reflux the mixture further for 1 hour (2) Polyvinyl alcohol formed due to the hydrolysis of PVAc slowly gets precipitated from the solution as a fine powder. Collect the product using a Buchner funnel and wash it with four 25 ml portions of anhydrous methanol. Dry the powder partially dried PVA to a 300 ml beaker containing 100 ml hot water (80°C). Allow the polymer to be dissolved over another 30 minutes (4) Add the PVA aqueous solution drop wise to a 1 litre beaker containing 200 ml of acetone. Stir the contents over a magnetic stirrer while addition is on. Collect the final precipitated powder and air dry it on a Buchner funnel before drying it to constant weight in a vacuum oven or desiccator at room temperature over 24 hours period. Weigh the product and report the yield

XIV Aim: Determination of melting point, storing time and gel time of phenolic resins

Requirements

Apparatus Melting point, capillary tubes, rubber bands, thermometer, beakers stop watch, spatula, hot plate (electric), crucible

Reagents and chemicals Paraffin, resin sample (commercial phenol formaldehyde rasin), hexamethylenetetramine

Procedure

(1) Grind finely a small amount of resin in a mortar and introduce it into a melting point capillary tube. Attach the resin filled capillary tubes to the bottom of thermometer and place it in a paraffin bath. Raise the temperature of the bath by using an electric hot plate or Bunsen burner at a rate of 1°C per minute. Note down the temperature at which melting starts and completes. Report this temperature range as melting temperature for the resin (2) Take 10 g of finely powdered resin sample in a petri dish. Place the petri dish on the surface of an oil bath (maintained at 135–150°C) with the help of tongs. Hold the dish firmly. Then gently stir the

molten resin with a spatula and note the time between the melting and hardening of the resin. This time is known as storing time (3) Place finely ground 1 g resin sample in a crucible and mix 0.1 g of finely powdered hexamethylenetetramine. Place the crucible on an electric hot plate at 150°C. Move the resin backwards and forwards with the spatula and note down the time interval between the melting and gelation. This time is known as gel time.

XV Aim: Determination of molecular weight of polymer by amine end group analysis

Requirements

Apparatus   100 ml three necked flasks with stoppers (glass), micro-burette, heating mantle and controllers, magnetic stirrer, stirring motor, glass condenser, small liquid nitrogen flask, small laboratory mill, aluminum weighing pan and balance capable of measuring 0.0001 g.

Reagents and chemicals   Commercially available or laboratory prepared polyhexamethylene sebacamide (nylon 6,10), reagent grade phenol crystals, dry methanol, 0.05, 0.1, 0.2, 0.15 or 1.0 N standardised hydrochloric acid, thymol blue indicator (0.1% in distilled water).

Procedure
(1) The polyamide sample must be broken and ground into fine pieces not longer than 1 mm in dimension. This is to be done in a laboratory ball mill. The procedure takes 2.5 to 3 hours. Hence the sample has to be made ahead of time. Polyamides and other reagents/chemicals used are extremely sensitive to moisture. Hence care must be taken to avoid the exposure of sample, reagents and chemicals to moisture or air. If needed the polyamide material should be cooled in nitrogen before it is crushed, grounded and milled (2) Place 35 g of phenol and 20 ml of methanol in each of the three 100 ml three necked flasks. Add finely ground and accurately weighed (1.5–2.0 ± 0.0001 g) three polyamide samples into each of the flasks. Fit the flasks with stoppers and condenser (3) Reflux the contents till all the sample is dissolved. Cool the flasks to room temperature and replace the condenser with stirrer (4) Add 0.2 ml of thymol blue indicator solution to each flask. Titrate the contents of the flask to a pink end point by fine addition of 0.2 ml of standardised HCl solution. The concentration of HCl solution has to be selected in such a way that finite titer volumes are obtained. Calculate the amount of HCl added (5) Calculate the molecular weight

$$\overline{M}_n = (\text{sample wt. (g)} \times 1000)/(\text{titre in ml} \times \text{normality})$$

Degree of polymerisation $= \overline{M}_n/M_0$
where M_0 is the mol. wt. of repeat unit
   Tabulate your results and estimate a mean value of $\overline{M}_n$ for the sample from the three calculated values

XVI Aim: Estimation of number average molecular weight by hydroxyl end group analysis

Requirements

Apparatus   250 ml iodine flasks, 50 ml burette, 10 ml pipettes, hot plate with magnetic stirrer, aluminium weighing pan, balance for accurate weight measurement (0.0001 g).

Reagents and chemicals   Polyethylene glycol or any other hydroxy terminated polymer, potassiumhydrogenphthalate, acetic anhydride (reagent grade), reagent grade pyridine, phenolphthalein, cresol red and thymol blue indicators and n-butanol.

Procedure

Hydroxy equivalent: (1) Prepare ahead of time 0.4 N methanolic sodium hydroxide (one litter quantity). Standardise the solution against potassium hydrogen phthalate (0.5 N) using phenolphthalein indicator. Prepare the acetylating agent by dissolving the acetic anhydride in pyridine. For a 250 ml of total volume, dissolve about $68.85 \times N$ ml, where N = the normality of methanolic sodium hydroxide and the rest is pyridine. Prepare the mixed indicator by adding 1 part 0.1% aqueous cresol red to 3 parts 0.1% aqueous thymol blue (both these indicators have to be neutralized with NaOH) (2) Weigh the 250 ml iodine flasks. Accurately weigh the polymer sample into 250 ml iodine flasks. The weight of the polymer sample to be taken depends upon the molecular weight of the sample. Following weights of sample are recommended for appropriate molecular weights:

$\overline{M}_n$	Weight (g)
400	1.2
800	1.8
1200	2.0
1600	2.3
2000	3.0

(3) Carry out the following addition to the two flasks containing polymer sample and also two other flasks with no sample (to be considered as blank) (i) 10 ml of acetylating reagent (stopper immediately), (ii) 10 ml of distilled water and 10 ml of pyridine after placing the magnetic bar into the flask and placing the flasks on a stirring hot plate, (iii) Heat the flasks for 5 minutes at approximately 100°C. Remove the flasks from the stirring hot plates and cool to room temperatures (iv) Add 10 ml of n-butanol to each of the cooled flasks. Add 5–6 drops of mixed indicator. (v) Titrate the contents to a neutral end point with 0.04 N NaOH. Record the titer values

Acid equivalent: (1) Accurately weigh 2–3 g of polymer sample into each of the iodine flasks. Consider the other two flasks as blanks (2) Add 25 ml of pyridine and place a magnetic bar in each of the flasks (3) Heat each of the flasks on a hot plate at 105–110°C until the sample is dissolved. Add 10 ml of distilled water and heat for further 5 minutes. Cool the flasks to room temperature (4) Add 10 ml of n-butanol and 5 drops of mixed indicator to each of the flasks (5) Carry out the titration of the contents of all the flasks with 0.4 N NaOH to the first blue end point. This end point is sensitive and fades rapidly (6) Calculate the hydroxyl equivalent, acid equivalent and the average molecular weight of the sample by using the following relations,

$$\text{Hydroxyl equivalent} = \frac{(\text{blank reading} - \text{sample reading}) \times \text{normality of [NaOH]}}{\text{Sample weight (g)}}$$

$$\text{Number average molecular weight} = \frac{2000}{\text{hydroxyl equivalent} + 2 \, (\text{acid equivalent})}$$

XVII Aim: Determination of viscosity average molecular weight of polystyrene in toluene by dilute solution viscosity method

Requirements

Apparatus Constant temperature thermostatic bath, Ubbelohde viscometer with efflux time greater than 120 seconds at 25°C, stop watch capable of registering flow time up to ± 0.1 sec. Sintered G4 and G5 glass funnels, rubber suction bulb, 100 ml volumetric flasks with standard joints, 10 and 25 ml stoppered volumetric flasks and thermometer etc.

Reagents and chemicals Polystyrene, reagent grade toluene, chromic acid solution (prepared as saturated potassium dichromate in conc. sulfuric acid)

Procedure

(1) Prepare a stock of 100 ml of polystyrene solution in dry toluene, first by adding 75 ml of toluene to accurately weighed (about 2 g) polystyrene. Stopper the flasks and agitate it gently. Allow about 24 hrs. time for complete dissolution with agitation. Once the polymer is fully dissolved, make up the total volume to 100 ml. Magnetic stirring and heating facilitates complete and fast dissolution (2) Filter the solution with sintered G4 funnel. Prepare 5 more solutions with concentrations namely 1.5, 1.0, 0.75, 0.5 and 0.25% (w/v) by diluting the stock solution with toluene. These diluted solutions are to be prepared in 25 ml quantity of each (3) Clean the Ubbelohde viscometer by filling it with chromic acid over night and repeated rinsing with distilled water. Dry it in hot oven. Deliver about 20 ml of dry toluene into viscometer. Place the viscometer in a constant temperature water bath maintained electronically at 25 ± 0.01°C. Allow the equilibrium to be attained for 15 minutes. By using the suction rubber bulb and closing the pressure relieve tube with a finger, allow the solvent to be sucked through capillary into upper flow bulbs of viscometer. Then release the pressure and allow the solvent to flow through. Start the stop watch when the bottom of the liquid meniscus passes the upper timing mark. Stop the stop watch when bottom of the meniscus touches the lower timing mark. Repeat this for three times and take the mean value as the flow time of solvent, t_0 (sec) (4) Deliver the lowest concentration

solution of polystyrene into viscometer and measure the flow time three times as described previously for the solvent. Take out the viscometer every time out to deliver a fresh solution and measure its flow time, t. One can also do direct dilutions in the viscometer using modified Ubbelohde versions (5) Calculate various viscosity terms by using the flow times of solvent and each polymer solutions by: $\eta_r = t/t_0$, $\eta_{sp} = \eta_r - 1$, $\eta_{red} = \eta_{sp}/C$. The intrinsic viscosity $[\eta]$ of polymer solution is obtained by graphs using the Huggin's relation:

$$\eta_{sp}/C = [\eta] + k' [\eta^2] C$$

(6) Tabulate the results listing the concentration, C, average flow time, t, η_r, η_{sp} and η_{red}. Plot the graph of η_{sp}/C versus C and obtain $[\eta]$ and k' from the intercepts and slopes by extrapolating to zero concentration (7) Estimate the viscosity average molecular weight of polystyrene using Mark-Houwink relation:

$$[\eta] = K \, \overline{M}_v{}^{\alpha}$$

$$= 1.15 \times 10^{-4} \, \overline{M}_v{}^{0.72} \text{ for polystyrene in toluene}$$

Note Mol. wt. of other polymers can be similarly determined by dissolving it in a suitable solvent and measuring viscosity of dilute polymer solutions at fixed temperature. For some common polymers, the value of K and α are given in Table 3(1).

XVIII Aim: Calculation of dimensions of polystyrene coil assuming a freely jointed chain and from dilute solution viscosity data

Requirements

Apparatus Balance for accurate weight (0.0001g) measurement, constant temperature thermostatic bath, Ubbelohde viscometer with efflux time greater than 120 seconds at 25°C, stop watch capable of registering flow time up to ± 0.1 s. Sintered G4 and G5 glass funnels, rubber suction bulb, 100 ml volumetric flasks with standard joints, 10 and 25 ml stoppered volumetric flasks and thermometer

Reagents and chemicals Cyclohexane, benzene, polystyrene, reagent grade toluene, chromic acid solution (prepared as saturated potassium dichromate in conc. sulfuric acid) etc.

Procedure

(1) Prepare 2% (w/v) stock polystyrene solutions in benzene (25°C) and cyclohexane (34°C) (i.e. θ-temperature). Dilute the stock solution to required concentration in the same manner as described in Exp. XVII. Care must be taken to keep the polystyrene solutions at 34°C because it defines θ-conditions for polystyrene (2) Measure the flow times for benzene (25°C), cyclohexane (34°C) and polystyrene solutions in benzene (25°C) and polystyrene solutions in cyclohexane (34°C) by the same procedure as Exp. XVII (3) Calculate η_r, η_{sp}, η_{red} and $[\eta]$ for both the sets (4) Dimensions of polystyrene chain: (i) calculate

chain expansion factor, 'α' from the relation $\alpha^2 = [\eta]_{ben} / [\eta]_{cy(theta)}$, where $[\eta]_{ben}$ = intrinsic viscosity of polystyrene in benzene or non θ-solvent conditions, $[\eta]_{cy(theta)}$ = intrinsic viscosity in cyclohexane (34°C) or in θ conditions. (ii) Evaluate the viscosity average molecular weight of polystyrene from intrinsic viscosities under θ and non θ conditions by:

$$[\eta]_{ben} = 1.06 \times 10^{-4} \ \overline{M}_v{}^{0.72}$$

$$[\eta]_{cy(theta)} = 0.82 \times 10^{-3} \ \overline{M}_v{}^{0.5}$$

(iii) From the M_v values in θ and non θ conditions calculate the degree of polymerisation, DP as DP = mol. wt. of polystyrene/mol. wt. styrene repeat unit (iv) Calculate number of bonds N in freely jointed chain by N = 2 DP. Then evaluate the root mean square end-to-end distances $<\bar{r}_\theta^2>^{1/2}$ and $<\bar{r}_{non\theta}^2>^{1/2}$ by equation,

$$<\bar{r}^2>^{1/2} = b \, (2N)^{1/2}$$

where b = bond length = 1.54 Å for C–C bond (v) the characteristic ratio of dimensions, C_N is calculated as

$$C_N = <\bar{r}^2>_\theta^{1/2} / <\bar{r}^2>_{non\theta}^{1/2}$$

(vi) Finally calculate the steric or hindrance factor σ, which is the measure of thermodynamic chain flexibility of chain by using the equation

$$C_N = \sigma^2(1 - \cos \tau)/(1 + \cos \tau)$$

where τ = valence of angle for a C–C single bond., i.e. 109.5°

XIX Aim: To qualitatively identify an unknown polymer by thermogravimetry polymer finger printing

Requirements

Apparatus Thermogravimetric analyser, data acquisition system and recorder plotter, Dewar flask, gas flow meter, calibrated weight standards, microbalance (weight measurement accurate to ± 0.0001 g)

Reagents and chemicals Source of dry and purified nitrogen, polymer samples of polyethylene, high density polyethylene (HDPE), low density polyethylene (LDPE), polystyrene, polymethyl methacrylate of syndiotactic and isotactic type

Procedure
(1) Consult the operational manual for the handling of instrument. Take the help of specialist/analyst, if available (2) Weigh the pan and record the weight. Add 10–20 mg of polystyrene to the pan and reweigh. Record the weight (3) Run the polystyrene sample on the instrument following a linear programmed temperature usually 5, 10 or 20°C/min till a final temperature of 550–600°C (4) Get the recorded TG traces as weight % loss against temperature (This is thermogram). Once the first run is completed, allow the instrument to cool to room temperature,

clean the pan and sample holders. Repeat the runs for other polymers i.e. polyethylene and polymethyl methacrylate. Obtain the TG runs of as many samples of these three polymers as possible. Get all the traces on a single graph (5) Repeat the whole procedure with an unknown polymer sample freshly and obtain the TG curve. By comparing the unknown TG trace with known plots, identify the polymer. Attempt to analyse the thermal behaviour of polymer sample.

XX Aim: Determination of melting temperature, T_m, glass transition temperatures, T_g and heat of fusion, ΔH_f of a given polymer sample using differential scanning calorimeter (DSC)

Requirements

Apparatus   DSC instrument and all the other accessories needed for the gas flow, accurate weight measurements and cooling system etc.

Reagents and chemicals   Polystyrene, polyvinyl chloride and polymethyl methacrylate

Procedure

(1) Weight out exactly 10–15 mg polymer sample onto the DSC sample container. Place the sample cell on the plotform with an empty reference cell on the reference platform. Consult the instrument manual for its operation, data acquisition and plotting of results. If necessary, take the help of specialist technician (2) Run the sample with an heating rate of 25 or 30°C/min. Obtain the DSC thermogram (3) Look for the sudden change in the heat flow with respect to base line and identify the exo or endotherms and their transition temperatures. Get the area under the peak either by the data analyzer or manually by using a planimeter (4) Calculate the heat of fusion (ΔH_f) using the relation, $\Delta H_f /(\text{J/g}) = (A/m)$ (BE Δq_s), where A = peak area in cm^2, m = sample mass in mg, B = time scale in min./cm, E = cell calibration coefficient at the temperature of measurement, Δq_s = Y axis scaling in mW/cm.

Some useful references for more laboratory experiments

A simple inexpensive molecular weight measurement of water soluble polymers using microemulsions, *J. Chem. Edu.*, 62, 545 (1985).

Determination of acrylonitrile/methyl methacrylate copolymer composition by infrared spectroscopy, *J. Chem. Edu.*, 60, 431, (1983).

Polyethylene glycol – A poor chemists crown, *J. Chem. Edu.*, 60, 77, (1983).

Sulphonation of polystyrene-orientation and characterisation of an ion exchange resin in organic chemistry, *J. Chem. Edu.*, 66, 613 (1989).

Viscometric determination of isoelectric point of a protein, *J. Chem. Edu.*, 40, 468 (1963).

Suggested Further Readings

Bran, B.J., and Billmeyer, Jr. F.W., *Techniques of Polymer Synthesis and Characterisation*, New York: Wiley, 1972.

Brydson, J.A., and Saunders, K.J., *Experimental Plastics Technology*, Methven Education Ltd., 1970.

Collins. E.A., Bares. J., and Billmeyer Jr. F.W., *Experiments in Polymer Science*, New York: Wiley, 1973.

Ke, B., *Newer Methods of Polymer Characterisation*, New York: Interscience Pub., 1964.

Mc Caffery E.M., *Laboratory Preparation for Macromolecular Chemistry*, New York: McGraw-Hill, 1970.

Pearce, E.M., Wright, C.E. and Bordoloi, B.K., *Laboratory Experiments in Polymer Synthesis and Characterisation*, Pennsylvania State University, Univ. Park, Pa, 1982.

Sandler, S.R., Karo, W., Bonesteel, J.A. and Pearce, E.M., *Polymer Synthesis and Characterisation—A Laboratory Manual*, New York: Academic Press, 1990.

Sorenson, W.R., and T.W. Campbell, *Preparation Methods in Polymer Chemistry*, New York: Wiley, 1968.

Appendix B

Pioneers in Polymer Science

Modern polymer science owes its development to present level because of original, innovative and highly intellectual thoughts of many scientists and technologists. The polymer science has seen many breakthroughs over the last five decades. The Nobel prize in chemistry, which is considered as the top most recognition of original and path breaking contribution, has gone five times to distinguished scientists for their contribution in polymer science. The Nobel prize for the year 2000, once again is bagged by polymer scientists for inventing conducting plastics. A brief biography of these scientists along with a sketch of their contribution are mentioned. Besides the Nobel laureates, there are many other popular names associated with different facets of polymer science. The sketches of some of these pioneers are also given for the benefit of readers. (Nobel laureates for their work on biological macromolecules are not included here).

Nobel Laureates

Hermann Staudinger (1881–1965)
Hermann Staudinger was honored with the Nobel Prize in chemistry for the year 1953. He was the first who coined the word 'Macromolecular' in 1920.

He put forwarded the concept of existence of giant molecules, which behave different than then well-known small sized simple molecules. Staudinger published great number of scientific papers. He published approximately 500 research papers on macromolecular compounds about 120 of these on cellulose and about 50 on rubber and isoprene. It is during this period, out of his first hand experience of working with the cellulose and rubber, he propagated that these compounds have molecules of a giant size and are different from simpler ones. His idea was rejected then by the academia. In fact it is said that an academician had advised him by saying "Dear colleague, leave the concept of large molecules well alone ... there can be no such thing as a macromolecule", after his major lecture devoted in favor of macromolecular concept. This happened in 1925. Today, however, it seems impossible to believe that this violent opposition not only to an original but also totally a new idea or innovative intellectual imagination, has existed relatively in recent times. The present day polymer

Hermann Staudinger

science thus owes a lot to Hermann Staudinger, who can be called as 'Father of Macromolecules'.

Staudinger was a prolific author and had written number of books. Some of these are Die ketene (Ketenes) (Enke, Stuttgart, 1912), Anleitung zur organischen qualitaten Analyse (introduction to organic qualitative analysis) (Springer, Berlin, 1923 (1st ed.) and 1955 (6th ed.), Tabellen Zu den Vorlessungen uber allgemine und anorganische Chemie (Tables for the lectures on general and inorganic chemistry) (Braum, Karlsruhe, 1947 (5th ed.), Die Hochmolekularen Organische Verbindungen, Kautschak und Cellulose (The high molecular organic compounds, rubber and cellulose) (Springer, Berlin, 1932), Organische Kolloid Chemie (Organic Colloid Chemistry) (Viewweg, Braunschweig, 1950 (3rd ed.), Fortschritte der Chemie, Physics und Technik der Makromolekularen Stritte (Progress of the Chemistry, Physics and technique of the macromolecular substances, jointly with Professor Viewweg and Professor Rohrs, volume I (1939) and (1942) (Lehmann, Munich) and Makromolekulare Chemie and Biologie (Macromolecular Chemistry and Biology) (Wepf and Co., Basle, 1947). Since 1947 staudinger has edited the journal Die Makromolekulare Chemie (Macromolecular Chemistry), published by Dr. A. Huthig, Heidelberg and Wepf & Co., Basle, Switzerland.

Hermann Staudinger was born in Worms, Germany on 23rd March 1981. He was educated in Worms, at Halle University and later at Darmstadt and Munich. He served many institutes as Professor, (Institute of Chemistry of the Techmische Hochschule, Karlsruhe, Germany for fourteen years from 1912), University of Freiburg, Germany (1912–1926). He had also served as principal of the Research

Institute of Macromolecular Chemistry, Freiburg, and as Head of the State Research Institute of Macromolecular Chemistry.

He received many honours and awards; and degree viz. Dr. Ing. H.C. of the Technische Hochschule, Karlsruhe, Dr. Rer. Nat. H.C. of University of Mainz, Dr. (c) H.C. of University of Salamanca and Dr. H.C. of University of Torino. Staudinger married Magda Woit, who was for many years his coworker and coauthor of numerous publications.

Giulio Natta (1903–1979)

Giulio Natta shared the Nobel Prize in chemistry for the year 1963 with Karl Ziegler. G. Natta began his career with a study of solids by means of x-rays and electron diffraction. He then used the same methods for studying the catalysts and structure of some high polymers. His kinetic research on methanol synthesis, on selective hydrogenation of unsaturated organic compounds and on oxosynthesis led to an understanding of the mechanism of these reactions and to an improvement in the selectivity of catalysts. In 1938 he began to investigate polymerisation of olefins and the kinetics of subsequent concurrent reactions. In 1953, he extended research work conducted by Ziegler on organometallic catalysts to the stereospecific polymerisation. This led to the discovery of new classes of

Giulio Natta

polymers with sterically ordered structures viz. isotactic, syndiotactic and diisotactic polymers and linear branched olefinic polymers and copolymers with an atactic structure. These studies led to the development of new important thermoplastic material, isotactic polypropylene at industrial scale.

By x-ray diffraction studies, Prof. Natta determined the exact arrangement of chains in the lattice of new crystalline polymers. He also synthesized new elastomers based on butadiene, and copolymers of ethylene with propylene. Prof. Natta's work also led to the synthesis of optimally active macromolecules from optimally inactive monomers and synthesis of crystalline and sterically ordered alternating copolymers from different copolymer couples.

Prof. Natta's scientific and technical activity is documented in 700 published papers out of which 500 concern stereoregular polymers, and by a large number of patents in many different countries.

Giulio Natta was born at Imperia, Italy on 26th February 1903. He graduated in chemical engineering at polytechnic of Milan in 1924. He joined as a faculty staff member of Pavia University as a full professor in 1933. He also served as a director of Institute of General Chemistry at this university before he was appointed as full-time professor of physical chemistry at University of Rome. He had association with the industrial chemistry departments of polytechnic of Turin and Milan polytechnic in similar capacity.

Prof. Natta was honorary member of several international chemical societies of Austria (1968), Switzerland (1963), Belgium (1962) and (New York Academy of Sciences (1958). He received several gold medals out of which one from the town of Milan (1960) is worth mentioning because it was given by the President of Italian Republic exclusively to meritorious personalities. Several international universities awarded honorary doctorate degree to Prof. Natta (Turin University and University of Mainz).

Karl Ziegler (1898–1973)

Prof. Karl Ziegler shared the 1963 Nobel chemistry prize with Prof. Natta. He carried out extensive work over the reactions in chemistry of organoaluminium compounds. Prof. Ziegler's Nobel Prize winning discovery, however, has been the organic mixed catalysts for the polymerisation of olefins (e.g. the synthesis of high density polyethylene).

Karl Ziegler was born at Helsa near Kassel in Germany on 26th November 1998. He graduated from University of Marburg/Lahn and served at universities of Frankfurt/Main and Heidelberg. Prof. Ziegler was the Director of the Max Planck Institute for Kohlen forschunrg in Mulheim/Ruhr from 1943 till 1969. He was also the founder director of Gesellschaft Deutscher Chemiker (German Chemical Society). He held honorary doctorate degrees from Technische Hochschulen, Hannover and Darmstadt, Universities of Heidelberg and Giessen. He had been bestowed with many medals out of all the most prominent being the award of distinguished order by the German Federal Government.

Karl Ziegler

Karl Ziegler was married to Maria Kurtz and has five grand children by his daughter and five by his son.

Paul J Flory (1910–1985)

Prof. Paul J. Flory was honoured for his fundamental work on polymers, with an award of Nobel Prize in chemistry for the year 1974. Prof. Flory had an unique distinction as Nobel laureate because unlike most laureates who are recognised primarily for either theoretical or experimental work, Flory was cited for achievements in both areas. Prof. Flory contributed almost to every branch of polymer chemistry including studies of polymerisation mechanism and structures, physical and mechanical properties both in bulk and in solution, elasticity of fibrous proteins and crystallisation of polymers from solution. Flory developed simple models to solve problems that had puzzled the scientists for many years. He himself had said once that neither simplicity nor complexity is inherent in object of inquiry but that both depend greatly on the point of view or perhaps insight of the observer.

Flory's interest in science was kindled by his remarkable teacher Carl W. Holl, Professor of Chemistry at Manchester College from which he was graduated in 1931. He then moved on to graduate school of Ohio State University and completed his dissertation in the field of photochemistry and spectroscopy in

Paul J Flory

1934. He joined Du Pont Company and assigned to small group headed by Dr. W.H. Carothers who invented nylon and neoprene. Flory during his stay at Du Pont explored the fundamentals of polymerisation and other novel aspects of polymeric substances. Flory also participated in the research and development programme on synthetic rubber, which was short of supply due to second World War. Flory then joined Cornell University and was offered a professorship in 1948. He then had a productive and satisfying period of research and teaching, which culminated into one of the originally written books in polymer chemistry "Principles of Polymer Chemistry", published by Cornell University Press in 1953.

The well known contribution of Flory has been that his models help in (i) understanding the polymer chain statistics, (ii) configuration of polymer chains, (iii) rational interpretation of physical measurements in dilute polymer solution and (iv) quantitative characterisation of macromolecules. The work on all these aspects led to the formulation of the hydrodynamic constant called theta and recognition of theta point at which excluded volume interactions are neutralised. Experimental methods for monitoring the theta conditions have been proposed. Very soon many laboratories all over the world confirmed the ideal behaviour of polymers (natural and synthetic) under theta conditions.

Flory futher concentrated his work on spatial configuration of chain molecules and the treatment of configurational dependent properties by rigorous mathematical models. Prof. Flory had a stint at Standord University from 1961 and another fundamental book titled "Statistical Mechanisms of Chain Molecules" was authored by him. This book summarises the theory and its applications.

Flory got married to Emily Catherine Tabor in 1936 and has three children (two daughters and a son) and four grand children. His son was named as Paul John Flory Jr.

Bruce Merrifield

Bruce Merrifield's name is attached to solid phase peptide synthesis. He was honoured by Nobel Prize in Chemistry for the year 1984 for his work on peptide synthesis.

Bruce Merrifield

Merrifield had college training at Pasadena Junior College and then moved to University of California at Los Angeles. After graduation he worked at the Phillip R. Park Research Foundation. He worked as an assistant for Dr. D.W. Woolley at Rockfeller university on a di-nucleotide and nucleotide growth factors. He persistently pursued his interests in the subject, which eventually led to the idea of solid state peptide synthesis in 1959.

Peptides act as proteins which are biocatalysts and control, accelerate or even inhibit the biological and metabolical processes. Merrifield's contribution led to the synthesis of peptides with desired sequences of amino acids in the laboratory conditions. Merrifield was a Nobel guest professor at Uppsala University in 1968 and was elected a member of the U.S. National Academy of Sciences in 1972. He had received several awards (Lasker award for basic medical research (1969), the Gairdner award (1970), the intrascience award (1970) and American Chemical Society award (1972), to name a few. He also received honorary degrees from several universities (University of Colarado, Uppsala University, Yale Unversity and Colgate University).

Merrifield was born in Fort Worth, Texas on 15th July 1921. He married Elizabeth Furlong and has six children. Mrs. Merrifield, a biologist, spent considerable time at home for upbringing of the children and later joined her husband at the Merrifield Laboratory at Rockfeller University.

Pierre-Gilles de Gennes

Pierre-Gilles de Gennes discovered that the methods developed for studying ordered phenomenon in simple systems can be generalised to more complex forms of matter, in particular liquid crystals and polymers. de Gennes began working initially on magnetic phase transitions, but turned his attention during the period 1960–1980 to other more complicated phenomenon such as (i) transition from:

Pierre-Gilles de Gennes

(a) conducting to superconducting state in certain materials and (b) from an ordered state to discovered state in liquid crystals, (ii) regularities in the geometrical arrangement and movement of polymer chains, and (iii) conditions that define the stability in microemulsions etc. de Gennes treated some of the complex systems, which eluded simple explanations based on generalised mathematical models that are already in place for studying the ordered and disordered phenomenon in simple systems.

de Gennes, made outstanding contribution in perceiving common features in ordered phenomenon in very widely differing physical systems such as magnets, superconductors, liquid crystals, polymer solutions, gels, porous media and other soft systems. The proposal by de Gennes, that "scaling laws" can be used for predicting the dynamics of ordered to disordered transitions or vice versa in the above mentioned complex systems, stimulated a great deal of theoretical and experimental work. Of all his studies, understanding of phase transitions in liquid crystals and dynamics in polymer chains, led to technical and commercial exploitation especially of liquid crystals in daily used items such as pocket calculators, wrist watches, flat TV screens and other display systems using colored digital numbers. de Gennes modelled the behaviour of polymer chains in dilute to more concentrated solutions and even in very high concentrations that are encountered in polymer melts. A new branch of polymer physics i.e. polymer dynamics has emerged due to the extensive experimental and theoretical studies by the School of Polymer Physics, founded and chaired by de Gennes under STRASACOL, a joint project with physicists and chemists from Strassbourg, Saclay and the College de France. de Gennes proposed "blob" model, which states that a certain typical segment of a polymer chain can move as if it were free, even in more concentrated solution and "reptation" model, which describes the segmental motion in polymer chain in terms of serpentine movement within a tangle or loop of surrounding polymer chains.

One of the most important discoveries of de Gennes is that "order in disorder" exists in the arrangement of polymer chains in dilute to concentrated solutions or even in polymeric melts, under some applied conditions identical to those under whose influence, a system of magnetic movements moves from order to disorder. De Gennes book titled "The Physics of Liquid Crystals" (1974) and "Scaling Concepts in Polymer Physics" (1979) became standard and authoritative publications in the respective subjects.

P.G. de Gennes was born in Paris, France in 1932. He had his education from Ecole Normale from 1955 to 1959. He was with Atomic Energy Center (Saclay, France) working mainly on neutron scattering and magnetism under the guidance of A. Herpin, A. Abragam and J. Friedel. He then worked as post-doctoral scientist with C. Kittel at Berkeley, USA in 1959 and returned to France to serve in French Navy. He left the navy in 1961 to take up an academic position as Assistant Professor in Orsay and founded and established Orsay group on Supraconductors. He became a Professor at the College de France, Paris. His interests also kept widening from supraconductors to liquid crystals, interfacial phenomenon, especially the dynamics of wetting and physical chemistry of adhesion etc. De Gennes received prizes and medals from several reputed academic societies. Of all these following can be cited: Holweck prize from the joint French and British Physical Society, the Ampere prize from the French Academy of Sciences, the Worlf prize from Israel, Gold Medal from the French CNRS, the Matteuci Medal from Italian Academy, the Lorentz Medal from Dutch Academy of Arts and Sciences and polymer awards from both American Chemical and American Physical Societies (ACS and APS). He has also been honoured with the award of memberships by several prestigious societies such as French Academy of Sciences, Dutch Academy of Arts and Sciences, the Royal Society, American Academy of Arts and Science and the National Academy of Sciences, USA.

Hideki Shirakawa, Alan G. Mac Diarmid and Alan J. Heeger

The Nobel prize in Chemistry for the year 2000 was shared by three scientists A.J. Heeger, A. G. Mac Diarmid and H. Shirakawa for their discovery of conducting plastics. The work of these three scientists has revolutionized the whole thinking about plastics or polymers. Because it is well known that polymers and plastics are used as excellent insulators i.e. chemical substances that cannot conduct the electricity. In fact, it is everybody's routine observation that insulating plastic material made up of polymers is coated around the copper or aluminum wires to prevent any leakage of electricity being carried by these metallic wires. Thus the pioneering work of these three scientists has changed the whole concept of electrically conducting materials, which so far have been confined mainly to metals.

It all started when Prof. Shirakawa and his co-workers synthesised polyacetylene way back in 1974, from acetylene by using Ziegler-Natta catalyst. (K. Ziegler and G. Natta shared Nobel Prize in chemistry for the year 1966 for their invention of insertion or stereo-specific polymerisation). It is worth mentioning that, in fact Ziegler and Natta prepared for the first time polyacetylene in 1958 by polymerising acetylene in n-hexane using a mixture of triethyl titanium/titanium tetrapropoxide as a catalyst. The resulted polyacetylene was black in appearance and highly crystalline with regular structure. The powder was very susceptible to air and was infusible and insoluble in almost all the solvents. However, it was not a good conductor.

Similarly, polyacetylene initially obtained by Shirakawa in 1974 had metallic appearance but was not a conductor. However, in 1977, Prof. Shirakawa, and Prof. Alan MacDiarmid, University of California at Santa Barbara, combinedly discovered that the conductivity of polyacetylene films can be improved almost by 10^9 times by oxidising the polyacetylene (PA) films with chlorine, bromine and iodine vapours. This treatment of PA films with halogens is called as doping in analogy to a similar process through which the conductivity of semiconductor materials is enhanced. The doped form of PA had a conductivity of 10^5 Siemens per meter. This value was higher than for any other conducting or semi-conducting polymers. Metals such as silver and copper have metallic conductivities of above

10^8 S m^{-1}. Thus for the first time, the three pioneering scientists have been successful in developing polymer plastic having conductivities of about 10^3 times closer to that of highly conducting metals.

The major contribution of Prof. Shirakawa is that he effected the polymerisation of acetylene to polyacetylene at the surface of the catalyst system, taken as concentrated solution in toluene and in inert atmosphere. Besides this modification, Prof. Shirakawa used a mixed catalyst system of triethylauminium/titanium tetrabutoxide.

The PA film, initially obtained by Prof. Shirakawa inside the walls of the reaction vessel had two forms, as expected in Ziegler-Natta polymerisation. The PA form, when polymerised in toluene under inert atmosphere and at –78°C, was copper colored and had all *cis* configuration (a *cis* content of about 95%). Similarly, when polymerisation was conducted using the same catalyst but in *n*-hexadecane as solvent media and at a temperature of 150°C, Prof. Shirakawa obtained PA films having all *trans* configuration. The *cis*- and *trans*- forms of PA are shown below:

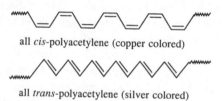

all *cis*-polyacetylene (copper colored)

all *trans*-polyacetylene (silver colored)

The conductivities of all *cis* - polyacetylene films were $10^{-8} – 10^{-7}$ Sm^{-1} and while *trans* - polyacetylene was a better conductor with $10^{-3} – 10^{-2}$ Sm^{-1}.

Meanwhile Prof. Alan Heeger and Prof. MacDiarmid were collaborating on the studies of metallic properties of covalent inorganic polymers. Prof. MacDiarmid met Prof. Shirakawa in Tokyo, and came to know about the synthesis of PA films in *cis*- as well as *trans*- forms. After the fruitful discussion and exchange of ideas between the two, it was suggested that Shirakawa's group refine the polymerisation process and Prof. MacDiarmid modify the PA films by doping process. These proposals were based on a previous observation by Shirakawa's group that PA films with all *trans*- configuration when treated with bromine or chlorine decreased IR transmission without changing the silvery color of the films. Prof. Shirakawa's group mean while refined their polymerisation process and could synthesize PA films with controlled *cis* / *trans* double bonds along the chain. Prof. MacDiarmid, using the PA films from Prof. Shirakawa's lab, had collaborated with Prof. Heeger's lab. Then many experiments went on. The doping of PA films was done with I_2 and AsF_5. The group has now established that doping of *cis*- polyacetylene with iodine and AsF_5 increases the conductivity of PA films by a factor of 10^{11} Sm^{-1}. This value is exceptionally very high and the investigations of these three scientists have opened up totally a new field of polymers i.e. conducting plastics. Besides polyacetylene, several other polymers are reported in the literature, which have potential to act as good conductors in doped form. These polymers, for example are polypyrrole, polythiophene (and its derivatives), polyphenylvinylene and polyaniline etc.

The application areas of conducting polymers are far and wide. One of the major positive commercial and technological aspects of the polymers is that transparent, homogeneous films of any thickness can be fabricated at very low costs because, such a processing is often done from solutions. The discovery by Shirakawa, MacDiarmid and Heeger of conducting plastics, have opened following major applications of conducting polymers- (i) As electromagnetic shields of electronic circuits, (ii) As anti-static coating materials on photographic films to prevent their damage from exposure to electric discharge, (iii) As hole injecting electrode material in light diode (LED) device, (iv) As candidates for electroluminiscent displays (for example in mobile telephone, remote control and other such devices), (v) As facilitators for full color video matrix displays, (vi) As electrochromic windows (smart windows), (vii) As sensing material in sensors, (viii) As field effect transistors and microwave absorbing screen coatings.

The brief biosketch of each of the Nobel laureates in chemistry for the year 2000 is as follows:

Hideki Shirakawa

Prof. Hideki Shirakawa, a professor of chemistry at the Institute of Material Science, University of Tsukuba, Japan, got his Ph. D. degree from Tokyo Institute to Technology (Titech) in 1966. Prof. Shirakawa has been a faculty member,

Hideki Shirakawa

holding different positions at Institute of Material Science, Tsukuba University, for more than last twenty years.

He started exploring polymer science unprecedently to a newer area of conducting polyacetylene (PA) films from electrically insulating polyacetylenes. (He had initiated this work, when he was at Tokyo Institute of Technology.) What is more heartening to note that Prof. Shirakawa's success in converting insulating PA to an electrical conducting one has resulted from an accidental experimental mistake. During laborious and lengthy laboratory experiments on refining the synthesis process of PA, a thousand fold too much catalyst was added in one of the experimental batch and the result was a fascinating silvery PA material, with all *trans* configuration of polymer chain. This beautiful silver film has been found to possess superior properties (hither to unheard to for other known polymers) close to metals.

Prof. MacDiarmid, University of Pennsylvania heard about the Prof. Shirakawa discovery of controlled synthesis of PA films having metal like properties, he had invited Prof. Shirakawa and worked together with Prof. Alan Heeger, University of California. The combined efforts of these three brilliant brains have led to the complete understanding of ways to modify the PA films from insulators to conductors. (Their seminal paper titled "Synthesis of electrically conducting organic polymers: halogen derivatives of polyacetylene $(CH)_n$" appeared in J. Chem. Soc., Comm., 579 (1977).) Dr. Shirakawa was born in 1936 in Tokyo and can be called as saint scientist, who dedicated his life to the cause of teaching and research in the frontier area of science and has blessed the humanity with his pioneering contributions, which may convert the silicon based circuits from polymer based electronics to nano scale integrated circuits.

Alan G. MacDiarmid

Prof. Alan MacDiarmid is the co-discoverer of the new and epoch making area of conducting polymers, more commonly known as synthetic metals. If prof. Shirakawa is responsible for the synthesis of hyperfine and superior quality of PA films, Prof. MacDiarmid handled these films and modified them with electrochemical doping by using all the expertise and innovation available with him and ultimately developed a prototype conducting polymer. Prof. MacDiarmid's scientific interest has been around the studies on and rediscovery of polyaniline, which is proving to be not only technologically important but also foremost industrial conducting polymer. His efforts have always been to improve the conductivity of polyaniline and its isomeric forms and enhance their mechanical properties. Other aspects that attracted his attention are reversible sensors and LED's based on polymers made from aniline and its isomeric forms.

Dr. MacDiarmid was born in Masterton, New Zealand in 1927 has been presently a Professor of chemistry at the University of Pennsylvania. He had obtained his higher education at the Universities of New Zealand, Wisconsin, Cambridge and joined the faculty of university of Pennsylvania way back in 1955.

MacDiarmid

Prof. MacDiarmid published approximately 600 research papers and claimed 20 patents exclusively on conducting polymers particularly on synthesis, chemistry, doping, electrochemistry, conductivity, magnetic and optical properties and processing of polyacetylene and polyanilines. Prof. MacDiarmid has been recipient of numerous awards and honorary degrees both nationally (in USA) and internationally.

Alan J. Heeger

Prof. Alan J. Heeger was the co-inventor and co-pioneer of new branch of polymer science i.e. conducting plastics. Prof. Heeger put his masterly expertise experience on semi conducting and metallic polymers and experimental facilities at the disposal of Prof. MacDiarmid and Prof. Shirakawa in characterising and analysing the modified or doped polyacetylene (PA) films.

The trio of polymer scientists, together proved beyond doubt that an insulator polymer can be modified to possess conducting and mechanical properties close to metals. Thus this trio of genius brains be called as father group, that invented electrically active polymers, which are going to revolutionize the electronic

Alan J. Heeger

industry in the coming years. Professor Alan Heeger and his collegues at University
of California, Santa Barbara, have been doing pioneering research in inventing
novel polymer materials that possess electrical and optical properties similar to
conventional metals. The beauty of Prof. Heeger research group is that their
experiments focussing on issues related to transport process in semi conducting
polymer devices, their fundamental electronic structure and light emission properties
have led to the new polymer materials, which can be easily processed and have
equally competing mechanical, optical and electrical properties similar to metals
Prof. Heeger's group can handle studies on LEDs, light-emitting electrochemical
cells (LECs) and lasers, all fabricated from semi conducting and conjugating
polymers. The characterisation of conducting polymers using broad range of
spectroscopic methods (including ultrafast femtosecond, and time resolved)
quantitative measurements of photo and electroluminescene, quantum efficiencies
and photo conductivity etc. is the hall mark Prof. Heeger's research group.

Dr. Heeger was born on 22nd January 1936, at Sioux city. He had his B.S.
with high distinction from University of Nebraska, and his Ph.D. from University
of California, Berkeley. His professional life was started in 1962 at University of
Pennsylvania as assistant professor and decorated the position of Professor at the
same place for the period 1967–1982. He has been a professor of physics since
1982 at University of California, Santa Barbara. He has received several national

and international honors in the form of fellow, prizes, awards and honorary doctorate. He had more than 650 scholarly publications to his credit and had filed more than 40 patents in the area of conducting polymers.

Source: The sketches of Nobel laureates are extracted from the official website of Nobel foundation, Stockholm, Sweden through internet.

Herman Francis Mark

Herman Francis Mark, popularly known as Herman Mark is one of the exceptional personalities in the international scientific/academic world. He had a multi-dimensional (faceted) personality, in which a deeply commited soldier, an honest and serious teacher, an innovative and intuitive researcher, accommodative and commanding administrator and finally a struggling man who had always converted odds against his own as well as his scientific community favor, all these qualities combined into one. Herman Mark was widely known as the father of polymer science and his enormous contributions as a researcher and inspiring teacher/

Herman Francis Mark

lecturer are mainly responsible for asserting the place, polymer science has presently taken in academic and industrial circles.

Another uniqueness of Herman Mark was that he was equally at home, at universities and industrial laboratories and his pioneering work and deeply committed efforts have a great influence in (i) phenomenal growth of polymer industry, (ii) design of curricula for polymer science in academic institutions for establishing polymers as a discrete branch of chemistry and (iii) laying the foundation of polymer research journals and specialized monographs in the field. The credit of being the chief architect of a polymer section of International Union of Pure and Applied chemistry (IUPAC), goes to Herman Mark. The hallmark of Herman Mark's personality was his informal and quite cheerful approach, which earned him the nick name "Geheimrat" among his collaborators and associates.

Herman Mark had to abruptly stop his studies and join Austrian army with the outbreak of the First World War. As a brave soldier, Mark fought from all fronts in the battle field, which won him not only fourteen gallantry awards for his bravery, but also wounded him several times. He was so spirited that, he persuaded his senior army officer to allow him to lead a counter attack on Italian army to recapture the Monte Ortigora, which Austrian army lost earlier. He was allowed to do so and emerged as victor in recapturing the lost mountain peaks. However, he was captured by the Italian army and kept as a prisoner of war in a convent near Bari, Italy. To keep himself away from the boredom of prison life, Herman Mark resumed his study of chemistry, which he had started two years earlier while recuperating from a battle wound. This unending enthusiasm for pursuing chemistry, in otherwise patriotic, and soldier Mark, has blessed the scientific community with one of the brilliant brains that has left indelible mark in the history of science especially polymer science.

Herman Mark worked with Wilhelm Schlenk for his doctoral thesis on synthesis and characterisation of the pentaphenylethyl free radicals. Mark impressed his thesis advisor so much that, Schlenk has invited him to join University of Berlin, where Dr. Schlenk took the chair previously occupied by Emil Fischer. Mark then took up the position as director, Institute for Fiber Research, upon an invitation from Fritz Haber, director of Kaiser Wilhelm Institute (KWI). Mark then unleashed his never ending enthusiasm and conducted research work on the detailed analysis of x-ray crystallographic studies on cellulose fiber as well as zinc wires. Mark and one of his associates Michael Polanyi have proved from x-ray diffraction studies the presence of crystallised and oriented regions along the fiber axis analogous to crystal orientation existed in metallic wires. The fruitfulness of Mark and Polanyi research work can be judged by the fact that, they could establish, (i) crystal structure of hexamethylene tetramine (1923), graphite (1924), (ii) dimerisation of oxalic acid (1924), (iii) birefringence pattern observed in the crystal structure of calomel (1926), (iv) the carbon-oxygen bond length from crystallographic study of carbon dioxide (1925, 26) and (v) structure similarities of ethane and diborane crystals etc.. Thus an originally trained organic chemist has directed his efforts to the problems in physics and physical chemistry. The voluminous x-ray crystallographic data was collected both on simple as well as high molecular weight substances by Mark and his associates has been used to solve the structural problems in polymeric solids even today.

The turning point in the Mark's career as well as to the development of modern polymer science took place in 1926, when Mark had attended the meeting of Gesselschaft Deutscher Naturforscher und Arzte in German city of Dusseldorf. He had delivered a lecture titled "Roentgenographic determination of structure of organic molecules, especially high molecular substances". He reviewed his whole work emphasizing the utility of important information that can be gained from the parameters of unit cells and space groups even if detailed molecular structure is not known. This was in the backdrop of severe skepticism by the scientists about the concept of large molecules, envisaged by Nobel laureate Herman Staudinger. Herman Mark was careful enough in his presentation and then did not commit to the presence of long chains but indicated that the crystallites of cellulose fibers behave like a large molecule.

The authority on the physics aspects of cellulose fibers, which Mark gained in Berlin, has prompted K.H. Meyer, director of I.G. Farbenindustrie (then Germany's largest chemical corporation), to invite Mark as director of the newly established research laboratory on high molecular compounds. Thus Mark moved to Ludwigschafen in 1926. The company, Mark joined was a prominent producer of rayon and cellulose acetate fibers. The conductive environment, Mr. Meyer offered, led Mark to continue his fundamental studies on elucidation of cellulose structure. Taking the clue from Mr. Polanyi's earlier observation that x-ray diffraction studies on cellulose fibers reveal presence of long chains, Mark and Meyer came with a pioneering research paper in 1928, solving the diffraction pattern from cellulose fibers and gave first experimental evidence of presence of a long chemical chain. Thus Mark and Meyer became the discoverers of first polymer crystal structure, that has survived the test of time.

After this, Mark has turned his attention to Hevea rubber (natural rubber) and solved not only its crystal structure but in the process settled for the first time the question of chemical constitution from the spatial point of view. The diffraction studies by Mark and Meyer have conclusively proved that the natural rubber i.c. polyisoprene has a *cis-* configuration around carbon - carbon double bond. Mark has also estimated the energy required to break its covalent bonds and based on his findings could predict the ultimate strength of cellulose fiber. Besides these studies Mark had collected enormous data from electron diffraction patterns of simple molecules such as carbon tetrachloride, benzene, cyclohexane, and *cis-* and *trans-* 1, 2-dichloroethylene. These studies led to the knowledge of bond lengths, bond angles and the angle of rotation. The Mark and Meyer's association was so fruitful that they together wrote a first monograph (in 1930) on "Der Aufbau der hochpolymeren organische Naturstoffe" ("The Structure of High Molecular Organic Natural Substances") and another authoritative book titled "Physik und Chemie der Zellulose (Physics and Chemistry of Cellulose) independently by Mark. With this work, Mark and Meyer had clearly established that the polymer molecules behave like flexible coils due to hindered rotation around the bonds along the polymer backbone. This finding was however severely contested by Herman Staudinger, who insisted that polymer molecules were stiff rods. Not only this, Staudinger bitterly attacked Mark and Meyer, for what he felt a kind of intrusion by the Mark and Meyer duo into his own original thoughts on macromolecules, which were almost rejected by scientists in 1926.

Meanwhile, as fate wished German government was taken over by Hitler's party and the management of I.G. industries had asked Herman Mark to leave the job and country at earliest, because Mark's father Herman Carl Mark was a jew. This brought Mark back to Vienna, Austria, as Professor of Chemistry at University of Vienna. Forgetting the bitter treatment he had received in Germany and keeping the great memories of his achievements in polymers in mind, Mark had straight away embarked on the design of curriculum in polymer science. He started his research afresh on the mechanism of polymerisation and viscosity of polymer solutions. He with E. Guth had developed a statistical theory of elasticity of a rubber molecule, which explained the elasticity shown by cross-linked rubber.

Once again unsettlement struck again, and Hitler's army occupied Austria and Mark was dismissed from professorship because of his close friendship with then Chancellor Dollfuss who was later murdered by Nazis in 1934. Herman Mark then managed through one of his close friends to get out of the jail and along with his wife and two young sons left Austria to Switzerland and proceeded to England enroute to Canada. He revived an early offer from director, International Paper Company, Hawkesbury, Ontario, Canada as a research director for two years concentrating on the improvement of manufacturing of wood pulp, cellulose acetate and viscose. The work on viscose was very important because Du Pont, USA has introduced the viscose fibers in tired cards. This collaboration helped Mark to go to USA, on a proposal by one of the board members of the Polytechnic Institute of Brooklyn, (poly) New York, where he was to work on a Du Pont sponsored consultancy work and also act as a faculty member at Poly. Meanwhile the war in Europe brought many refugees into USA and Mark was assigned to the Shellac bureau for testing and chemical characterisation of Shellac. Since this material was imported from Asia and the war situation demanded a synthetic substitute and Mark grabbed the challenge. His expanded team of associates namely A.V. Tobolsky, P.M. Doty, B.H. Zimm, S. Krimm and R.S. Stein, all of whom had been introduced to polymers by Mark and later became leading names in polymer research.

The intense polymer research activities of Herman Mark's group led to the foundation of Institute of Polymer Research, the first graduate program in USA in 1947. The enthusiastic leadership and enormous scientific contacts of Mark with scientists all over the world created a congenial atmosphere and forum for exchange and cross fertilization of ideas among the scientists of Institute with others. On every Saturday mornings, symposia on subjects related to rapidly growing advances in polymers, were held. Soon by 1944, Mark was able to establish one of the world's outstanding scientific centers in polymers. Mark meanwhile helped in establishing Weizmann Institute in Palestine and similar other centers in Soviet Union.

Herman Mark was not deterred by advancing age and continued to give his annual lectures on "What is new in polymers" till into his eighties. He had travelled the world widely in delivering lectures in Universities and Industries. He was at home at both the institutions. He acted as editor of Journal of Polymer Science and was also consultant to many polymer industries and U.S. Government. It is said that out of about 500 overseas visits by Mark, two of them were highly memorable to him. He was invited to Japan in 1962 to present a lecture to the Japanese Emperor, an honour exclusively and uniquely reserved for Nobel laureates, even through Mark did not belong to this elite club. Thus Mark's efforts in polymer research were honoured befittingly by Japanese scientists. Mark was one of the first two American scientists, who visited China in 1949 after the communist government seized the power.

Herman Mark was honoured by many awards, prizes and medals. The Universities of Leige, Berlin, Uppsala, Vienna, Madrid, Prague and Technion in Haifa conferred honorary doctorates. He was offered memberships of Royal Institute of Great Britain, National Academy of Sciences. He received following medals/prizes; Hertz medal (1928), Nichols medal (1960), Gibbs medal (1975), Humbold award (1978), Wolf prize (1979), Perkin medal (1980) and Michelson-Morley award (1989).

As always, fate wished, Mark's personal life was deeply affected by deaths, he lost his father, he lost his wife Mimi Schramek in 1970 after a long struggle with angina pictoris, his beloved son Peter in 1979. He lived his last two years of a long life with his second son Hans, who was Chancellor of University of Texas. Mark left to heavenly abode on 6th April 1992 thus ending his earthly life, which began on 3rd May 1895.

Stephanie Kwolek

Stephanie Kwolek has invented one of the modern world's most readily recognized and widely used material, Kevlar fiber. We are all aware that polymers are not hundred percent crystalline and hence it is normally expected that polymers with high mechanical strength are rare. But Stephanie Kwolek proved otherwise. Kwolek pioneered low temperature processes for the preparation of condensation polymers. Her determination and love for the work at hand were so great, that she had met with a string of success in the search for new and better polymers. In fact it all happened, when she took up a research position with Du Pont's textile fibers laboratory in Buffalo, New York, after completing her BS in chemistry (1946) from Carnegie Mellon University, Pittsburg. She had a deep interest in science and medicine as a child. She wanted to study medicine, but the weak financial position of her family did not allow her to enter a medical school. She has found her job at Du Pont very interesting and challenging and she could work with all the zeal and enthusiasm. Kwolek has synthesized hundreds of new polymers, including Kapton polyimide film and Nomex aramide polymer and fiber.

Not satisfied with her already history creating achievements, she has carried out her experiments to make stronger and stiffer fibers of tremendous rigidity and strength. In 1960's, her efforts led to the discovery of an entirely new branch of synthetic polymers i.e. liquid crystalline polymers. She had succeeded in

Stephanie Kwolek

synthesizing the first pure monomers to synthesize polybenzamide. But she found that these monomers were ultra-sensitive to the moisture and heat and too easily underwent hydrolysis and self polymerisation. Then she discovered an acceptable solvent and created appropriate low temperature polymerisation conditions for the intermediate monomers.

Not knowing that, she was to make a new discovery in synthesizing a polymer that would change the world, she began her work at the superpolymers lab as usual in one morning in 1964. Under mild and earlier set up polymerisation conditions, Kwolek obtained an aramid polymer, which was in a fluid and cloudy form. No other scientists would have realised that this new polymer could produce fibers with much greater strength and stiffness. But acting on her own instinct, Kwolek asked for spinning out the product and the result was astonishing not only for her, and for her company but for the whole world. The new synthetic fiber produced was very thin but at the same time five times stronger than steel and had superior properties than any other fibers available. So it is obvious that, Du Pont was delighted at Kwolek's discovery and took up the studies for standardizing the conditions for producing the commercial form of Kevlar fiber,

Kevlar fiber, as we all know now, is used to produce bulletproof vests that protect the lives of policemen on duty. It has been used as a shield to protect the important personalities from impending shooting attacks. Kevlar fibers have many other applications including radial tyres, brake pads (replacing asbestos), racing sails, fiber optic cables, water-, air- and spacecraft shells and mooring and suspension bridge cables, skis, safety helmets, hiking and camping gear, suits for fire fighters, cut resistant gloves and tennis rackets etc. Thus Kwolek's invention has ensured the sale of hundreds of millions of dollars per year worldwide.

Kevlar was not the Kwolek's only invention. She had held about 17 patents on synthetic polymers by the time she had retired in 1986. One of the patents was for the feasible spinning methods for the production of commercial aramide fibers. She has been acting as a part time consultant for Du Pont, even after retirement. Keeping all the commercial success aside, Kwolek has received numerous awards including Kilby award and the National Medal of technology. She has been honoured for outstanding achievements in science by inventor's Hall of Fame. In recognition of her own pioneering achievements and for her encouragement of young generation of scientists, Stephanie Kwolek has been honoured in 1999 by Lemelson-MIT lifetime achievement award. In April 1999, San Francisco's exploratium had honoured her at a special ceremony.

Santi Ranjan Palit (1912–1981)

Santi Ranjan Palit is regarded as father of polymer research in India. Santi Ranjan was born in 1912 at Calcutta, had his university education in Calcutta. He had started his professional carrier as a lecturer in Vidyasagar College and became a very successful teacher. During this time he had authored a book "Elementary Physical Chemistry" which was accepted by teachers and students alike. He joined indian Lac Research Institute at Ranchi as a research assistant

Santi Ranjan Palit

and wrote significant publications. He earned a D. Sc. degree from Calcutta University and on invitation from Prof. McBain of Stanford University moved to U.S.A.

He worked on detergents and solubilisation of soaps. Santi Ranjan then joined Prof. Herman Mark at the Polytechnic Institute of Brooklyn, New York, U.S.A. He worked on the solvency of high polymers. His stay with Prof. mark left in him a strong desire to pursue polymer research work back in India. He joined first as planning officer and later as head of department of physical chemistry at Indian Association for Cultivation of Science, Calcutta. He had pursued his work in the physical chemistry of polymers rigorously and published several research papers, monographs and professional books. He had extensively visited several universities and laboratories for delivering subject specific and general lectures.

Santi Ranjan has six children and many grand children.

Leo Baekelund (1863–1944)

Baeklund was Dutch chemistry professor, inventor and enterpreneur. He was a rare personality having qualities of an·intellectual academic, innovative researcher and highly successful entrepreneur, all together. He was in search of a synthetic substitute for shellac, the purified secretions (lac) of the larvae of the insect 'Kerria laccan'. While working in his Bronx, New York, garage laboratory, he came upon a phehol-formaldehyde formulation of multiple properties. The resin once formed was very hard and intractable, could be cut and machined.

It was an excellent electrical insulator, highly resistant to water and other

Leo Baekelund

solvent attacks. Baekelund immediately recognised its potential applications. He refined his process of synthesis and production and maximised its unique properties. He kept the secret of formulation to himself and marketed his invention in 1909 under the trade name "Bakelite", chosen after himself. Thus, the whole credit of development of modern commercial plastics goes to Baekelund. Bakelite however is not the first commercial plastic material and this credit is still with the "celluloid". The extent of enterpreneurship visioned by Baekelund can be gazed by the fact that, Bakelite is still manufactured commercially and used at a level of multimillion dollars per year worldwide.

Wallace Hume Carothers (1896–1937)

Wallace Hume Carothers began his industrial career in 1927 at the DuPont Company. He was instrumental in launching a research program on fundamentals of polymer synthesis. He was able to synthesise a large number of macromolecules with interesting properties. Carothers synthesised the polymers step-wise using known condensation reactions of low molecular weight organic substances. He studied reaction mechanisms involved in their synthesis and the thermodynamics of the polymerisation processes. In fact, Carothers in 1931 gave a classification scheme for the types of polymcrisations. He distinguished between addition polymers and condensation polymers in the following manner. Addition polymers have composition of monomeric units that is identical with that of monomer molecules and addition polymers are formed by addition polymerisation.

Wallace Hume Carothers

Condensation polymers are those which have monomeric repeat units that are not identical with the monomers. These are formed by polycondensation in which small molecules are eliminated.

Carother's conclusions were entirely consistent and were complimentary to Staudinger's work in Germany. By careful selection of reactants or monomers, Carothers obtained polymers with molecules of great length compared to their cross section i.e. fiber forming polymers. After an extensive laboratory trial, Carothers and his research group were able to develop the polymerisation process for the production of first commercially successful synthetic rubber and synthetic fiber nylon. Thus, Carothers is regarded as grand daddy of all modern day engineering plastics. Unfortunately, the brilliant successful career of Wallace Hume Carothers came to a tragic end in the form of a suicide in 1937 at the age of just 41.

References for Further Reading

Seymour, R.B., ed. *History of Polymer Science*, Washington: Am. Chem. Soc., 1982.

Morawetz, H., *Polymers, The Origin and Growth of a Science*, New York: Wiley-Interscience, 1985.

Seymour, R.B., ed., *Pioneers in Polymer Science*, Boston: Kluwer, 1989.

Mark, H.F., *From Small Organic Molecules to Large: A Century of Progress*, Washigton DC: Am. Chem. Soc., 1993.

Appendix C

Questions

Questions and Clues Based on Conceptual Understanding

1. Polyacrylonitrile is non-elastomeric. Why?

   Strong electron withdrawing ability of the cyano group (—CN) produces negative charge on the pendant group with a counter balancing positive charge located on the backbone. This results into dipolar interaction and close packing hence crystallinity.

2. Some polymers can be cross-linked by high energy radiation. Why?

   A polymer when irradiated with high energy radiation, creates radical sites on the chains which form cross-links with other chains (PVC is cross-linked this way, HDPE is difficult to cross-link).

3. Some polymers are transparent while others are opaque. Wholly amorphous polymers may be expected to be transparent. Why?

   Optical transparency requires that the material must not be made (at the microscopic level) of phases of significantly different refractive indexes. Crystalline and amorphous regions in a polymer have different refractive indices. Hence when polymer is wholly amorphous, it is opaque.

4. LDPE is a branched polymer. Why?

   Due to the internal chain transfer, where the growing end of the chain has tendency to abstract a H-atom from a —CH_2—group from the same chain generates free radical where the new chains grow, adding branches (back biting).

5. Thermal decomposition of polymer of formaldehyde is called unzipping. Why?

   The decomposition is reverse of propagation and produces formaldehyde.

6. Why is the termination process not preferred through coupling in cationic polymerization?

   Same charge on macrocations does not allow coupling.

7. Vinyl chloride-vinyl acetate coplymer is more flexible and can be more easily processed than PVC, Why?

   Presence of vinyl acetate units in the chain decreases the regularity in PVC chains (decreases intermolecular forces) thus lowers T_g and consequently produces flexibility making the copolymers easily processable (*internal plasticization*).

8. Isotactic polypropylene(PP) films are less permeable to gas as compared to atactic polypropylene, Why?

   Isotactic PP is more crystalline, possesses high density and low permeability.

9. Why do aramides possesses high melting point?

   Polyamides due to extensive intermolecular H-bonding are highly crystalline.

Aromatic polyamides (aramides) due to the planar benzene ring have even more favorable effect on T_m because the polymer chains become highly rigid.

10. Polybutylene terephthalate is more flexible than polyethylene terephthalate. Why?
Decrease in the magnitude of inter molecular hydrogen bonding in the polyester can be expected for the diol (butylene glycol) as compared to ethylene glycol. Thus, polybutylene terephthalate is more flexible.

11. Ubbelohde capillary viscometer is often preferred over Ostwald viscometer in polymer mol. wt. determination. Why?
In the former, the bottom end of the capillary is above the level of the liquid and so the pressure head is independent of the volume of the liquid. Thus, successive dilutions can be done in the Ubblehode viscometer.

12. Specific heats of polymers are higher as compared to metals. Why?
Polymers are usually poor conductors of heat.

13. Cold cracked rubber does not contract easily. Why?
In cold conditions, the rubber is a glassy polymer without any rubber-like flexibility. Rubbery state is seen only at $T > T_g$.

14. Ethylene is more easily polymerised by free radicals than isobutylene. Why?
Due to absence of electron donating group, ethylene is less polar than isobutylene.

15. Why does solution polymerisation often result to low mol. wt. polymer?
Due to the chaintransfer through solvent.

16. Why does the T_g of a polymer increase in the presence of a filler?
The intermolecular attraction impedes the segmental mobility and thus T_g increases.

17. Benzoyl peroxide is usually stored as solution. Why?
Due to its explosive nature.

18. Zimm plot constructed from light scattering results on polymer solutions provides usuful information about the characterization and solution behaviour-elaborate
Zimm plot is the double extrapolation method to zero angle and zero concentration and provides $\overline{M}_w$. The conc. dependence at zero angle gives second virial coefficient while angle dependence at zero concentration gives particle scattering function which gives radius of gyration.

19. Polymers obtained by condensation polymerization are more crystalline than addition polymers. Why?
This is due to the presence of highly polar functional groups, condensation polymers, e.g. nylon, terylene are highly crystalline and usually do not dissolve below melting temperature.

20. Polymers like polyacrylonitrile, polyacrylates are often prepared by anionic poymerization. Why?
Electron withdrawing (–CN,–COOR) groups reduce double bond electron density and thus favour anionic polymerization.

21. Aqueous solution of polyethylene glycol (PEG) shows decreasing polymer-water interaction with increase in temperature (inverse solubility relationship). Why?
Though several explanations are provided, a simple one could be the disruption of H-bonds between etherial oxygen of PEG and water molecules with increase in temperature.

22. The solubility parameter value (δ) for the solvents n $C_8H_{17}OH$, nC_4H_9OH and CH_3OH show a increasing trend. Why?
n-Octanol ($\delta = 10.3$), *n*-butanol ($\delta = 11.4$) and methanol ($\delta = 14.5$) show the order because the contribution of polar group becomes less significant as the nonpolar chain increases.

23. Kevlar (aramide) is more rigid than nylons. Why?
This is due to the delocalised electrons on the benzene ring that make the ring

rigid. The high electron density in the chains result in stronger intermolecular attraction between the chains.

24. Thiokols (polysulphides) do not make hard rubbers. Why?
Due to not being vulcanized.

25. The melt viscosity increases fast with increase in the mol.wt. of polymer. Why?
The melt viscosity is proportional to the 3.4 power of molecular weight.

26. Though difficult in processing, ultrahigh mol. wt. (UHMW) polyethylene is used in making some durable items.
Very high mol. wt. gives toughness to polyethylene.

27. Why are dust free solutions necessary for the determination of mol. wt. of polymers by light scattering method?
A single dust particle due to its large size would result in very large scattering leading to erroneous result.

28. Ethyl benzene is soluble in methanol but polystyrene is not. Why?
The low entropy of polymer solute reduces its solubility.

29. Natural rubber on heating (masticating) with about 30% sulphur becomes a hard plastic ebonite. Why?
Extensive cross-linking makes polyisoprene behave like an influsible plastic.

30. Ionomers and polyelectrolytes both contain charged pendant groups but behave differently. Why?
Ionomers have low content of charged groups and are usually insoluble in water. Polyelectrolytes are highly charged water soluble polymers.

31. Only proteins can be considered as monodispersed polymers. No synthetic polymer is mono-dispersed. Why?
Proteins are built up of α-amino acids where a fixed number of α-amino acids are arranged in a definite sequence leading to a macromolecule of a definite molecular weight.

32. Methanolic solution of polyvinyl acetate when treated with NaOH precipitates out. Why?
The polymer gets hydrolysed to polyvinyl alcohol which is insoluble in methanol.

33. Addition polymers are normally not crystalline Why?
Due to the presence of asymmetric carbon atom in the chain.

34. π/C is extrapolated to zero concentration in the membrane osmometric method to obtain $\overline{M}_n$. Why?
Even the dilute polymer solutions behave nonideally and therefore the equation $\pi/C = RT/\overline{M}_n$ holds only at infinite dilution.

35. Polystyrene develops yellow colour on uv irradiation. Why?
Abstraction of a hydrogen atom from C atom adjacent to the radical centre is energetically favourable and results in unsaturation in conjugation with the pendant aromatic ring. This leads to yellow colour.

36. Cotton is used as a filler to impart mechanical strength to polymers. Why? Cotton being a fibrous material is quite strong and thus provides mechanical strength to the polymer.

37. Why the threshold molecular weight for polyethylene is higher than that for nylons?
Strong H-bonding in nylon molecules develops mechanical strength in the material at low mol. wt.

38. Polyisobutylene does not show stereo-regularity whereas as polypropylene does. Why?
Polypropylene has asymmetric centres which are absent in polyisobutylene.

39. Carbon tetrachloride and disulphides like RSSH are efficient chain transfer agents. Why?

C—Cl bond in CCl_4 is weak and the resultant free radical gets resonance stabilized. In bisulphides C—S bond is weak and breaks easily.

$$—M^* + RSSH \rightarrow — MSH + RS^*$$

40. Paraffin wax has structure similar to HDPE but lacks in strength. Why?
    The critical chain length for the onset of chain entanglement is not achieved for paraffin wax due to its low mol. wt.

41. The contraction of an expanded rubber band on heating is a spontaneous process. Why?
    Large negative contribution from the entropy factor in the free energy change makes the process spontaneous.

42. The critical chain length required for the onset of chain entanglement for PMMA, PS and PIB (no. of monomers in the chain are 208, 730 and 610, respectively) are different. Why?
    The cirtical chain length depends on the polarity and shape of the polymer chain.

43. Polyethylene terephthalate is stiffer and has higher m. pt. than polyethylene-adipate and polybutylene terpthalate. Why?
    Presence of more methylene groups makes the molecule flexible and the phenyl group makes it stiff.

44. T_g and T_m of polystyrene are higher than HDPE. PVC has still higher T_g and T_m. Why?
    The presence of bulky pendant phenyl groups in polystyrene restricts rotation of chain segments. Polar Cl atoms of PVC show dipole-dipole attraction between chains and thus it has high T_g and T_m.

45. Polyvinyl alcohol crystallises readily but not polyvinyl acetate. Why?
    Ordered polymer molecules with small pendant groups crystallise more readily than those with bulky groups.

46. The refractive index (dn/dc) increment if not measured precisely, causes serious error in $\overline{M}_w$ by light scattering. Why?
    The error in dn/dc is magnified as the light scattering equation uses this term as square $(dn/dc)^2$.

47. Polymer chemistry is considered as a relatively new branch of chemistry. Why?
    Before 1920 when Hermann Staudinger for the first time gave the concept of macromolecules, no one believed that a compound can have molecular weight as high as a few thousands. Hence the development of this branch had really taken place since 1920's.

48. Interpenetrating polymer networks (IPNs) form polymer mixtures. Why?
    IPNs are obtained by polymerising a monomer in the presence of a polymer. This gives rise to a multiphase system similar to a polyblend which may be described as an intimate combination of two polymers.

49. Ionomers are superior to LDPE. Why?
    Polyethylene containing methacrylic acid in the chain (or ethylene - methacrylic acid copolymer with low methacrylic acid content) is an ionomer. Introduction of methacrylic acid decreases the crystallinity and improves the toughness and adhesive action. The salt of the copolymer (ionomer) sets as cross-linked polymer at ordinary temperatures.

50. Small amount of divinylbenzene (DVB) is added in the polymerisation of styrene for use as an ion exchange resin. Why?
    DBV contains two double bonds and thus cross-links polystyrene.

51. Halogen containing polymers should not be burnt. Why?
    The emission of HCl (for example by burning PVC) is harmful to living beings.

52. Secondary cellulose acetate is more widely used than tertiary cellulose acetate. Why?

Due to its solubility in less expensive solvents like acetone.

53. Coatings containing less solvent or those which are water borne are preferred. Why?

To minimise the environmental pollution due to the evaporation of nonaqueous solvents.

54. Polyurethanes and epoxies are good adhesives. Why?

The presence of strong polar groups makes them good adhesives.

55. Phenol-formaldehyde resin (PF), like melamine resin is not used for dinner ware. Why?

Bakelite (PF resin) is dark brown in color and it is difficult to impart attractive colors to it.

56. Plasticised PVC is considered toxic. Why?

Due to the presence of plasticizer and residual monomer.

57. The solubility parameter of aliphatic polar solvents decreases with increase in a homologous series. Why?

Increase in alkyl chain weakens the contribution of polar group.

58. Styrene when copolymerised with maleic anhydride gives alternating copolymer. Why?

The reactivity ratio $r_1(k_{11}/k_{12})$ is 1 while $r_2(k_{22}/k_{21})$ is zero. Thus, a monomer free radical adds to other comonomer.

59. Suspension and emulsion polymerization have no problem of heat dissipation (polymerization is exothermic). Why?

The inert dispersion medium does not allow the viscosity to increase.

60. Head to tail configuration is preferred over head to head configuration. Why?

Due to the steric factors.

61. Block and graft copolymers are very good stabilizers for colloidal dispersions. Why?

Due to the presence of two distinct moeities in the same molecule, these resemble to surfactants (soaps and detergents) and thus adsorb efficiently.

62. Single crystals of some polymers can be obtained from crystallization in dilute solution but not from melt. Why?

If crystallization is achieved from a very dilute solution, single, nearly perfect crystals may be obtained from many polymers. This revolutionary discovery stimulated much research on the behaviour of crystalline polymers. In single crystals the long molecules fold back on themselves every 10 nanometers or so, so that more or less flat platelets are formed, with the molecules lying perpendicular to the plane of the crystal. Crystallization from the melt leads to much less regular structures, for it is very difficult to disentangle long molecules from each other, and a given molecule may participate in several crystalline units. A melt - crystallized polymer typically contains regions that are rather well ordered (crystallites), tied together by uncrystallised segments of molecules.

63. Teflon is an inert polymer. Why?

The most electronegative fluorine atoms provide strong attraction in the molecules thus make it resistant to chemicals.

64. Chain stiffness depends on chemical structure. Why?

Strongly polar groups, such as those containing oxygen, nitrogen, sulfur, and other polar atoms exert the strongest attractive forces. Bulky side groups attached to the main chains stiffen the chains and also by their bulk may prevent a close fit between chains. Long side chains tend to act as internal plasticizers.

65. Light scattering is a highly useful technique. Why?
    The fact that the amount of light scattered by a solution is a function of the molecular weight of the dissolved polymer molecules has provided a means for the determination of weight-average molecular weights up to several million. The light scattering measurement can also yield valuable information regarding the shape of the molecule in solution.

66. In what ways does the reinforcement affect the properties of rubbers and plastics?
    With rubbers, strength and modulus may be increased up to a point by reinforcement or increased cross-linking at the expense of elongation, while with plastics, the modulus is increased but the strength and toughness are often decreased by these factors.

67. Plasticizers make the polymer flexible and rubbery. Why?
    The addition of plasticizers to a polymeric material makes it to be softer and more rubbery in character. Plasticizer molecules are held in association with the polymer chains by secondary valence forces. They separate the segments/molecule of polymer, thus reducing the effective intermolecular attractive forces.

68. *cis*-1, 4 Polydienes make very flexible chains and are rubbery, soft and non-brittle. Why?
    The flexibility of linear polymer chains and of the segments between cross-linked (cured) products is decreased by the presence of polar group and regularity in the molecular structure. A non-polar irregular chain should be the most flexible. Products containing highly flexible chains are rubbery, soft, and non brittle, with relatively high resistance to impact and to tear.

69. Osmometric method for mol. wt. determination is preferred over viscosity and GPC methods. Why?
    Osmometry is an absolute method and does not require calibration. Automated high speed osmometers are commonly used. However, osmometry does not provide any information on polydispersity.

70. Osmometry provides number average mol wt. of polymers. Why?
    Osmotic pressure is a colligative property i.e. depends only on the number of molecules and not on their size or value.

71. The exponent α of Mark-Houwink equation is close to unity for aromatic polyamides. Why?
    The polymer molecules of polyamides behave as rigid rod-like because of the presence of ring structures followed by semi-conjugation along the chain.

72. Above the ceiling temperature, depolymerization is favourable process. Why?
    The ΔG for polymerization is negative as long as ΔH (exothermic, $\Delta H = -ve$) is greater than $T \Delta S (\Delta G > 0)$ at higher temperature and thus the reverse (depolymerization) process becomes favourable.

73. Softening temperature of nylon 6, 6, is higher than that of nylon 6, 10. Why?
    Increase in the methylene groups along the chain (sebacic acid containing 10 carbon atoms making nylon 6, 10 with hexamethylene diamine, and adipic acid containing 6 carbon atoms for nylon 6, 6) weakens H-bonding between molecules and thus nylon 6, 10 has lower softening temperature compared to nylon 6, 6.

74. Chain structure of elastomers differs from that of thermoplastics and thermosetts. Elastomers are usually described as loosely linked network where that intermolecular linkages are few (thermosetts are fully cross-linked network polymers whereas thermoplasts are linear) which allows considerable extension without breaking the chemical cross-linking on applying force (elastomeric deformation).

75. The reduced viscosity versus concentration plot is extrapolated to zero concentration to obtain intrinsic viscosity needed in the mol. wt determination. Why?

Extrapolation eliminates the effect of polymer concentration on viscosity of solution.
76. Under what condition the intrinisic viscosity is considered independent of molecular weight?
When the polymer behaves as a hard sphere. $[\eta] = KM^{\alpha} = K$ for ($\alpha = 0$). This is the case for all globular protein.
77. A useful range of mol. wt. varies from 20,000 to 2,00,000 Why?
Below 20,000 the polymer does not possess enough strength (polymers usually have threshold mol. wt. when their mechanical strength becomes maximum and constant. Very high mol. wt. polymers (>200,000) are not preferred as they possess very high melt viscosity which creates problems in processing.
78. T_g of polyethylene, polypropylene and polystyrene are in the order PE < PP < PS. Why?
Presence of bulky pendant groups hinders segmental motion and thus increases T_g.
79. T_g of polyethylene is low as compared to PVC and PAN Why?
The polar Cl and CN groups in PVC and PAN hinder the segmental motion and thus increase T_g.
80. Polyethylene oxide has low T_g as compared to polyethylene. Why?
The insertion of an oxygen atom between the methylene groups in the polymer backbone increases internal rotation (thus segmental motion) giving a lower value of T_g for PEO.
81. The T_g of polystyrene is less than T_g of poly α-methylstyrene and that of polymethylacrylate is less than PMMA. Why?
Additional methyl group hinders the segmental motion and thus increases T_g.
82. T_g of nylon 6, 6 is about the double to that of polyester of comparable chain. Why?
Strong interaction (H-bonding) is present in nylon 6,6 but absent in polyester, in which only dipole-dipole forces operate. This increases T_g remarkably.
83. Teflon is a linear thermoplast polymer yet it has characteristics of thermosett materials. Why?
Due to the presence of most electronegative F atoms in regular configuration.
84. Natural rubber needs vulcanization. Why?
Linear polyisoprene is tacky polymer with poor strength, poor resistance to solvents and little durability as the chains can slip past. Cross-linking develops strength, makes the rubber hard and easily processable.

Questions Based on Multiple Choice

Set I

1. Hardening of plastics often involves cross-linking. This process is called
   (a) Curing
   (b) Vulcanisation
   (c) Compounding
   (d) Plasticization
2. Which of the following classes belongs to polymer:
   (a) Phospholipids
   (b) Steroids
   (c) Enzymes
   (d) Vitamins
3. In order to obtain a high mol. wt. polymer by polycondensation
   (a) The byproducts should be removed during polycondensation
   (b) Excess of one monomer in a monomer pair should be avoided
   (c) Monomers should be highly pure
   (d) Each of the above is essential

4. The synthesis of polymers of low molecular mass using chain transfer agents is often described as
   (a) Chain polymerisation (b) Step polymerisation
   (c) Emulsion polymerisation (d) Telomerisation

5. The final product of emulsion polymerisation is latex. The polymer can be obtained from this latex by
   (a) Coagulation (b) Non-solvents
   (c) Evaporation (d) Crystallisation

6. The mixing of two polymers yields
   (a) Block copolymer (b) Alternating copolymer
   (c) Polyblend or polymer alloy (c) None of the above

7. Which of the following natural polymers are polysaccharides
   (a) Casein, gelatin, collagen, keratins
   (b) Gums, agar, heparin, alginates
   (c) Urease, α-chymotrypsin, amylase, lipase
   (d) Insulin, oxytocin, lipase, ribonuclease

8. Gutta percha is
   (a) *Cis* 1, 4-polyisoprene (b) *trans* 1, 4-polyisoprene
   (c) Vulcanised polyisoprene (d) Ebonite

9. The polymerisation of a vinyl monomer often predominatly involves the
   (a) Head-to-tail arrangement (b) Head-to-head arrangement
   (c) Both the above (d) Both the above are identical

10. Polyethylene oxide can be called as
    (a) Polyethylene glycol (b) Polyoxirane
    (c) Polyoxyethylene (d) All of the above

11. Which of the following is a polar polymer
    (a) Polyvinyl alcohol (b) Polystyrene
    (c) Polyethylene (d) All of the above

12. Which of the following is a non-polar polymer
    (a) Polyacrylic acid (b) Polyvinyl alcohol
    (c) Polypropylene (d) Polyacrylonitrile

13. Which of the following is a polyelectrolyte
    (a) Polyvinyl sulphate (b) Polyvinyl alcohol
    (c) Polymethyl acrylate (d) Polyacrylonitrile

14. Block copolymer molecule in selective solvent (good solvent for one block, precipitant for the other) associate and thus behave like
    (a) Surfactants (b) Dyes
    (c) Proteins (d) Drugs

15. The concentration of polymers in solution is often expressed as
    (a) g/dl or wt% (b) Molarity
    (c) Molality (d) Vol %

16. Plasticisers do not affect
    (a) Modulus (b) T_g
    (c) Dielectric loss (d) T_m

17. Excess entropy of polymer solutions is
    (a) Zero (b) Highly positive
    (c) Negative (d) Either positive or negative

18. Solution polymerisation in CCl_4 often leads to
    (a) Low mol. wt. polymers (b) High mol. wt. polymers
    (c) Cross-linked polymers (d) Polydispersed polymers

19. Relaxation is not affected by

(a) Tacticity   (b) Crystallinity

(c) Plasticizers   (d) Molecular weight

20. The lower limit of mol. wt. of about 10,000 for membrane osmometry is due to
    (a) The assumption of infinite dilution
    (b) Polymer-polymer interactions
    (c) Leakage through the membrane
    (d) All of the above are true

21. The upper limit of about 5,000 mol. wt. for vapour phase osmometry is due to
    (a) The small effect on the solvent activity for higher mol. wt. polymers
    (b) The low volatility of higher mol. wt. polymers
    (c) Approximations in the theory
    (d) Inefficiencies in heat transfer

22. Which of the following is the initiator for suspension polymerisation
    (a) Potassium persulphate   (b) Benzoyl peroxide
    (c) AIBN   (d) Lewis acids

23. In order to see the fluctuations in scattered light intensity in quasi-elastic light scattering
    (a) A laser must be used as a light source
    (b) Solute concentration must be very small
    (c) The particle size should be less than 1/20 of the wavelength
    (d) The scattering volume must be very small

24. Which of the following is a thermoplastic
    (a) Ebonite   (b) Vulcanised rubber
    (c) Bakelite   (d) HDPE

25. Pyrolysis GC is not generally applicable to polymers
    (a) Thermoset (cross-linked, insoluble)
    (b) Block copolymers
    (c) Terpolymers
    (d) Polymers with heteroatoms (O, N, Cl) in the backbone

26. A "good solvent" is one in which
    (a) The second virial coefficient is negative
    (b) The polymer will always dissolve rapidly
    (c) The solubility parameter of the solvent differs from that of the polymer by less than 30%
    (d) Different polymer molecules do not interact strongly with each other

27. At which temperature will the polymer coil be larger in poor solvent
    (a) At theta temperature   (b) Below theta temperature
    (c) Above theta temperature   (d) None of the above

28. Which of the following techniques yields a number-average molecular weight
    (a) Viscometry   (d) Osmometry
    (c) Light scattering   (d) Ultracentrifugation

29. Which of the following is a branched chain polymer
    (a) HDPE   (b) Isotatic polypropylene
    (c) LDPE   (d) Amylose (starch)

30. In copolymerisation, one does not usually oserve that the product of the reactivity ratios is very large ($r_1/r_2 \gg 1$). In this case,
    (a) no polymerisation takes palce
    (b) A perfectly alternating copolymer is formed
    (c) k_{11} is small compared to k_{22}
    (d) No copolymer is formed

31. Which one of the following is not a requirement for using a wet chemical (titration) method to determine the amount of some reactive group in a polymer

(a) The reaction should be rapid

(b) The reagent needs to be a small molecule so it can have access to buried groups in the interior of the polymer coil in solution.

(c) The reaction should be quantitative

(d) It is necessary to be able to monitor either the product formed or the reagent consumed.

32. The T_g of a polymer can be determined by
    (a) IR   (b) NMR
    (c) Dilatometry   (d) TGA

33. Which would be the better solvent for polystyrene
    (a) *n*-pentane   (b) benzene
    (c) acetonitrile   (d) methanol

34. In which of the following polymers will hydrogen bonding predominate
    (a) Natural rubber   (b) HDPE
    (c) Cellulose   (d) Cellulose nitrate

35. Which of the following is a monodisperse polymer
    (a) Natural rubber   (b) Cellulose
    (c) Casein   (d) Nylon

36. Which of the following is thermosett or cross-linked polymer
    (a) Cellulose   (b) Unvulcanised rubber
    (c) Cellulose nitrate   (d) Bakelite

37. Which one of the following polymers can be prepared by anionic, cationic and free radical polymerization?
    (a) Polyethylene oxide   (b) Polyvinyl chloride
    (c) Poly(vinyl methyl ether)   (d) Polystyrene

38. Which of the following will yield a cross-linked polyester when condensed with adipic acid
    (a) Ethyl alcohol   (b) Ethylene glycol
    (c) Glycerol   (d) Hexamethylene diamine

39. Tetrafluoroethylene is a monomer for
    (a) Neoprene   (b) Butyl rubber
    (c) Teflon   (d) Glyptals

40. The intermolecular forces operating only in crystalline polymers like terylene, nylons, cellulose are
    (a) H-bonding   (b) Dipole-dipole interaction
    (c) Hydrophobic interaction (d) London dispersion forces

41. End group analysis gives
    (a) $\overline{M}_n$ (b) $\overline{M}_w$
    (c) $\overline{M}_v$ (d) $\overline{M}_z$

42. Which of the following is thermoplastic
    (a) Terylene   (b) Nylon
    (c) Polypropylene   (d) All of the above

43. Wool and silk are natural polymers. These are basically
    (a) Proteins   (b) Polysaccharides
    (c) Polyesters   (d) Polyethers

44. Vulcanisation of natural rubber by sulphur makes it
    (a) Water soluble   (b) Soft
    (c) Hard   (d) Less elastic

45. The polymer Buna N is a copolymer of
    (a) Butadiene and acrylonitrile   (b) Butadiene and styrene
    (c) Butadiene and ethylene   (d) Butadiene and isoprene

46. Which of the following will not dissolve in a solvent or melt on heating
    (a) PVC   (b) Teflon
    (c) Bakelite   (d) PMMA
47. HDPE differs from LDPE in
    (a) Degree of crystallinity   (b) Molecular weight
    (c) Melting temperature   (d) All of the above
48. Camphor, dibutyl phthalate, tricresyl phosphate are all examples of
    (a) Antioxidants   (b) Plasticizers
    (c) Curing agents   (d) UV stabilizers
49. The glass transition is a second order thermal transition in amorphous polymers due to
    (a) Segmental motion   (b) Molecular motion
    (c) Both   (d) None of the above
50. Nylon 6 is obtained from
    (a) Caprolactam
    (b) Hexamethylene diamine and adipic acid
    (c) Amino acids
    (d) Chloroprene

Set II

1. Terylene is a
   (a) Polyester   (b) Polyamide
   (c) Vinyl polymer   (d) Acrylic polymer
2. Polydispersity index is often expressed as the ratio
   (a) T_g/T_m (b) $\overline{M}_w/\overline{M}_n$
   (c) Both   (d) None of the above
3. No synthetic polymer can be monodispersed. Nature's monodispered polymer is
   (a) Starch   (b) Cellulose
   (c) Insulin   (d) Natural rubber
4. GPC is a chromatographic technique which separates polymer molecules in a sample according to their size and provides
   (a) Fractionation   (b) Mol. wt. distribution
   (c) Molecular weight   (d) All of the above
5. Which of the following thermal methods cannot be used to determine T_g
   (a) TGA   (b) DTA
   (c) TMA   (d) DSC
6. Living anionic polymerisation can be conveniently used to produce
   (a) Block copolymers   (b) Homopolymers
   (c) Thermosetts   (d) Thermoplasts
7. Separation of different low mol. wt. (MWD) distribution samples from a highly polydispersed polymer is called fractionation which can be done by
   (a) Partial dissolution method   (b) Partial precipitation method
   (c) GPC   (d) All of the above
8. Which of the following polymers are often highly crystalline
   (a) Fibres   (b) Plastics
   (c) Elastomers   (d) Surface coating materials
9. At temperature above T_g but below T_m, a polymer may be considered as a
   (a) Crystalline solid   (b) Super cooled crystal
   (c) Glassy solid   (d) None of the above
10. Which of the following can be used for the synthesis of addition polymers
    (a) Free radical polymerisation (b) Ionic polymerisation
    (c) Coordination polymerisation (d) All of the above

11. Nylon 6, 6 can be conveniently produced directly as yarn by
    (a) Melt polycondensation
    (b) Solution polycondensation
    (c) Interfacial polycondensation
    (d) All of the above

12. Addition polymerisation involves the breaking of a π bond and formation of 2-σ bonds and is a/an
    (a) Exothermic process
    (b) Endothermic process
    (c) Reversible process
    (d) None of the above

13. Polycondensation often involves the elimination of small molecules like H_2, CH_3OH, HCl gas, etc. and these are removed to prevent reversible reaction by
    (a) Distillation
    (b) Fractionation
    (c) Crystallization
    (d) Sublimation

14. The imperfect crystalline regions made from bundles/aggregates of ordered chain are called
    (a) Spherulites
    (b) Crystallites
    (c) Single crystals
    (d) Dendrites

15. The value of α in Mark-Houwink equation $[\eta] = KM^{\alpha}$ in theta solvent is
    (a) 1.0
    (b) 0
    (c) 0.5
    (d) 0.8

16. The value of second varial coefficent (A_2) in the equation $\pi/C = [RT/M] (1 + A_2C)$ for polymer in θ-solvent is
    (a) 0
    (b) 1.0
    (c) 0.5
    (d) None of the above

17. Which of the following most known polymer scientists did not receive Nobel Prize
    (a) Hermann Staudinger
    (b) Karl Ziegler and Giulio Natta
    (c) Herman Mark
    (d) Paul Flory

18. Bakelite was the 1st synthetic polymer made by
    (a) Henry Baekelund
    (b) Charles Goodyear
    (c) William Carothers
    (d) Hyatt

19. At any time when the chain polymerisation progresses, the reaction vessel contains
    (a) Only monomer
    (b) monomer + polymer
    (c) Monomer + dimer + trimer + tetramer, etc.
    (d) All of the above

20. Polycondensation involving step-wise addition of monomer/monomer pair is also called
    (a) Step polymerisation
    (b) Condensation polymerisation
    (c) Step-addition polymerisation
    (d) All of the above

21. Which is not true for polymers
    (a) A small amount of dissolved polymer increases the viscosity of liquid
    (b) Polymer dissolution is a slow process involving swelling
    (c) Polymers do not show sharp m.pt. and liquid polymers cannot be changed into gaseous form on heaing
    (d) Polymers sublime on heating

22. Which of the following additives is added during the polymerisation
    (a) Plasticizers
    (b) Antioxidants
    (c) Thermal stabilizers
    (d) Chain transfer agents

23. What is not true for thermosetts
    (a) They are cross-linked polymers
    (b) They do not melt/soften on heating
    (c) They do not dissolve in any liquid
    (d) They do not have very high molecular mass

24. The property of a homopolymer can be improved by
    (a) Copolymerisation with some other monomers
    (b) Forming polymer alloys
    (c) Adding plasticizers
    (d) All of the above
25. Polymer solutions can be considered colloidal solutions in that they show
    (a) Electrophoresis   (b) Light scattering
    (c) Coagulation   (d) Peptisation
26. Polymer solutions can be grouped into
    (a) Association colloids   (b) Molecular colloids
    (c) Lyophobic colloids   (d) All of the above
27. Starch, cellulose, glycogen are all the polymers of
    (a) Glucose   (b) Sucrose
    (c) Glucose + Fructose   (d) None of the above
28. All enzymes are made of
    (a) Proteins   (b) Proteins and Polysaccharides
    (c) Nucleic acids   (d) None of the above
29. DP = 1/1–p is the equation for polycondensation that relates degree of polymerisation (DP) to the extent of reaction, p. It is known as
    (a) Carother's equation   (b) Huggin's equation
    (c) Mark-Houwink equation   (d) Copolymerisation equation
30. In polycondensation, the extent of conversion, p, approaches unity, the heterogenity index $(\overline{M}_w/\overline{M}_n)$ in the polymer is
    (a) 1   (b) 2
    (c) 100   (d) 0
31. The high increase in viscosity of the monomer as the polymerisation proceeds is often described as
    (a) Autoacceleration   (b) Trommsdorff effect
    (c) Gel effect   (d) All of the above are same
32. The kinetics of addition polymerisation shows molecular weight as
    (a) Proportional to monomer concentration
    (b) Inversely proportional to (Initiator conc.)
    (c) Both
    (d) None of the above
33. 1, 4 Polyisoprene shows
    (a) *Cis-trans* isomerism   (b) Tacticity
    (c) Both of the above   (d) None of the above
34. Polypropylene shows
    (a) *Cis-trans* isomerism   (b) Stereoisomerism
    (c) Structural isomerism   (d) Functional isomerism
35. Polystyrene and poly (p-xylene) are
    (a) *Cis-trans* isomers   (b) Structural isomers
    (c) Stereoisomers   (d) Functional isomers
36. At least two functional groups are required in the monomer/monomer pair. The functionality of phenol—a monomer for Bakelite is
    (a) 1   (b) 2
    (c) 3   (d) 4
37. The coil dimension $(\overline{r}^2)^{1/2}$ of a polymer molecule is related to the unperturbed dimension $(\overline{r}_0^2)^{1/2}$ by $(\overline{r}^2)^{1/2} = \alpha(\overline{r}_0^2)^{1/2}$. Under θ-condition the value of chain expansion factor, α, is

(a) 0   (b) 0.5

(c) 1.0   (d) 2.0

38. In good solvents where solvent molecules processes an affinity for polymer molecule, the polymer coil
    (a) Expands   (b) Contracts
    (c) Expands or contracts   (d) Does not change size

39. The Hildebrand's solubility parameter is equal to cohesive energy density and has the unit
    (a) $(cal/cm^3)^{1/2}$ (b) cal/cm^3
    (c) J/m^3 (d) None of the above

40. For the dissolution of amorphous polymers in non-polar solvents, solubility parameter values for solvent and polymers should be
    (a) Nearly equal   (b) Zero
    (c) Largely different   (d) None of the above is true

41. What is true for membrane osmometry for the determination of mol. wt of polymers?
    (a) It gives $\overline{M}_n$
    (b) Dynamic osmometers provide quick determination of osmotic pressure
    (c) The solvent should not be a very good solvent
    (d) All are correct

42. Which of the following method only will give reliable estimate of $\overline{M}_n$ of a polymer with mol. wt. >1,00,000?
    (a) Ebbluiometry   (b) Cryoscopy
    (c) End of group analysis   (d) Membrane osmometry

43. The determination of $\overline{M}_w$ from light scattering involves a double extrapolation on the same graph. This grid like figure is called
    (a) Zimm plot   (b) Chromatogram
    (c) Turbidity plot   (d) Mol.wt. distribution curve

44. Ultracentrifuge utilizes the determination of mol. wt. through the measurement of sedimentation constant and diffusion coefficient utilising equation called
    (a) Carother's equation
    (b) Huggins equation
    (c) Mark-Houwink-Sakurada equation
    (d) Svedberg equation

45. In GPC, the stationary and mobile phases are
    (a) Solid and liquid
    (b) Liquid and liquid
    (c) The single liquid acts as both the phases
    (d) Solid and gas

46. A thin film of melt crystallized polymer on viewing in optical microscope between crossed polars, show characteristic structure. These are called
    (a) Crystallites   (b) Spherulites
    (c) Single crystals   (d) Dendrimers

47. DTA can be utilized in polymer research for the determination of
    (a) T_g/T_m (b) Degree of crystallinity
    (c) T_c (d) All of the above

48. A polymer with high mol. wt. and high degree of crystallinity would behave as
    (a) Soft and waxy   (b) Hard and flexible
    (c) Hard and tough   (d) Soft and flexible

49. The reactivity ratios of monomer 1 and monomer 2 are r_1 and r_2. If $r_1 = r_2 = 0$
    (a) A block copolymer will form
    (b) An alternate copolymer will form

(c) A random copolymer will form

(d) Two homopolymers will form

50. Termination under control condition sometimes does not occur in:

(a) Anionic polymerisation

(b) Cationic polymerisation

(c) Free radical polymerisation

(d) Coordination polymerisation

Set III

1. Nylon 6, Nylon 6,6 and Proteins all are

(a) Polyamides

(b) Polyethers

(c) Polyesters

(d) Polyolefins

2. Natural rubber is a

(a) *Cis* 1, 4 Polyisoprene

(b) *Trns* 1,4 Polyisoprene

(c) 1, 2 Polyisoprene

(d) 1, 3 Polyisoprene

3. Which of the natural polymer is not a condensation polymer

(a) Nucleic acids

(b) Polysaccharides

(c) Proteins

(d) Natural rubber

4. Which of the following is not a polysaccharide

(a) Starch

(b) Cellulose

(c) Heparin

(d) Amylase

5. Which is considered as a first synthetic polymer

(a) Bakelite

(b) Teflon

(c) Neoprene

(d) Nylon 6

6. Which of the following polymers possesses elastomeric properties

(a) Thiokol

(b) Plypropylene

(c) Polymethyl methacrylate

(d) Polyvinylidene chloride

7. Which of the following is not a fiber

(a) Polypropylene

(b) Polyethylene terephthalate

(c) polyacrylonitrile

(d) Polyurethane

8. Which of the following is not a cellulose derivative

(a) Cellophane

(b) Celluloid

(c) Gum cotton

(d) Cellobiose

9. Which of the following is not a copolymer

(a) SBR

(b) Butyl rubber

(c) Saran

(d) Teflon

10. The concept of macromolecules was propounded by the Nobel laureate

(a) Staudinger

(b) Natta

(c) Flory

(d) Sanger

11. Solid phase synthesis of polymer was first demonstrated by the Nobel laureate

(a) Ziegler

(b) Merrifield

(c) Herman Mark

(d) Paul Flory

12. Bisphenol-A is a monomer used in the preparation of

(a) Polycarbonates and epoxy resins

(b) PF and UF resins

(c) Polyurethanes

(d) Alkyd resins

13. Which of the following is not an initiator for free radical addition polymerisation

(a) Benzoyl peroxide

(b) Azobis isobutyronitrile

(c) Lewis acids

(d) Persulphates

14. The vulcanization process, i.e. mastication of rubber with sulphur to make it useful is due to:

(a) Carothers

(b) Goodyear

(c) Schonbein

(d) Hyatt

15. Which of the following is a thermosett
    (a) UF polymer
    (b) Teflon
    (c) PMMA
    (d) SBR
16. Plasticisers are added to polymers to decrease its
    (a) T_g
    (b) T_m
    (c) Solubility
    (d) Crystallinity
17. The T_g of a polymer can be determined by
    (a) Dilatometry
    (b) DTA/DSC
    (c) TMA
    (d) both (a) and (b)
18. Light scattering provides molecular weight, which is
    (a) Weight average
    (b) Number average
    (c) Z-average
    (d) Viscosity average
19. Which of the following techniques is not based on colligative properties
    (a) Ebbuleometry and Cryoscopy
    (b) Vapour pressure osmometry
    (c) Membrane osmometry
    (d) Viscometry
20. The light scattering intensity and sedimentation velocity of particles depends on their
    (a) Sizes
    (b) Number
    (c) Colour
    (d) None of the above
21. Which of the following would you expect to be a crystalline polymer
    (a) Polybutadiene
    (b) SBR
    (c) Nylon 6, 6
    (d) LDPE
22. Which of the following will have highest value of T_g
    (a) Polyethylene
    (b) Polypropylene
    (c) Polystyrene
    (d) Polymethyl styrene
23. Which of the following polymers is difficult to crystallize
    (a) Polyesters
    (b) Polyethylene
    (c) Polypropylene
    (d) Polyvinyl carbazole
24. Carbon fibre is obtained by the cyclisation followed by dehydrogenation of polyacrylonitrile. It is a
    (a) Ladder like polymer
    (b) Star like polymer
    (c) Comb like polymer
    (d) Branched polymer
25. The monomer added in small amount in the polymerisation of styrene to cross-link it for use as an ion exchange resin or as column material for GPC is
    (a) Divinylbenzene
    (b) Butadiene
    (c) Propylene
    (d) None of the above
26. In emulsion polymerisation, the initiator is
    (a) Soluble in water
    (b) Soluble in monomer
    (c) Insoluble in both
    (d) Soluble in both
27. Which of the following polymerisation techniques offers problem of heat dissipation
    (a) Solution polymerisation
    (b) Bulk polymerisation
    (c) Suspension polymerisation
    (d) Emulsion polymerisation
28. Which of the following is not polymerised alone but is copolymerised with styrene to yield alternating copolymer
    (a) Butadiene
    (b) Maleic anhydride
    (c) Acrylonitrile
    (d) Vinyl acetate
29. Polyvinyl alcohol is obtained by the hydrolysis of polyvinyl acetate but not by the polymerisation of vinyl alcohol monomer because the monomer is
    (a) Difficult to purify
    (b) Converts into acetaldehyde
    (c) Difficult to polymerise
    (d) Water soluble
30. The T_g of polymers behaving like rubber should be

(a) Less than room temperature   (b) Higher than room temperature
(c) Equal to room temperature   (d) Any of the above

31. Which of the folowing abbreviating forms is a condensation polymer
    (a) PAN   (b) PMMA
    (c) SBR   (d) PET
32. Which of the following materials is used in the production of terylene
    (a) DMT   (b) VAM
    (c) Bisphenol   (d) Glycerol
33. The polycondensation of a monomer pair each having bifunctionality always gives
    (a) Linear polymer   (b) Branched polymer
    (c) Cross-linked polymer   (d) Any of the above
34. Which of the following is not a thermosett
    (a) Glyptals   (b) Bakelite
    (c) Epoxies   (d) Butyl rubber
35. The most inert polymer, used in non-sticking kitchen ware is
    (a) Teflon   (b) Melamine resin
    (c) PVC   (d) PMMA
36. The monomers in a polymer molecule are joined through
    (a) H-bond   (b) Covalent bonds
    (c) Electrovalent bonds   (d) Dipole-dipole interaction
37. The polymer used in making buckets, mugs, storage tanks, TV cabinets, etc. is
    (a) HDPE   (b) Polypropylene
    (c) PVC   (d) Polystyrene

Answers

Set I

1. (a)	2. (c)	3. (a)	4. (d)	5. (a)
6. (c)	7. (b)	8. (b)	9. (a)	10. (d)
11. (a)	12. (c)	13. (a)	14. (a)	15. (a)
16. (d)	17. (a)	18. (a)	19. (c)	20. (c)
21. (a)	22. (c)	23. (a)	24. (d)	25. (a)
26. (c)	27. (c)	28. (b)	29. (c)	30. (c)
31. (c)	32. (c)	33. (d)	34. (c)	35. (c)
36. (d)	37. (d)	38. (c)	39. (c)	40. (b)
41. (a)	42. (d)	43. (a)	44. (c)	45. (a)
46. (c)	47. (a)	48. (b)	49. (a)	50. (a)

Set II

1. (a)	2. (b)	3. (c)	4. (b)	5. (d)
6. (a)	7. (d)	8. (a)	9. (d)	10. (d)
11. (a)	12. (a)	13. (a)	14. (a)	15. (c)
16. (a)	17. (c)	18. (a)	19. (b)	20. (d)
21. (d)	22. (d)	23. (d)	24. (d)	25. (b)
26. (b)	27. (a)	28. (a)	29. (a)	30. (b)
31. (d)	32. (a)	33. (c)	34. (b)	35. (b)
36. (c)	37. (c)	38. (a)	39. (a)	40. (a)
41. (d)	42. (d)	43. (a)	44. (d)	45. (c)
46. (b)	47. (d)	48. (a)	49. (a)	50. (a)

Set III

1. (a)	2. (a)	3. (d)	4. (c)	5. (a)
6. (a)	7. (b)	8. (d)	9. (d)	10. (a)
11. (b)	12. (a)	13. (c)	14. (b)	15. (a)
16. (a)	17. (d)	18. (a)	19. (b)	20. (a)
21. (c)	22. (d)	23. (d)	24. (a)	25. (a)
26. (a)	27. (b)	28. (b)	29. (b)	30. (a)
31. (d)	32. (a)	33. (a)	34. (d)	35. (a)
36. (b)	37. (b)			

Fill in the blanks

Set I

1. The term 'polymer' was first used by the Swede _____
2. The structure that repeats in a polymer molecule is called _____
3. Network copolymers require one of the monomers with functionality _____
4. Intrinsic viscosity is also known as _____
5. Membrane osmometry, unlike vapour pressure osmometry, viscosity and GPC, does not require _____
6. A thermoplastic with regular chain amenable to crystallization on stretching to a temperature between T_g and T_m, improves mechanical property. This process is called _____
7. A highly regular type of graft polymer where the grafts are separated by uniform lengths in the backbone can be called _____ polymer
8. Liquid crystals show _____ behaviour
9. Poly (phenylene terephthalamide) or Kevlar shows _____
10. H-bonding raises _____ of polymers due to the hindered segmental rotation
11. For several common polymers T_g is approximately equal to _____ of T_m
12. In solution polymerization the solvent simply acts as _____ for the dissolved species like monomer and initiator etc.
13. _____ equation predicts the polycondensation reaction
14. _____ is a linear polymer of β-glucose with crystalline structure
15. A glassy polymer on increasing the temperature changes to a soft, flexible, rubbery mass at _____
16. The first synthetic polymer made by Henry Baekelund is _____
17. Depending upon the nature of the α-amino acids, a _____ molecule can be globular (water soluble) or fibrillar (water insoluble)
18. Polysaccharides, polyamides, polynucleotides and polyisoprene can be grouped as _____
19. Polyacrylonitrile on heating followed by dehydrogenation forms a ladder like polymer called _____
20. _____ polymers can be obtained by polymerising a monomer in the presence of a polymer
21. _____ polymers have structural resemblance to surfactants and thus form micelles and adsorb from solution
22. _____ can be conveniently used to determine $\overline{M}_n$ of polymers with fairly high value of molecular weight
23. _____ is used as a coating material in cooking frying pans due to its inert nature
24. Nature can produce mono-dispersed polymers (no perfectly monodispersed

polymers are made in the laboratory). The Nature's mono-dispersed polymers are
_____

25. *Cis*-1, 4 polyisoprene is a gummy polymer obtained from the latex of rubber trees. It could be used commercially only after the discovery of vulcanisation process first used by _____

26. Based on the cohesive energy density that measures intermolecular attraction, Hilderbrand introduced _____ that helps in predicting solubility of a polymer in a solvent

27. Styrene-maleic anhydride copolymer has charged carboxylic groups and can be called _____

28. _____ can be made simply by physical mixing of two homopolymers and sometimes possess better properties than the constituting homopolymers

29. At any time during the course of polymer formation, the reaction vessel contains only monomer and polymer in _____ process

30. _____ are low mol. wt. and low volatile materials that often decrease glass transition temperature of polymers and thus facilitate in processing and give desired properties to polymer

31. _____ anionic polymerization process is very suitable method for obtaining block copolymers with controlled mol. wt. and molecular composition/structure

32. The value of $\alpha = 0.5$ (in the equation $[\eta] = KM^{\alpha}$) and second virial coefficient A_2 = 0 (in the equation $\pi/C = [RT/\overline{M}_n](1 + A_2C)$ for any polymer in _____ solvents

33. The _____ of the monomers for a particular monomer pair are important in predicting the composition of a copolymer

34. An _____ copolymer is one when the concentration of the monomers in the feed (in the polymerization vessel) is same as that in the copolymer formed

35. _____ was the first to give the concept of macromolecules in 1920 (much criticized then by the scientists) and was awarded Nobel Chemistry Prize in 1953

36. The polydispersity index of a polymer sample is often expressed as the ratio _____ where it is one for a monodispersed polymer

37. Stereoregular polypropylene is a highly _____ and can be used as fibre

38. The polyethylene produced from the polymerisation of ethene by Ziegler's catalyst at mild temperature/pressure conditions gives _____, a polymer with high degree of crystallinity

39. Poly (sodium acrylate) is an example of a _____ which dissociates in water to give a macroanion and sodium ions

40. Polystyrene (cross-linked) made by copolymerisation of styrene with small amount of divinylbenzene on sulphonation gives a polymer which is highly useful as _____ in water treatment

41. Polycondensation of two pure reactants each with a bifunctionality taken in exactly equimolar ratio may lead to an extremely _____ polymer

42. A polystyrene ball thrown on the floor will shatter like glass into pieces but the same ball when heated to 100°C prior to throwing will show _____

43. The common polymers like PE, PP, PS, PVC, terylene, nylon, can be recycled from waste by _____ polymers are not reusable

44. The dissolution of a polymer first involves swelling. A polymer showing _____ swelling in a solvent will not form solution

45. A polymer will be a useful rubber if its T_g is less than room temperature. However, for excellent elasticity the polymer is desired to have a few _____

46. Polyethylene oxide, polyvinyl alcohol, poly (sodium acrylate), polystyrene sulphonate, polyvinyl methyl ether are the examples of _____ polymers

47. Maleic acid does not polymerize by itself although it contains unsaturation as well as two functional groups. However, styrene and maleic anhydride can be copolymerised to give an _____ copolymer

48. Chain transfer agents are often added during the polymerisation if low mol. wt. polymer is desired. The chain transfer can by itself take place in _____ polymerisation technique

49. T_g of a polymer can be decreased by _____ plasticization (without adding any plasticizer) using copolymerization

50. _____ is a quick, reliable and efficient technique for the fractionation of a polydispersed polymer as well as for molecular weight distribution analysis

51. The hereditary material DNA, is a polynucleotide containing long repeating units of base-sugar-phosphate. The DNA molecule exists as a double helix, this understanding about the structure of DNA is due to investigations by _____

52. Glyptals, epoxies, alkyds, formaldehyde resins can all be grouped together as _____

53. PVC, nylons, polyesters, cellulose, PAN can all be considered having strong inter molecular forces broadly known as _____

54. In polycondensation, the molecular weight builds up slowly (unlike chain polymerisation) but the reaction goes to completion yielding high mol. wt. polymer, if the both the bifunctional reactants are present in equal amount. The DP can be controled by using a _____

55. Benzoyl peroxide, AIBN, persulphates and redox systems e.g. ferrous ions and H_2O_2 generate _____ which initiate the polymerisation of unsaturated monomers

56. The important condition for a low mol. wt. substance to form a polymer is that it must contain atleast bifunctionality. This may arise from the presence of functional groups or _____

57. Polymerisation is an exothermic process. Bulk polymerisation technique often has the problem of heat dissipation. No such problem is present in solution polymerisation, but chain transfer by solvent gives, low mol. wt. polymer. _____ and emulsion polymerisation are thus more advantageous techniques for polymerisation

58. Polycondensation can be done by melt, solution and _____ techniques. The latter technique is quite useful for making nylon 6, 6 directly in the form of continuous filament

59. Polybutadiene and polyisoprene are not useful as rubbers but their cross-linked products like vulcanised rubber or _____ like butyl rubber, SBR, NBR are highly commercialized elastomers

60. A chromatographic technique useful in the identification and detection of monomers, polymer additives and polymers is called _____

61. Most monomers and polymers may be qualitatively and quantitatively identified by _____

62. A technique commonly employed to obtain information on the crystal structure of polymers is _____

63. When two monomers are polymerised together, the process is called _____

64. Dextran, dextrin, heparin, starch, cellulose, glycogen are all biopolymers from the class _____

65. Two polymers or their solutions when mixed and when noncompatible exhibit _____

66. _____ equation named in honour of the ultracentrifuge inventor and is used in the determination of mol. wt. of polymers by this instrument

67. The double chain heat resistant polymers are called _____

68. _____ is used in the preparation of epoxy resins and polycarbonates
69. Certain polymers particularly thermoplastics in the glassy state undergo a localized deformation called _____
70. Trommsdorff-Norris effect or gel effect is also known as _____
71. The copolymers of ethylene and methacrylic acid are often known by their generic name _____
72. An allotropic form of carbon other than diamond and graphite is called _____
73. Polyvinyl acetate on _____ gives PVA which is a useful adhesive
74. Crystalline polymers are often dissolved in solvents above their _____
75. Polymers with filler materials, particularly fibres with improved strength are called _____ plastics
76. An additive that readily undergoes chain transfer with a macroradical is called _____
77. Cross-linked polymers do not form solution and can undergo only _____
78. Polymers prepared by polycondensation of phenol with formaldehyde under acidic conditions are called _____
79. A measure of mechanical strength of materials expressed as the ratio of stress to strain is called _____
80. A thermal analytical method that is often used for the determination of T_g of polymers is _____
81. Degradation of polymers from chain ends successively is called _____
82. Jute, wood flour, carbon black, glass, talc, sand and silica products are polymer additives called _____
83. Natural polymeric biocatalysts which possess high substrate specificity are called _____
84. A kind of termination step in addition polymerization other than coupling is _____
85. _____ polypropylene has all the pendant methyl groups on one side of the chain
86. Latex from the rubber tree is _____ to extract natural rubber
87. Polymer formation takes place within the _____ micelles in emulsion polymerisation
88. Pyrolysed polyacrylonitrile is used as a reagent for _____ reactions in organic synthesis
89. The OH stretching vibrations can be seen in the IR spectrum of PVA at _____
90. Polymer microstructures can be examined by _____
91. Formaldehyde resins (e.g. PF, UF) are processed by _____
92. Colloidal dispersions of solid polymer in a dispersion medium are called _____
93. Epoxies, alkyd resins, PVA, Polycyano-acrylates are all examples of _____
94. Polyamides, aromatic polyamide, carbon fibres are examples of _____ polymers
95. Number average molecular weight of a polymer can be conveniently determined by _____
96. Low molecular weight polymers with end functional groups can be characterised for mol. wt. by _____
97. Polymer spherulites can best be examined by _____
98. Water soluble charged polymers are called _____
99. A polymerisation technique that involves surfactant micelles in a heterogeneous mixutre is _____
100. A minimum requirement for a monomer to form a polymer is the _____ in its structure

Set II

1. Linear polyethylene made using Ziegler's catalyst is often abbreviated as _____
2. The formation of gel often leading to explosion in the bulk polymerisation of unsaturated monomers like acrylonitrile is called _____
3. The first synthetic polymer made by Henry Backeland is _____
4. The useful elastomeric material which is chiefly polyisoprene (with little isobutylene as a comonomer) is called _____
5. In solution polymerisation, solvent often acts as _____ leading to low mol. wt. polymer
6. _____ are often added to make the polymer soft and flexible which interact with polymer and weaken intermacromolecular attraction and thus decrease T_g
7. Emulsion polymerisation often results into _____ from which polymer is not separated by is used as such, e.g. in emulsion paints
8. On cooling a molten polymer, its _____ viscosity shoots up
9. The milk protein that has been used to make plastics in olden days is _____
10. Polyacrylonitrile when cyclised followed by dehydrogenation forms _____
11. Polymerisation of a monomer in a medium consisting of a polymer leads to a multiphase system called _____
12. _____ can be determined experimentally using any colligative property like osmotic pressure, lowering in vapour pressure, etc.
13. _____ are the monomers in abbreviated forms needed for producing polyester terylene
14. Natural rubber creeps of its own weight but when cross-linked with sulphur (a process invented by Charles Goodyear) it becomes useful. The process is commonly called _____
15. The hardening of plastics, often associated with cross-linked and is called _____
16. A highly inert polymer that is used in kitchen-ware and was discovered accidentally by Plunkett is _____
17. _____ provide important information on the composition of copolymers formed from the two comonomers
18. _____ is the solvent, when the polymer molecule exists in its unperturbed dimension at a definite temperature
19. The polysulphides often used as elastomers are known by the trade name _____
20. _____ gives rise to the statistically regular or irregular arrangement of pendant groups (like methyl group in polypropylene)
21. _____ are the polymers (like polyacetylenes) that conduct electricity
22. _____ is a very useful technique employed for the preparation of tailor-made block copolymers
23. A polymer like ABS built up of 3 different monomer units can be called a _____
24. _____ is a technique for processing polymers into continuous films or sheets
25. Polyethylene containing a few methacrylic acid monomers along the chain can be called an _____
26. A process where the cellulose is regenerated for use in making rayon is called _____
27. Aromatic polyamides are thermally stable polymers commercially known as _____
28. The T_g of polystyrene can be decreased by copolymerising styrene with little butadiene. This process of decreasing T_g is called _____
29. Polymerisation above a temperature, which depends on monomer, is not thermodynamically feasible. This temperatue is called _____

30. The fully extended length of a polymer molecule based on bond distance and bond length is expressed as _____

31. An elastomer (e.g. natural rubber) flow of its own weight. This is called _____

32. Natural rubber when cross-linked with 30% S produces a hard plastic like substance called _____

33. The organised malteese like structures observed in polymers under polarised microscope are called _____

34. The _____ and _____ polymerisation in a monomer $CH_2 = CHX$ often lead to stereopolymers

35. The German scientist who in 1920 gave the concept of macromolecules (which was highly criticised at that time but later won him Nobel Chemistry 1953 prize) was _____

36. _____ undergoes ring-opening polymerisation to produce nylon 6

37. The Nobel laureate _____ is known for his solid phase synthesis of macromolecules

38. A linear polymer chain exhibits coiling due to _____ arising mainly as a result of internal rotation around a single bond

39. The Nobel physical chemist who is known for the advancement in the theory of polymer solution is _____

40. Cotton clothes are first dipped in NaOH solution to avoid shrinkage was a useful process developed by John Mercer and is known as _____

41. Regular coiling of flexible chains of many biopolymers forms _____ structures

42. The dimension of a polymer coil in solution particularly for branched polymers is often expressed as _____

43. The viscosity average mol. wt. becomes number average mol. wt. when the empirical constant of Mark-Houwink equation, α, becomes _____

44. Small hole like machine through which a molten polymer or its solution is passed to obtain fibres is called _____

45. Styrene-butadiene block copolymers can be grouped _____ elastomers

46. Polymers of silicon compounds are collectively called _____

47. _____ first prepared linear polyethylene using low pressure technique

48. The high degree of crystallinity in nylon fibres arises as a result of intermolecular _____

49. _____ is experimentally evaluated in dilute solution viscosity method and is related to molecular weight

50. _____ is the breakdown of a polymer into simpler components by means of microorganisms

51. HDPE can be obtained as _____ by crystallisation from dilute solutions

52. PMMA and polycarbonates possess high _____

53. Nylons are the synthetic polymers, which possess _____ structures

54. To obtain high mol. wt. polymer by step-polymerisation, _____ of the monomer pair must be maintained

55. While the mechanical properties of a polymer become independent of mol. wt. above a critical value, a sharp increase in _____ restricts the processing of very high mol. wt. polymers

56. In emulsion technique, the polymerisaton takes place within the _____ of the emulsifier

57. _____ is often observed in solution polymerisation and results into low mol. wt. polymers

58. _____ are obtained when the two comonomers have same reactivity ratios

59. A lower critical solution temperature observed in PEO, PPO, PVME due to their inverse solubility relationship is often called _____

60. _____ possess structural similarities with surfactants and thus form micelles and exhibit efficient adsorption characteristics

61. All polymer solutions exhibit light scattering and thus can be called _____ solutions

62. The double extrapolation of light scattering experimental data to obtain mol. wt. of polymers yield a grid which is called _____

63. _____ is the most widely used moulding technique for thermoplast involves the passage of the molten polymer from a nozzle

64. _____ is determined by dilatometry and provides a convenient determination of glass transition temperature

65. The techniques based on colligative properties, e.g. _____ and _____ are not convenient for the determination of mol. wt. of polymers

66. _____ in linear polymers makes them thermosetting plastics

67. Glyptals are made by the condensation of glycerol with _____

68. _____ can be obtained from osmometry or light scattering and provides significant information about polymer-solvent interaction

69. Flory's _____ provides information about the polymer-solvent interaction

70. The branched polymer component present in starch is called _____

71. Proteins undergo _____ on heating, with pH changes or when alcohol, urea, etc. are added to their aqueous solutions

72. _____ synthesised and characterised linear polyesters and polyamides as early as in 1930

73. Segmented polyurethane fibres are commonly known as _____

74. A short plastic tube, which is heated and expanded by air in the blow moulding process, is called _____

75. _____ is a cellulose derivative known for its use as thickener and antisoil redeposition agent in detergent formulations

76. Polyimides are _____ polymers

77. _____ are additives that are added to polymers to retard their oxidative degradation

78. A filled or reinforced plastic is often called _____

79. Cross-linked polymers with ionisable groups are used as _____

80. Continuous fibers of nylon 6, 6 can be made by _____

81. BF_3, $AlCl_3$ are the catalysts used in _____

82. Reacting the end groups to produce a stable polymer is called _____

83. Cross-linked resin prepared from trihydric alcohols and dibasic acids are called _____

84. Pyrolysed PAN (carbon fibre) is an example of _____ polymers

85. The polymer coil in solution under theta-conditions assumes _____ dimensions

86. Resistance to bending of a material is measured as _____

87. Hardness of materials is often measured in _____ scale

88. _____ are the substances formed by 3–10 monomer molecules

89. The point on a stress-strain curve below which there is reversible recovery is called _____

90. Fluids which exhibit lamellar flow can be called _____

91. The solubility parameter concept was given by _____

92. The property when a material possesses viscosity of liquid and elasticity of rubbers is called _____

93. The phenyl group in polystyrene and chloro group in PVC can be called _____ groups

94. Chain _____ make a polymer more rigid and thermally stable

95. The *trans*-1, 4 polyisoprene is called _____
96. Fibre reinforced materials are the major type of commercial _____
97. The Nobel chemistry prize for the year 2000 has been awarded for the polymers known as _____
98. Polymers containing other than carbon atoms in the main chain are called _____
99. Polymeric gels, solid polymer electrolytes and photoconducting polymers are called _____ polymers
100. Cellulose is water _____

Answers

Set I

1. Berzelius, 2. CRU, 3. More than two, 4. Viscosity number, 5. Calibration, 6. Crystallisation, 7. Comb, 8. Mesomorphic, 9. Highest mechanical strength, 10. T_g, 11. 1/2, 12. Inert liquid, 13. Carothers, 14. Cellulose, 15. T_g, 16. Bakelite, 17. Protein, 18. Biopolymers, 19. Carbon fiber, 20. IPN, 21. Block and Graft, 22. Membrane osmometry, 23. Teflon, 24. Proteins, 25. Charles Goodyear, 26. Solublility parameter concept, 27. Ionomer, 28. Polyblends, 29. Chain polymerisation, 30. Plasticizers, 31. Living, 32. Theta, 33. Reactivity ratio, 34. Ideal, 35. Staudinger, 36. $\overline{M}_w / \overline{M}_n$, 37. Crystalline, 38. HDPE, 39. Polyelectrolyte, 40. Ion exchange resin, 41. High molecular weight, 42. Elastic behaviour, 43. Thermosett, 44. Limited, 45. Cross-links, 46. Water soluble, 47. Alternate, 48. Solution, 49. Internal, 50. GPC, 51. Watson and Crick, 52. Thermosett resins, 53. Dipole-dipole force, 54. Monofunctional monomer, 55. Free radicals, 56. Double bond, 57. Suspension, 58. Interfacial, 59. Copolymers, 60. Pyrolysis gas chromatography, 61. IR spectroscopy, 62. X-ray diffraction, 63. Copolymerisation, 64. Polysaccharides, 65. Phase separation 66. Svedberg equation, 67. Ladder polymers, 68. Bisphenol A, 69. Crazing, 70. Autoaccelaration, 71. Ionomers, 72. Fullerene, 73. Hydrolysis, 74. Melting temperature, 75. Reinforced plastics, 76. Telogen, 77. Limited swelling 78. Novolacs, 79. Young modulus, 80. DSC, 81. Unzipping, 82. Fillers, 83. Enzymes, 84. Disproportionation, 85. Isotactic, 86. Coagulated, 87. Surfactant, 88. Dehydrogenation, 89. 3200–3550 cm^{-1}, 90. NMR, 91. Compression moulding, 92. Polymer colloids, 93. Adhesives, 94. Thermally stable, 95. Osmometry, 96. End group analysis, 97. Polarised microscope, 98. Polyelectrolytes, 99. Emulsion polymerisation 100. Bifunctionality

Set II

1. HDPE, 2. Tromsdorff effect 3. Bakelite, 4. Butyl rubber, 5. Chain transfer agents, 6. Plasticizers, 7. Latex, 8. Bulk, 9. Casein, 10. Carbon fibers, 11. IPN, 12. Molecular weight, 13. EG and PTA, 14. Cross-linking, 15. Mastication, 16. Teflon, 17. Reactivity ratio, 18. Theta solvent, 19. Thiokol, 20. Tacticity, 21. Conducting polymers, 22. Living polymerisation, 23. Terpolymer, 24. Calendering, 25. Ionomers, 26. Viscose, 27. Kevlar, 28. Internal plasticization, 29. Ceiling temperature, 30. Contour length, 31. Creeping, 32. Ebonite, 33. Spherulites, 34. Head to head and head to tail, 35. Staudinger, 36. Caprolactum, 37. Merrifield, 38. Chain flexibility, 39. P.J. Flory, 40. Mercerization, 41. Helical, 42. End to end distance, 43. One, 44. Spinneret, 45. Thermoplastic, 46. Silicones, 47. Karl Ziegler, 48. Bonding, 49. Intrinsic viscosity, 50. Biodegradation, 51. Single crystals, 52. Transparency, 53. Protein like, 54. Stoichiometry, 55. Melt viscosity, 56. Micelles, 57. Chain transfer, 58. Azeotropic copolymers, 59. Cloud point, 60. Block copolymers. 61. Colloidal, 62. Zimm plot, 63. Injection moulding, 64. Partial specific volume, 65. Ebulliometry and Cryometry, 66. Cross-linking, 67. Phthalic acid, 68. Second virial coefficient, 69. Interaction parameter, 70. Amylopectin, 71. Denaturation, 72. Carothers, 73. Spandex, 74. Parison,

75. Carboxymethylcellulose, 76. Thermally stable, 77. Antioxidants, 78. Composite, 79. Ion exchange resin, 80. Interfacial polycondensation, 81. Cationic polymerisation, 82. Capping, 83. Alkyd resins 84. Ladder, 85. Unperturbed, 86. Flexural strength, 87. Mhos, 88. Oligomers, 89. Yield point, 90. Newtonian fluid, 91. Hilderbrand, 92. Viscoelasticity, 93. Pendant, 94. Stiffening groups, 95. Gutta percha, 96. Composites, 97. Conducting plastics, 98. Inorganic polymers, 99. Specialty, 100. Insoluble

Appendix D

Glossary

ABS   Copolymer of acrylonitrile, butadiene and styrene

Acetate Rayon   A fiber made from cellulose diacetate

Acrilan   Fiber based polyacrylonitrile

Addition polymerisation   Polymerisation involving chain reactions in which chain carrier may be a free radical or an ion

AIBN   2, 2' -azo*bis*isobutyronitrile—a free radical initiator for polymerisation

Alternating copolymer   An ordered copolymer in which two comonomers are arranged alternatively in the polymer chain$-(-ABAB-)-_n$

Amino resins   Resins obtained from polycondensation of urea (or melamine) and formaldehyde

Amorphous regions   Regions of the polymer sample where the chains are tangled or disordered

Amylopectin   Highly branched starch molecule insoluble in water

Amylose   Linear starch molecule soluble in water

Anionic Polymerisation   A polymerisation initiated by an anion

Anisotropic   Properties directionally dependent (crystalline solids)

Antioxidant   A stabilizer which retards the oxidative degradation of a polymer

Antistat   An additive that reduces the static charges

Aramides   Aromatic polymides e.g. Kevlar that are thermally stable polymers

Atactic   A polymer with random arrangement of pendant groups on each side of the polymer chain

Average mol. wt.   Number, weight and Z averages defined as the first, second and the third powers of av. mol. wt, respectively, in a polydispersed polymer

Azeotropic copolymer   A copolymer in which the composition of monomers is the same as that of the feed

Beta cellulose   Cellulose soluble in 17.5% but insoluble in 8% aq NaOH

Biopolymers   Naturally occurring macromolecular substances e.g. proteins, polysaccharides, nucleic acids and rubber

Bingham Plastic   A plastic that doesn't flow until the external stress exceeds a critical threshold value

Block copolymer   A linear copolymer with constituent monomer units arranged in blocks (long sequences) in the chain

Bulk density	Apparent density that usually refers to powders and small particles
Bulk polymerisation	Polymerisation method of an unsaturated monomer that contains monomer and the initiator only
Butyl rubber	Copolymer of isoprene and isobutene as a useful elastomer
Calender	Equipment with a series of hollow rolls used for fabrication of films and sheets
Capping	The reaction of polymer end groups to prevent the dissociation of a polymer into monomer
Carbon fiber	A highly flame retardant polymer obtained from the cyclization and dehydration of polyacrylonitrile on heating
Casting	The solidification of a liquid resinous composition in a mould without the use of pressure
Cationic polymerisation	A polymerisation, which is initiated by a cation and subsequently, propagated by carbonium ion
Ceiling temperature	A threshold temperature above which a specific polymer can not be made
Cellophane	Regenerated cellulosic sheet produced from cellulose xanthate (viscous) solution
Cellulose	A natural polymer consisting of repeating beta glucose unit
Cellulose acetate	An ester formed by reaction of cellulose and acetic acid
Cellulose nitrate	Obtained after nitration of cellulose and classified as primary, secondary and tertiary depending on number of groups in each repeating anhydro glucose unit in cellulose are nitrated
Cellulose xanthate	A product of soda cellulose and CS_2
Chain stiffening groups	Groups in the polymer backbone which decrease the segmental motion e.g. amide
Chain transfer agent	A substance that destroys the activity of the growing chain
Chitin	A polymer of acetylated glucosamine present in the exoskeleton of shell fish
Cold drawing	Process of stretchting a polymer such that its crystalline regions become aligned or oriented in the direction of applied stress
Composite	A filled or reinforced plastic
Compounding	Process of mixing a polymer with additives
Contour length	The fully extended length of a polymer chain
Copolymer	A polymer made of more than two monomers
Coupling agent	A compound which improves interaction between the filler and a resin
Creep	Cold flow
Cross-link density	A polymer network formed when separate linear or branched chains are joined together along the chains by cross-links
Crystalline polymer	A polymer with ordered structure
Crystalline	Regions of crystallinity
Curing	Formation of extensive three dimensional cross-links in the final stage in the production of a resin
Curing agent	A substance that causes curing

Degradation	Decomposition of polymer chain by heat, light, oxygen, water, bacteria etc.
Degree of polymerisation	Number of the repeat units present in a polymer molecule
Dilatometry	A technique that measures changes in specific volume, often used to determine T_g
DNA	Deoxyribonucleic acid, a biopolymer responsible for the hereditary character
DSC	Differential scanning calorimetry-a technique that measures differences in change in enthalpy of a heated polymer and a reference standard based on power input
DTA	Differential thermal analysis—a thermal analytical method often used to monitor thermal transitions in polymers
Dry spinning	Extrusion of polymer solution through the holes of a spinneret into a hot chamber
Elastomers	Synthetic polymers with rubber-like properties
Elastic range	The range on stress-strain curve below yield point
Emulsion	The colloidal dispersion of one liquid into another which are imiscible with each other
Emulsion polymerisation	A technique to prepare a polymer, often in the form of a latex; the monomer is dispersed in a liquid (usually water) that contains the initiator. A surfactant emulsifies the monomer and its micelles are the focus of polymerisation
End group analysis	A method for the determination of number average mol. wt. of polymers if it contains functional end groups that can be determined (say by titration)
End to end distance	Shortest distance between the chain ends in a polymer molecule
Epoxy resin	A polymer obtained by the condensation of epichlorohydrin and bisphenol A
Extender	An inexpensive filler
Extrusion	A processing technique in which viscous plastic mass is continuously forced through an orifice to produce articles of various profiles such as tubes, rods etc.
Fiber	A crystalline polymer that can be used as a fiber
Filler	A solid additive which resricts the mobility of polymer chains and may also serve as an extender or reinforcing agent for the polymer
Flame retardants	Additives that reduce the flamability of a polymer
Flexural strength	A measure of resistance to bending
Flory-Huggins theory	Mathematical treatment that predicts the equilibrium behaviour between liquid phases containing polymer
Fractionation	A method to produce several low polydispersed sample from a highly polydispersed polymer sample
Fringe micelle model	A model showing crystalline and amorphous domains present in polymer
Gel	A semi-solid mass that forms usually when concentrated colloidal substances are stored
Gel effect	An enormous rise in viscosity in bulk polymerisation that leads to a high mol. wt. polymer (Autoaccelaration or Tromsdorff effect)
Gel point	The point at which gelation (cross-linking) begins

Geotextiles	Polymeric mats, sheets and textiles employed in the control of soil, water etc. in geological applications
GPC	Gel permeation chromatography—a chromatographic technique useful in the determination of molecular weight distribution in polymer
Glass transition temperature	A characteristic temperature for a given polymer at which a polymer transforms from glassy state to rubbery state, designated by T_g
Glycogen	A highly branched polysaccharide which serves as the reserve carbohydrate in animals
Graft copolymer	A special type of a branched copolymer in which the branches consist of one type of monomer units while the main chain consists of another type of monomer units
Gutta percha	A naturally occurring polymer, *trans*-1, 4-polyisoprene—a stiff isomer of natural polymer
Hevea Brazilensis	Natural rubber (*cis*-1, 4-polyisoprene)
HDPE	High density polyethylene (also low pressure polyethylene)
Homopolymer	A polymer made of a single type of monomer e.g. polyethylene
Humic acid	Natural polymeric aromatic carboxylic acids
Impact strength	The resistance of a test sample to a sharp blow
Inhibitor	A chemical used to stop polymerisation
Initiation	The first step of chain polymerisation where the free radicals or ions generated from the initiator attack the monomer molecule to form new reactive species
Initiator	Substance that generates reactive species like free radicals or ions that participate in the reaction with monomer molecule
Initiator efficiency	The fraction of radical or decomposed initiators that lead to the formation of active centres
Injection moulding	A process in which a heat softened polymer is injected under pressure into a cool mould where it hardens to give the desired article
Interfacial polycondensation	The process of producing a polymer by the reaction of the functional group containing monomers at the oil/water interface for example in the production of nylon 6, 6
Internal plasticization	Polymer chain flexibility achieved by reduction of chain symmetry through copolymerisation
Interpenetrating network	A multiphase polymer system that is obtained when a monomer is polymerised in presence of a polymer
Intrinsic viscosity	Reduced viscosity at infinite dilution, a parameter that is experimentally obtained from dilute solution viscosity measurements and is related to the molecular weight of the polymer by an empirical Mark-Houwink equation
Ionomer	Polymer that contains only a few charged groups in its molecule and is insoluble in water. A generic name for copolymer of ethylene and methacrylic acid
Isotactic	Arrangement in space where all the symmetric centres

	in the molecule have identical (either with d or l configuration)
Kinetic chain length	The av. number of monomers that add to a growing chain before it terminates (rate of propagation = rate of termination)
Lacquers	Polymer solutions containing pigment—a surface coating material
Ladder polymer	A polymer in which the backbone consists of a double strand much like a ladder that makes the polymer thermally stable
Latex	Stable dispersion of a polymer in water. For example, polyisoprene is dispersed in water as the sap of the rubber tree in a latex
Lignin	Non-cellulosic resinous component of wood which acts as a binder for the cellulosic fiber of wood
Living polymers	The product formed at the end of an anionic polymerisation in which all monomer is consumed, but the active centres have not been terminated. Addition of monomer (same or different) will continue the polymerization further
Mark-Houwink equation	The equation that relates intrinisc viscosity to molecular weight of polymers. $[\eta] = KM^{\alpha}$ where K and α are the constants for a particular polymer-solvent pair at fixed temperaure
Melt spinning	A technique for fiber processing in which a molten polymer is pumped at a constant rate under high pressure through a plate (called spinneret) containing several holes
Melt viscosity	Viscosity of a molten polymer
Microstructure	The local organisation of the repeat units in a copolymer or partially atactic polymer, i.e. sequence distribution
Modulus	Ratio of stress (applied force) to strain (elongation); a measure of stiffness of a polymer
Monomer	The starting substance that has minimum bifunctionality from which a polymer is made
Network	A high degree of cross-linking in linear polymers
Newtonian flow	A fluid whose viscosity is proportional to the applied viscosity gradient
Novalac	Polymer made by the condensation of phenol and formaldehyde under acidic conditions
Oligomers	Substances usually made up of a few monomer units
Pendant groups	Groups attached to the main polymer chain e.g. Cl in PVC
Plasticizer	Substance added to polymer to reduce its glass transition temperature, thus facilitating segmental motion and making polymer flexible
Polyblend	Multiphase polymer system consisting of two polymers
Polycondensation	Formation of polymer by the reaction of monomers containing two or more functional groups (also called step polymerisation)
Polydispersity index	The ratio of weight average mol. wt to number average mol. wt. that is a measure of the distribution of molecules of different sizes present in a polymer

Polyelectrolyte	A polymer containing charged groups that make it water soluble where it ionises to form a macro ion and counter ions
Polymer	Large molecules built up of thousands of small repeat units (monomers) which are joined together by covalent bonds
Polymer pioneers	Those scientists who contributed to the advancement of polymer science and include Nobel laureates, Staudinger, Ziegler, Natta, Flory, Merrifield, de Gennes and others like Hermann Mark, Marvell, and Carothers
Polymerisation	Process of making polymers from low mol. wt. substances (chain polymerisation from unsaturated monomers and step polymerisation from functional group containing monomers)
Polypeptide	An oligomer made up of few α-amino acid units covalently liked through covalent bond
Pultrusion	A process in which filaments are dipped in a prepolymer, passed through a die, and cured
Radius of gyration	rms distances of a chain end from the centre of gravity of its polymer molecules
Random copolymers	A copolymer in which there is no definite ordering in the arrangement of the two monomers from which the polymer is made
Rayon	Regenerated cellulose filament
Reclaimed rubber	Recycled rubber
Retarder	A substance that competes with the monomer for free radical during radical polymerisation. Decreases both the rate of polymerisationm and DP and is less reactive than inhibitor
Shellac	Resin secreted by coccid insect
Solubility parameter	Numerical value equal to the square root of the cohesive energy density. It is used to predict polymer solubility through matching the solubility parameter of the polymer to that of solvent
Solution polymerisation	A polymerisation technique in which the monomer is dissolved in an inert solvent and then polymerised. It overcomes the problem of heat dissipation encountered in bulk polymerisation but gives low molecular weight polymer due to chain transfer
Spandex	Elastic polyurethane fibers
Spherulites	Aggregates of polymer crystallites
Spinneret	A metal plate with many small, uniformly sized minute holes (used in processing of films)
Styrofoam	Foamed polystyrene
Suspension polymerisation	A polymerisation technique in which monomer contains the dissolved initiator and is then suspended in water (also called bead or pearl polymerisation)
Syndiotactic	A polymer in which the pendant groups are arranged alternatively on each side of the polymer backbone e.g. methyl groups in polypropylene
Tacticity	Arrangement of pendant groups in space

Tanning	Cross-linking of leather (protein) with tannic acid
Tensile strength	Resistance to pulling stresses
Telomer	Low mol. wt. polymer resulting from chain transfer agents. The additive that causes chain transfer is called talogen
Tenacity	Measure of fiber strength
Terylene	Polyethylene terephathalate
Thermogravimetric analysis	Measurements of change (nomally loss) in weight when a material is heated
Thermoplasts	Polymers which can be softened and moulded on heating
Thermosetts	Polymer that can be cured by the application of heat. These are cross-linked infusible material
Theta solvent	A solvent in which a polymer exists as a statistical coil (i.e. in unperturbed dimensions)
Theta temperature	Temperature at which a polymer of infinite mol. wt. begins precipitating from solution
Thiokol	Polysulphide elastomers
Transfer moulding	A process in which a preheated briquette is forced through an orifice into a heated mould cavity
Ultracentrifugation	A high speed centrifuge process empolyed for separation of sub-microscopic particles such as polymer chains and provides a tool for measuring molecular weight of polymers
UV stabilizer	An additive that retards degradation of polymer by UV radiation
Vapour phase osmometry	A technique that gives average molecular weight of polymers (usually for low molecular weight range)
Viscoelasticity	A property having characteristic features of both solids and liquids
Viscose process	Regeneration of cellulose fibers by precipitation of the sodium salt of cellulose xanthate in acid
Vulcanisation	The process of introducing cross-links in natural rubber (usually by masticating it with sulphur)
Weight average mol. wt.	The second moment of average of a polydispersed polymer
Wet spinning	A process of obtaining fibers by precipitation of polymeric solution
Yield point	A point on stress-strain curve below which there is reversible recovery
Z average mol. wt.	The third power average of molecular weight in a polydispersed polymer (larger than weight average mol. wt.)
Ziegler-Natta catalyst	e.g. $TiCl_3$–R_3Al
Zimm plot	A method for recording light scattering intensities measured at different polymer concentrations and different measured angles on a graph that on double extrapolation to zero angle and zero concentration gives mol. wt. of polymers (also gives the radius of gyration on extrapolation to zero angle and second virial coefficient on extrapolation to zero angle)
Zone melting	Fractionation of a polydispersed polymeric system by heating over a long period of time

Appendix E

Trade Names for Some Common Polymers

Acrilan (Orlon)	Polyacrylonitrile based fiber
Araldite, Epan	Epoxy Resins
Bakelite	Phenol-formaldehyde resin
Balata	Same as gutta percha
Cellophane	Sheet of regenerated cellulose
Celluloid	Plasticized cellulose
Collagen	Protein of the connective tissue
Dacron (Terylene)	Trade name for polyester fibers
Dextran	High molecular weight branched polysaccharide synthesised from sucrose by bacteria
Dextrin	Degraded Starch
Dynel, Vinyon	Copolymer of vinyl chloride and acrylonitrile
Ebonite	Highly cross-linked rubber with 35% sulphur
Galalith	Plastics from moulded casein
Glyptals	Alkyd Resins
Gutta percha	Natural *trans*-1, 4 polyisoprene
Humic acid	Polymeric aromatic carboxylic acid present in lignin
Hytrel	Thermoplastic elastomer
Keratin	A fibrillar protein
Kraton	Styrene-butadiene-styrene block copolymer
Perspex, Lucite	Polymethyl methacrylate
Pluronic	Ethylene oxide-propylene oxide block copolymers
Saran	Trade name for polymers of vinylidene chloride
Spandex	Elastic polyurethane fibre
Styrofoam	Foamed polystyrene
Teflon	Polytetrafluoroethylene
Thiokol	Polysulphide rubbers
Tricel	Cellulose triacetate
Vinoflex	Copolymer of vinyl chloride and vinyl isobutyl ether
Vinylite	Copolymer of vinyl chloride and vinyl acetate

Appendix F

Abbreviated Forms of Some Common Initiators/Monomers/Polymers/ Additives etc.

AA	Acrylic acid
ABR	Acrylonitrile-butadiene copolymer
ABS	Acrylonitrile-butadiene-styrene copolymer
AIBN	Azo *bis* isobutyronitrile
AM	Acrylamide
ASTM	American society for testing of materials
BPO	Benzoyl peroxide
CMC	Carboxymethylcellulose
DMT	Dimethyl terephthalate
DOP	Dioctyl phthalate
EO	Ethylene oxide
HDPE	High density polyethylene
HIPS	High impact polystyrene
LDPE	Low density polyethylene
LLDPE	Linear low density polyethylene
MA	Methyl acrylate
MF resins	Melamine-formaldehyde resins
MMA	Methyl methacrylate
NBR	Nitrile butadiene rubber
PAN	Polyacrylonitrile
PEO	Polyethylene oxide
PET	Polyethylene terephthalate
PF and UF resins	Phenol-formaldehyde and urea-formaldehyde resins
PMMA	Polymethyl methacrylate
PO	Propylene oxide
PPO	Polypropylene oxide
PTA	tere-phthalic acid
PTFE	Polytetrafluoroethylene
PU	Polyurethane
PVA	Polyvinyl alcohol
PVAc	Polyvinyl acetate
PVC	Polyvinyl chloride

SBR	Styrene butadiene rubber
TCP	Tricresyl phosphate
TDI	Toluene diisocyanate
TPE	Thermoplastic rubber
UHMWPE	Ultrahigh molecular weight polyethylene
VA	Vinly acetate

Appendix G

Structures of Repeat Units of Some Common and Specialty Polymers

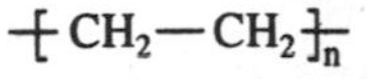

$\dashv CH_2-CH_2 \vdash_n$
Polyethylene

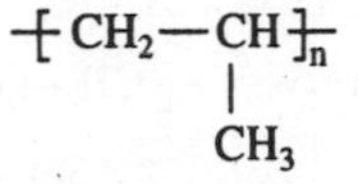

$\dashv CH_2-CH \vdash_n$
$\quad\quad\quad |$
$\quad\quad\quad CH_3$
Polypropylene

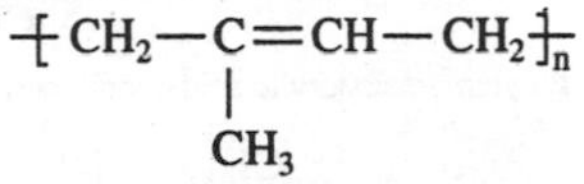

$\dashv CH_2-C=CH-CH_2 \vdash_n$
$\quad\quad\quad\quad |$
$\quad\quad\quad\quad CH_3$
1, 4-Polyisoprene

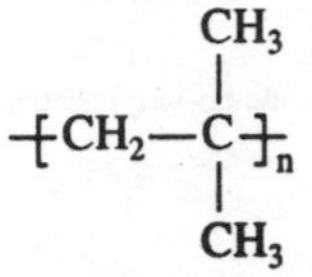

$\quad\quad CH_3$
$\quad\quad |$
$\dashv CH_2-C \vdash_n$
$\quad\quad |$
$\quad\quad CH_3$
Polyisobutylene

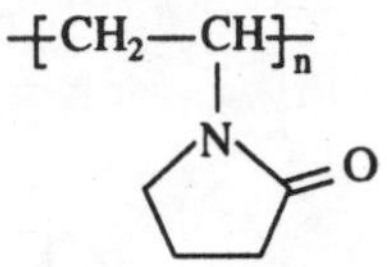

$\dashv CH_2-CH \vdash_n$
Polyvinyl pyrrolidone

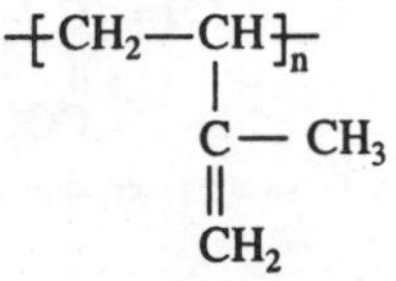

$\dashv CH_2-CH \vdash_n$
$\quad\quad\quad |$
$\quad\quad\quad C-CH_3$
$\quad\quad\quad ||$
$\quad\quad\quad CH_2$
3, 4-Polyisoprene

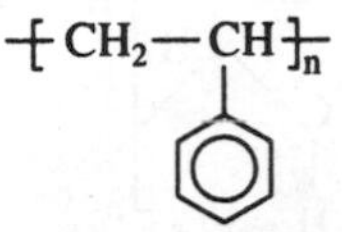

$\dashv CH_2-CH \vdash_n$
Polystyrene

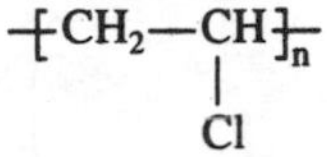

$\dashv CH_2-CH \vdash_n$
$\quad\quad\quad |$
$\quad\quad\quad Cl$
Polyvinyl chloride

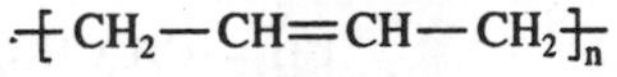

$\dashv CH_2-CH \vdash_n$
Polyacrylamide oxime

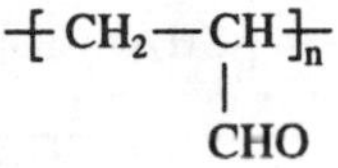

$\dashv CH_2-CH=CH-CH_2 \vdash_n$
Polybutadiene

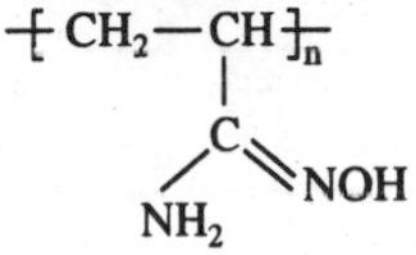

$\dashv CH_2-CH \vdash_n$
$\quad\quad\quad |$
$\quad\quad\quad CHO$
Polyacrolein

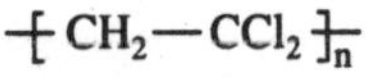

$\dashv CH_2-CCl_2 \vdash_n$
Polyvinylidene chloride

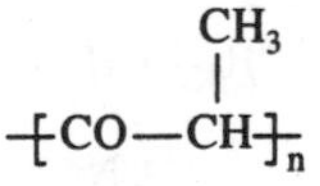

$\dashv CH_2-CH \vdash_n$
$\quad\quad\quad |$
$\quad\quad\quad CONH_2$
Polyacrylamide

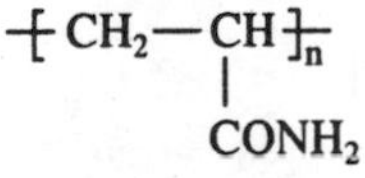

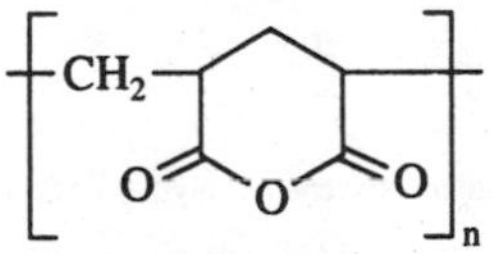

Polyacrylic anhydride

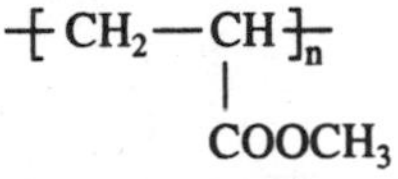

$\quad\quad CH_3$
$\quad\quad |$
$\dashv CO-CH \vdash_n$
Polyacetaldehyde

$\dashv CH_2-CH \vdash_n$
$\quad\quad\quad |$
$\quad\quad\quad COOCH_3$
Polymethyl acrylate

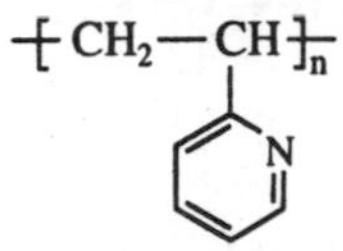

$\dashv CH_2-CH \vdash_n$
Polyvinyl pyridine

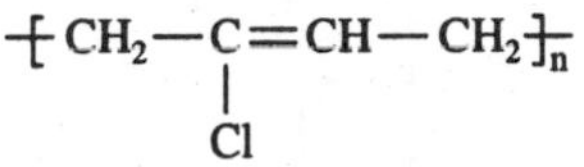

$\dashv CH_2-C=CH-CH_2 \vdash_n$
$\quad\quad\quad\quad |$
$\quad\quad\quad\quad Cl$
Polychloroprene (Neoprene)

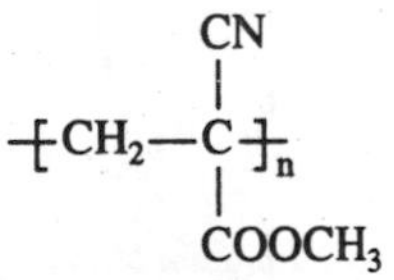

$\quad\quad CN$
$\quad\quad |$
$\dashv CH_2-C \vdash_n$
$\quad\quad |$
$\quad\quad COOCH_3$
Poly α-cyanomethacrylate

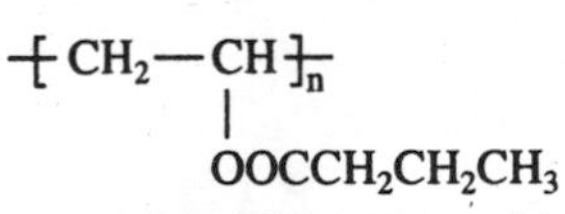

$\dashv CH_2-CH \vdash_n$
$\quad\quad\quad |$
$\quad\quad\quad OOCCH_2CH_2CH_3$
Polyvinyl butyrate

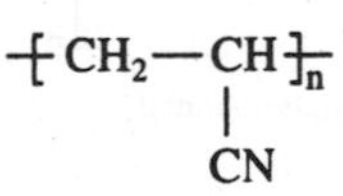

$\dashv CH_2-CH \vdash_n$
$\quad\quad\quad |$
$\quad\quad\quad CN$
Polyacrylonitrile

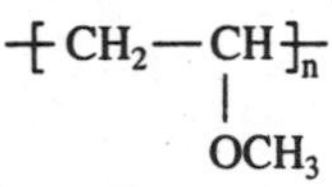

$\dashv CH_2-CH \vdash_n$
$\quad\quad\quad |$
$\quad\quad\quad OCH_3$
Polyvinyl methyl ether

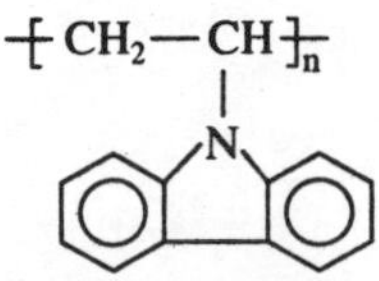

$\dashv CH_2-CH \vdash_n$
Polyvinyl carbazole

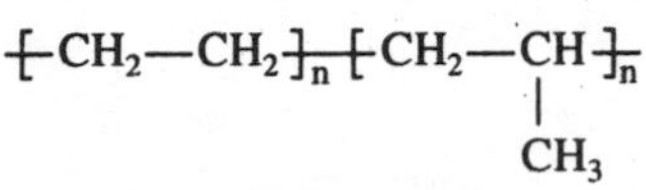

$\dashv CH_2-CH_2 \vdash_n \dashv CH_2-CH \vdash_n$
$\quad\quad\quad\quad\quad\quad\quad\quad |$
$\quad\quad\quad\quad\quad\quad\quad\quad CH_3$
Ethylene-propylene elastomer

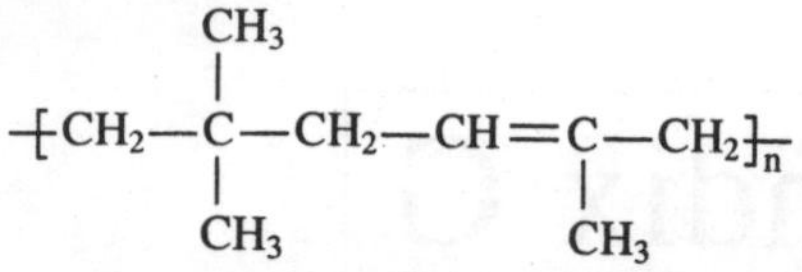

$$+CH_2-\underset{\underset{CH_3}{|}}{\overset{\overset{CH_3}{|}}{C}}-CH_2-CH=\underset{\underset{CH_3}{|}}{C}-CH_2\}_n$$

Butyl rubber

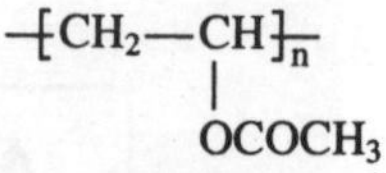

$$+CH_2-CH\}_n$$
$$\overset{|}{OCOCH_3}$$

Polyvinyl acetate

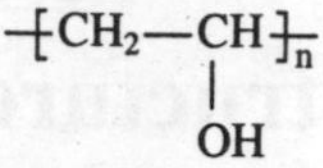

$$+CH_2-CH_2-CH_2-\underset{\underset{COO^{\ominus}}{|}}{\overset{\overset{CH_3}{|}}{C}}\}_n$$

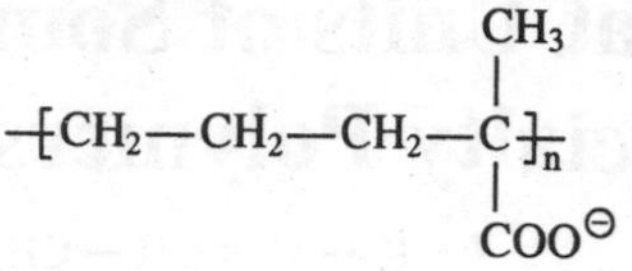

Ethylene-methacrylic acid copolymers (Ionomer)

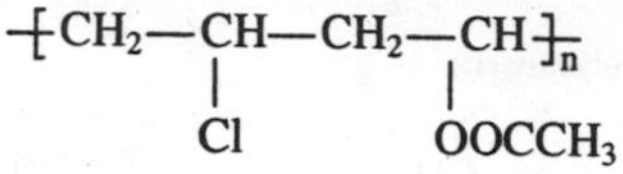

$$+CH_2-CH\}_n$$
$$\overset{|}{OH}$$

Polyvinyl alcohol (PVA)

$$+CH_2-\underset{\underset{COOCH_3}{|}}{\overset{\overset{CH_3}{|}}{C}}\}_n$$

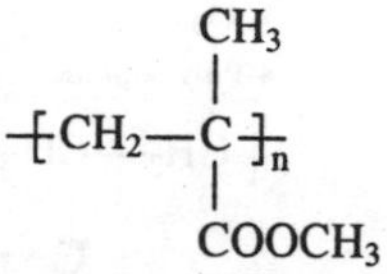

Polymethyl methacrylate (PMMA)

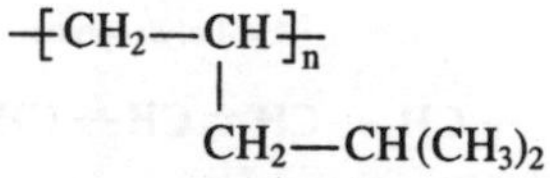

$$+CH_2-\underset{\underset{Cl}{|}}{CH}-CH_2-\underset{\underset{OOCCH_3}{|}}{CH}\}_n$$

Polyvinyl chloride-co-vinyl acetate

$$+CH_2-CH\}_n$$
$$\overset{|}{CH_2-CH(CH_3)_2}$$

Poly 4-methyl-1-pentene

$$+CH_2-CH\}_n$$
$$\overset{|}{COCH_3}$$

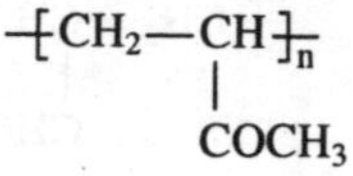

Polymethyl vinyl ketone

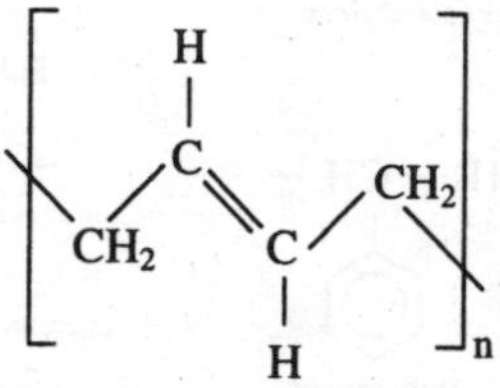

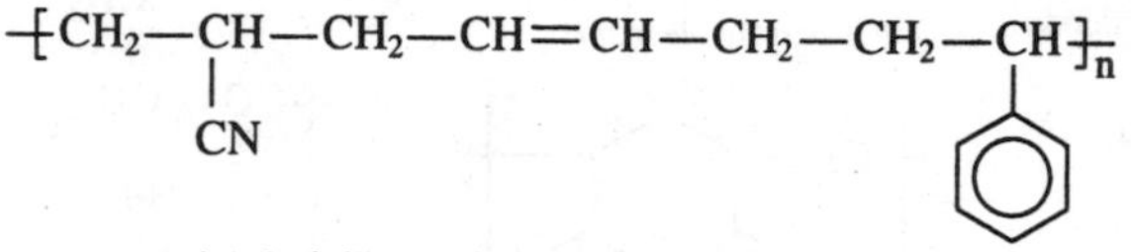

trans 1,4-polybutadiene

$$+CH_2-\underset{\underset{CN}{|}}{CH}-CH_2-CH=CH-CH_2-CH_2-CH\}_n$$

Acrylonitrile-butadiene-styrene terpolymer (ABS)

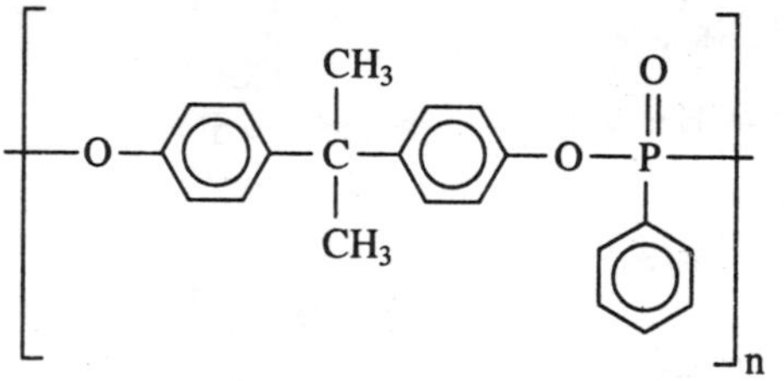

Poly 2,2-propane *bis*(4-penyl phenylphosphonate)

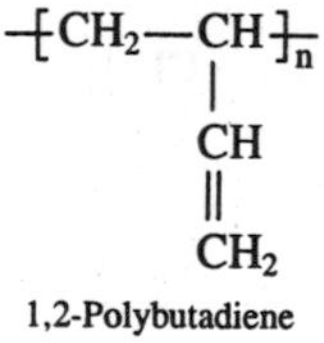

$$+CH_2-CH\}_n$$
$$\overset{|}{CH}$$
$$\overset{||}{CH_2}$$

1,2-Polybutadiene

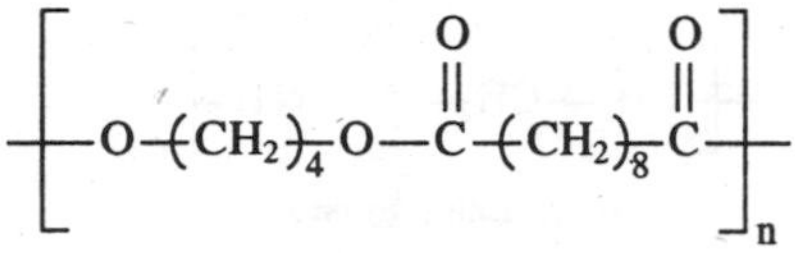

$$\left[O-(CH_2)_4-O-\overset{\overset{\displaystyle O}{\|}}{C}-(CH_2)_8-\overset{\overset{\displaystyle O}{\|}}{C} \right]_n$$

Polytetramethylene sebacate

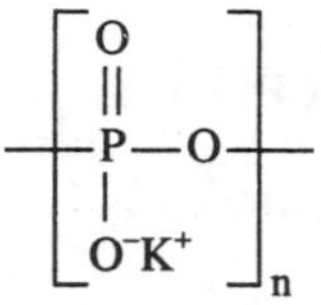

$$\left[\overset{\overset{\displaystyle O}{\|}}{\underset{\underset{\displaystyle O^-K^+}{|}}{P}}-O \right]_n$$

Potassium polymetaphosphate

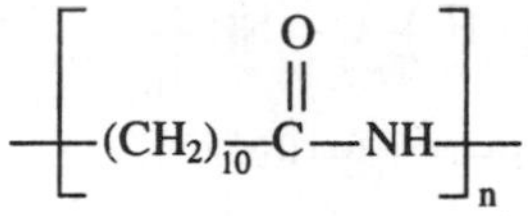

$$\left[(CH_2)_{10}-\overset{\overset{\displaystyle O}{\|}}{C}-NH \right]_n$$

Poly 11-undecanoamid (Nylon 11)

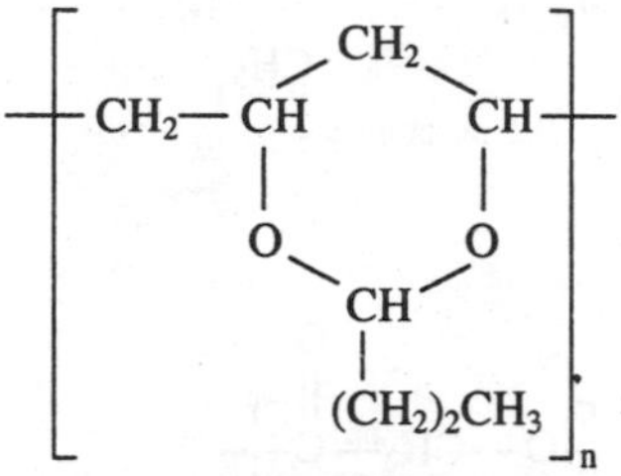

Polyvinyl butyral

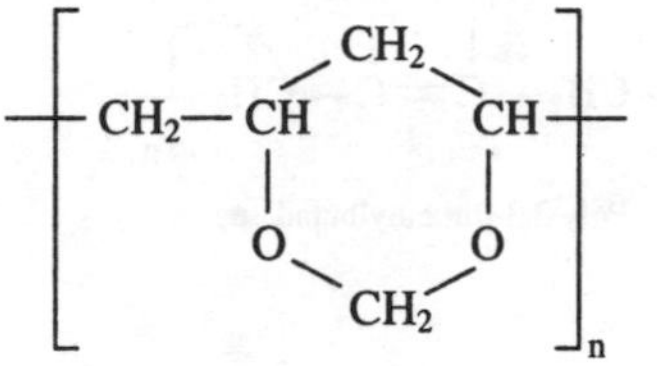

Polyvinyl formal

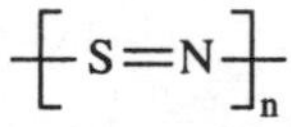

$$\left[S=N \right]_n$$

Polythiazyl

$$\left[CH_2-\bigcirc-CH_2-O-\overset{\overset{\displaystyle O}{\|}}{C}-CH_2-CH_2-S-CH_2-CH_2-\overset{\overset{\displaystyle O}{\|}}{C}-O \right]_n$$

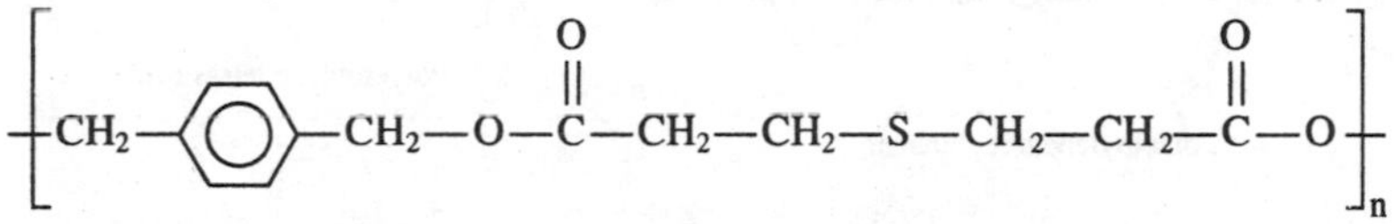

Poly p-xylylene thiodipropionate

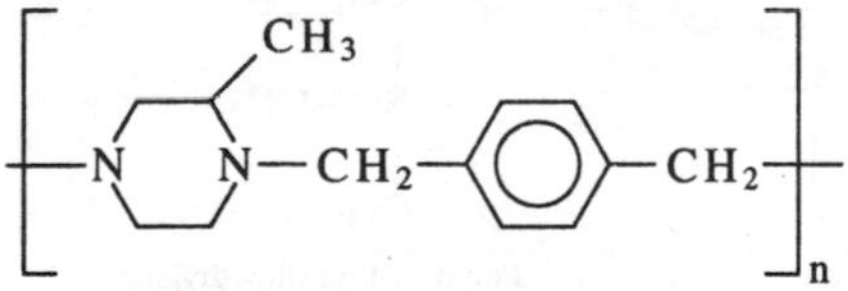

Poly 1,4-xylylenyl-2-methylpiperzine

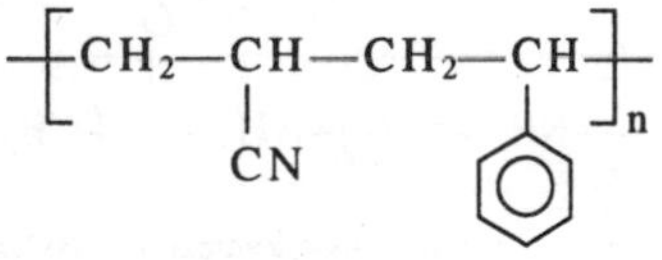

$$\left[CH_2-CH-CH_2-CH \right]_n$$
$$CN \qquad \bigcirc$$

Styrene-acrylonitrile copolymer

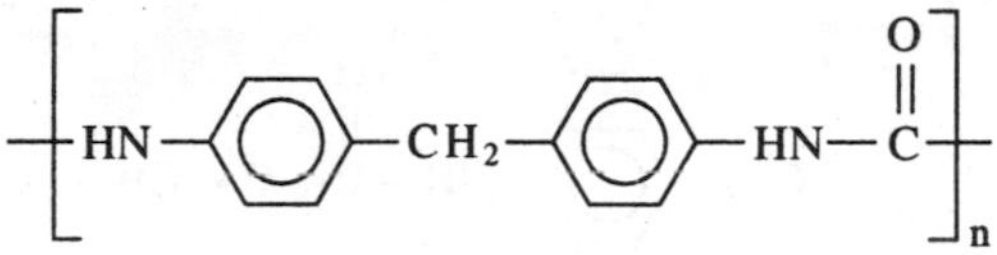

$$\left[HN-\bigcirc-CH_2-\bigcirc-HN-\overset{\overset{\displaystyle O}{\|}}{C} \right]_n$$

Polymethylene *bis* 4-phenylurea

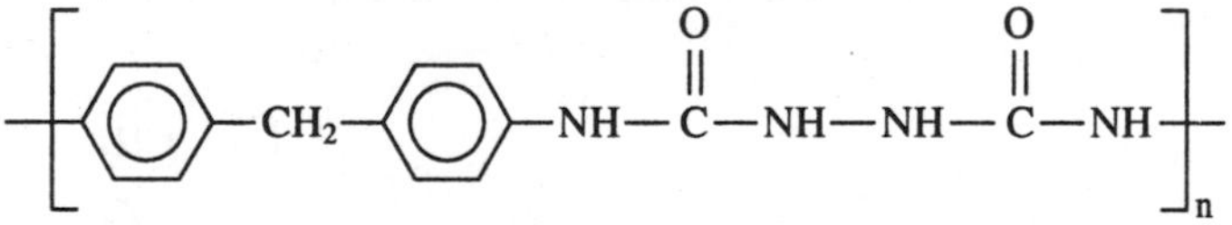

$$\left[\bigcirc-CH_2-\bigcirc-NH-\overset{\overset{\displaystyle O}{\|}}{C}-NH-NH-\overset{\overset{\displaystyle O}{\|}}{C}-NH \right]_n$$

Polymethylene *bis* 4-phenylurethylene

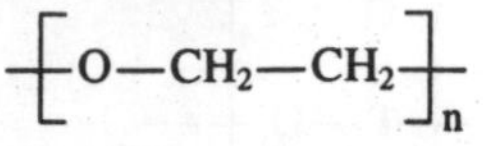

$$\left[O-CH_2-CH_2 \right]_n$$

Poly oxyethylene

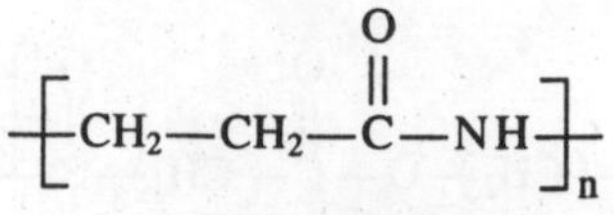

$$\left[CH_2-CH_2-\overset{\displaystyle O}{\overset{\|}{C}}-NH \right]_n$$

Poly β-alanine (nylon 3)

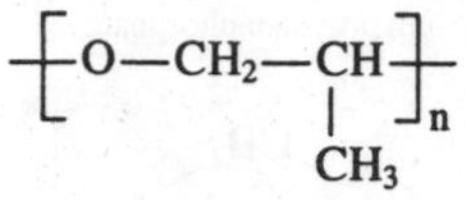

$$\left[O-CH_2-CH \right]_n \\ \qquad\qquad\quad CH_3$$

Polyoxypropylene

$$\left[(CH_2)_5-\overset{\displaystyle O}{\overset{\|}{C}}-NH \right]_n$$

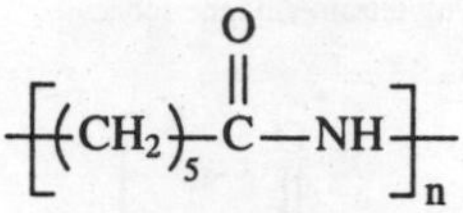

Polycaprolactam (Nylon 6)

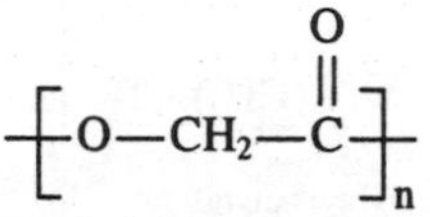

$$\left[O-CH_2-\overset{\displaystyle O}{\overset{\|}{C}} \right]_n$$

Polyglycollic ester

$$\left[CH_2-\overset{\displaystyle H_3C}{C}=\overset{\displaystyle CH_3}{C}-CH_2 \right]_n$$

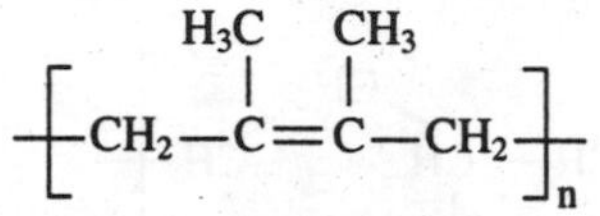

Poly 2,3-dimethylbutadiene

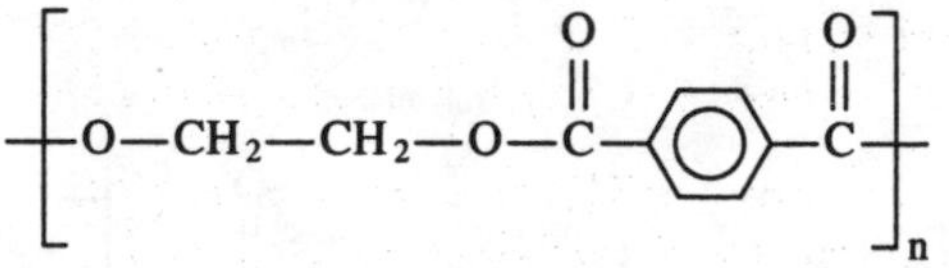

$$\left[O-CH_2-CH_2-O-\overset{\displaystyle O}{\overset{\|}{C}}-\bigcirc-\overset{\displaystyle O}{\overset{\|}{C}} \right]_n$$

Polyethylene terephthalate

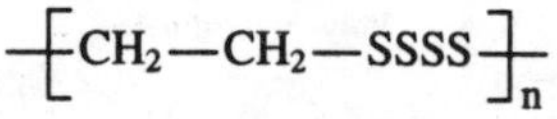

$$\left[CH_2-CH_2-SSSS \right]_n$$

Polyethylene tetrasulfide

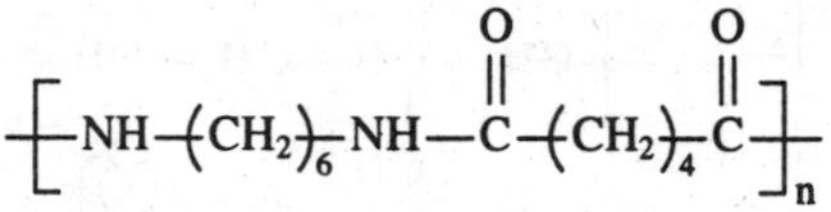

$$\left[NH-(CH_2)_6-NH-\overset{\displaystyle O}{\overset{\|}{C}}-(CH_2)_4-\overset{\displaystyle O}{\overset{\|}{C}} \right]_n$$

Poly hexamethyleneadipamide (Nylon 6,6)

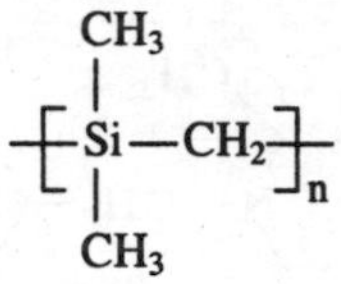

$$\left[\overset{\displaystyle CH_3}{\underset{\displaystyle CH_3}{Si}}-CH_2 \right]_n$$

Polydimethyl silmethylene

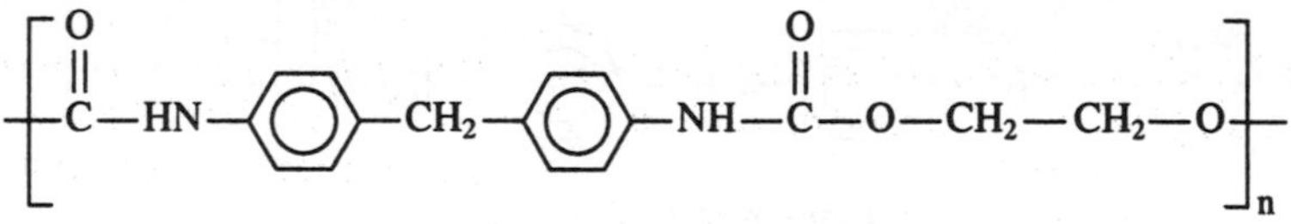

$$\left[\overset{\displaystyle O}{\overset{\|}{C}}-HN-\bigcirc-CH_2-\bigcirc-NH-\overset{\displaystyle O}{\overset{\|}{C}}-O-CH_2-CH_2-O \right]_n$$

Polyethylene methylene bis(4-phenylcarbamate)

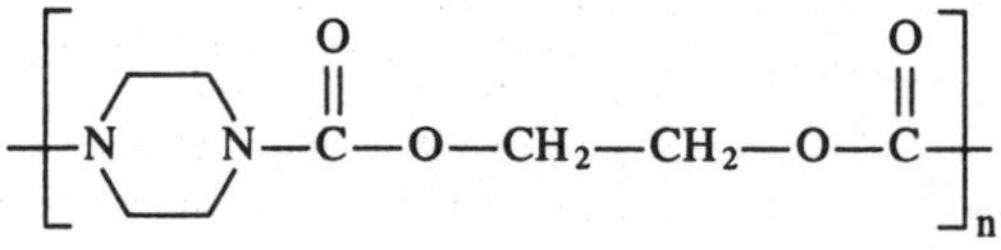

$$\left[N\bigcirc N-\overset{\displaystyle O}{\overset{\|}{C}}-O-CH_2-CH_2-O-\overset{\displaystyle O}{\overset{\|}{C}} \right]_n$$

Polyethylene N, N'-piperazinedicarboxylate

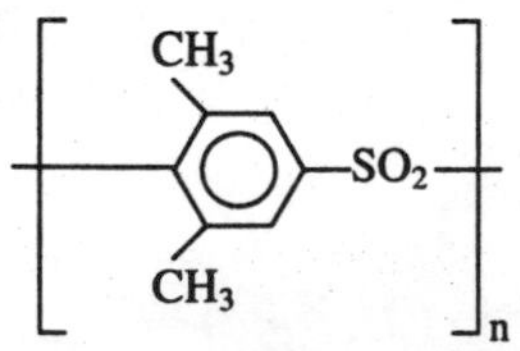

$$\left[\overset{\displaystyle CH_3}{\underset{\displaystyle CH_3}{\bigcirc}}-SO_2 \right]_n$$

Poly 3,5-dimethyl-1,4-phenylene sulphone

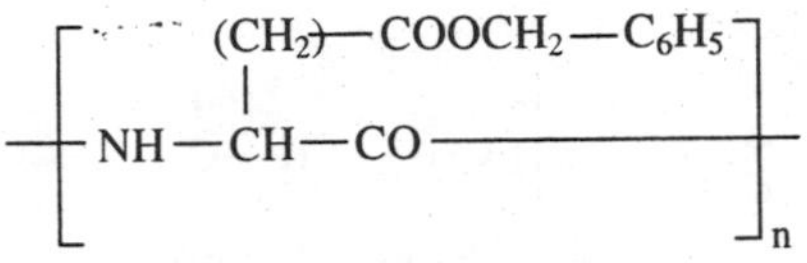

Poly γ-benzyl-L-glutamate

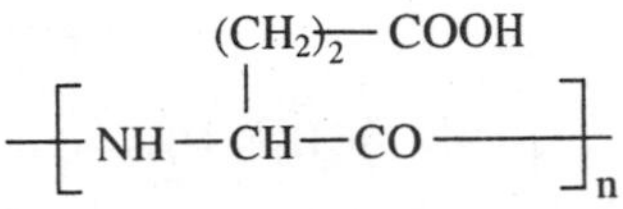

Poly α-L-glutamic acid

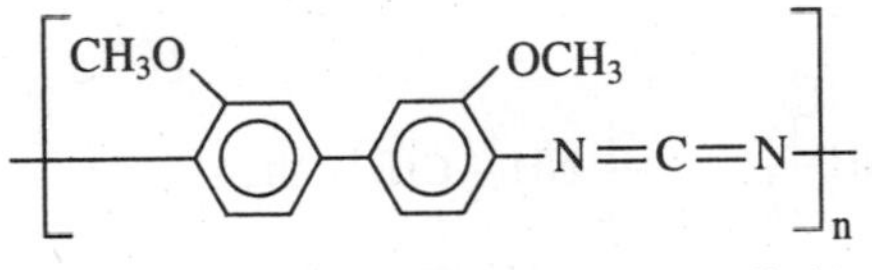

Poly 3,3'-dimethoxy-4,4'-biphenylene carbodiimide

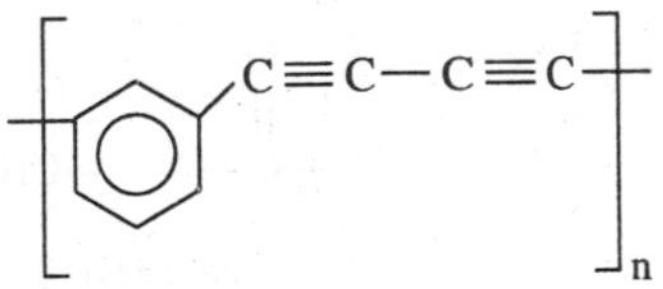

Polydiethynylbenzene

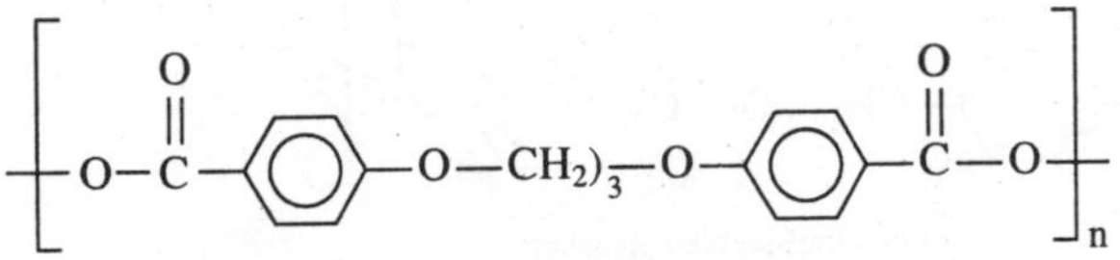

Poly 1,3-bis p-carboxyphenoxypropane anhydride

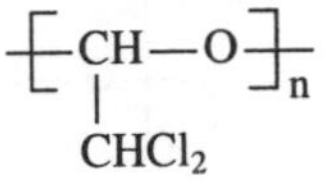

Polydichloroacetaldehyde

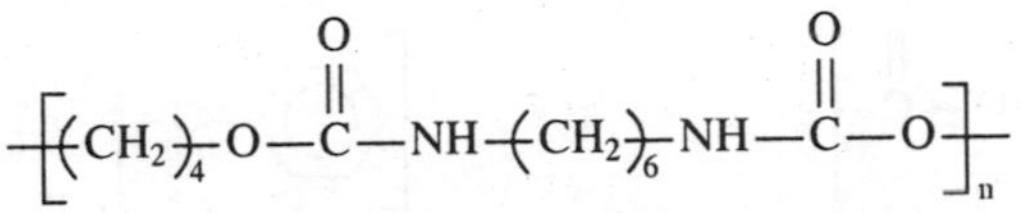

Polytetramethylene hexamethylenedicarbamate

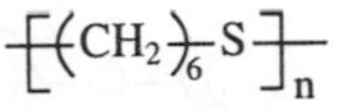

Polyhexamethylene

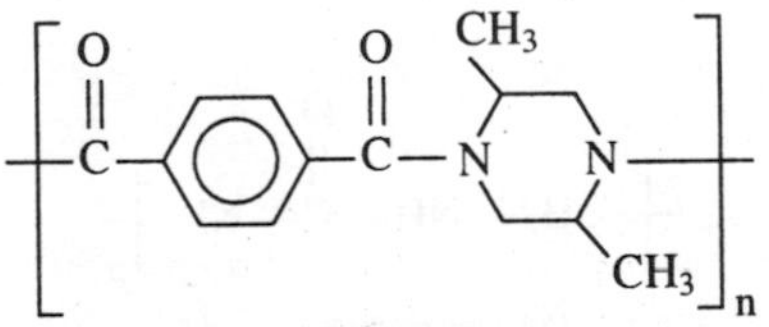

Poly 2,5-dimethylpiperazine terephthalamide

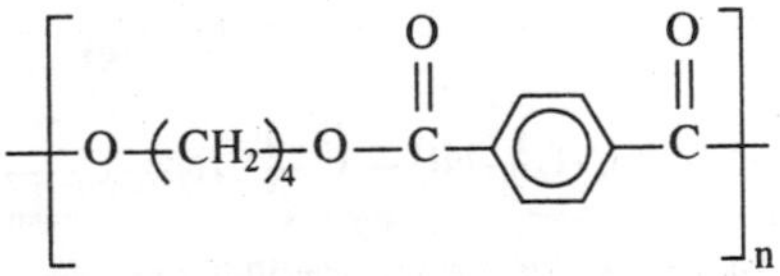

Polybutylene terephthalate

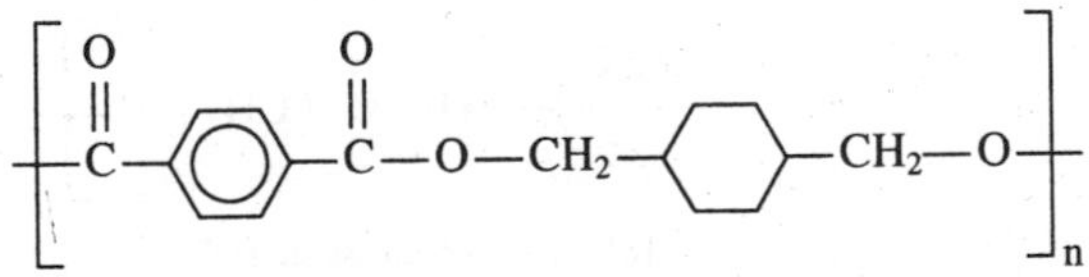

Polydihydroxy methylcyclohexyl terephthalate

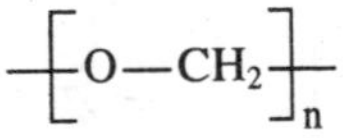

Polyoxymethylene

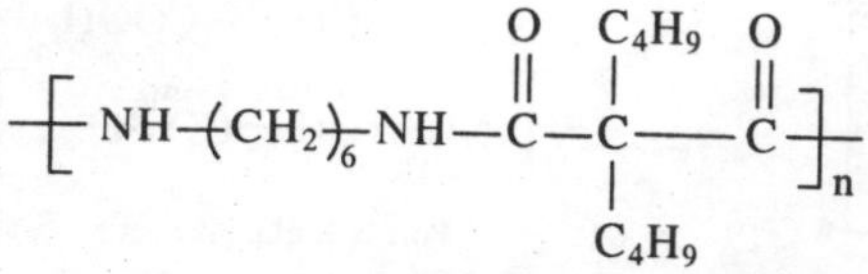

Polyhexamethylenedi-n-butylmalonamide

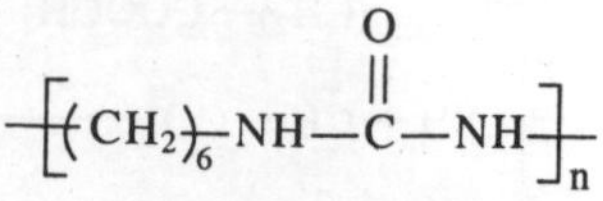

Polyhexamethylene urea

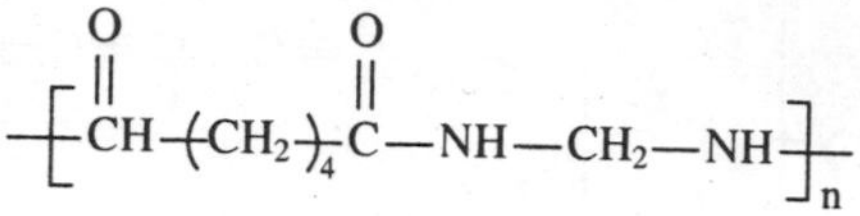

Polymethyleneadipamide (1,6 Nylon)

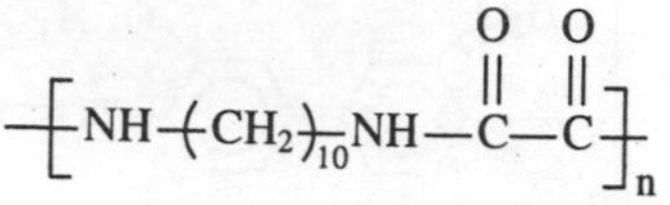

Polydecamethyleneoxamide

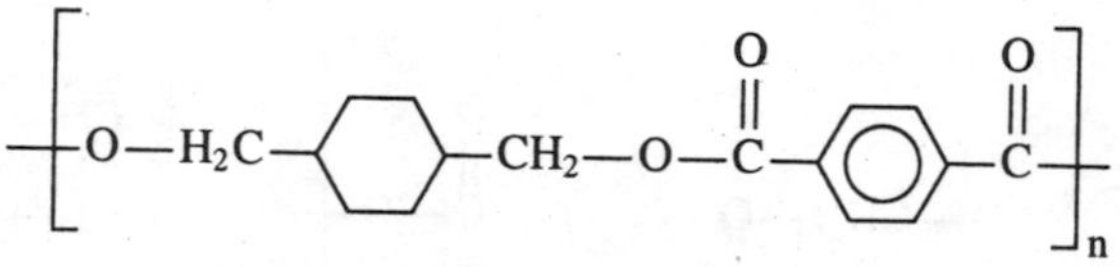

Poly 1,4-cyclohexanedicarbinyl terephthalate

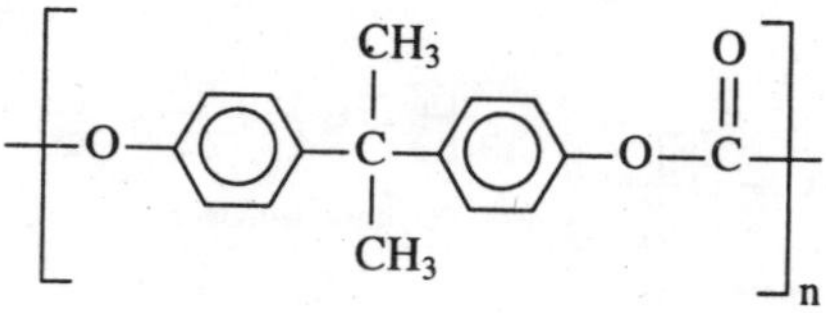

Polycarbonate

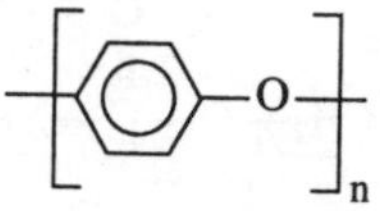

Polyphenylene oxide

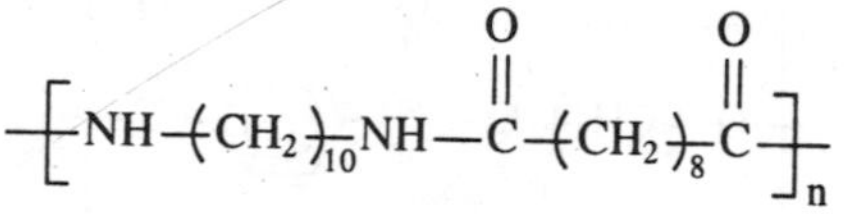

Polydecamethylenesebacamide (10,10 polyamide)

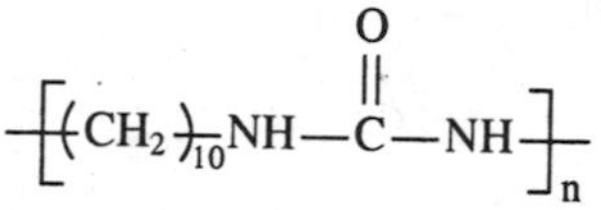

Polydecamethylene urea

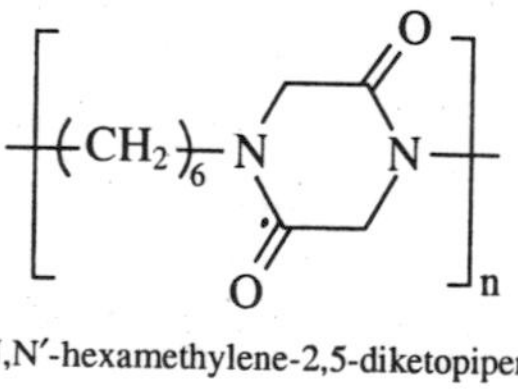

Poly N,N'-hexamethylene-2,5-diketopiperazine

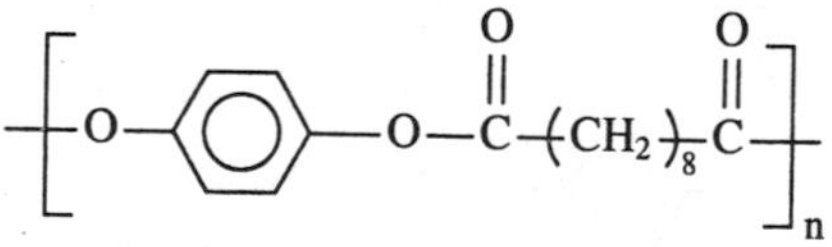

Poly 1,4-phenylene sebacate

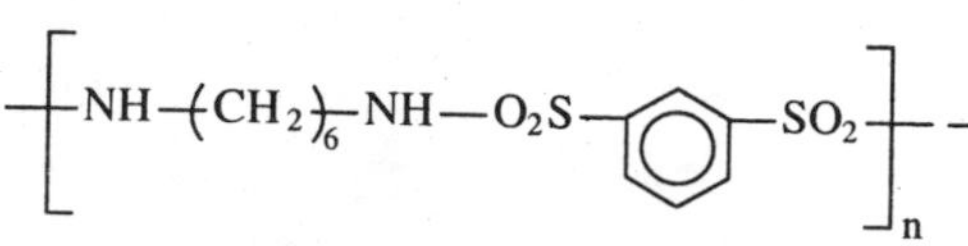

Polyhexamethylene m-bezenedisulfonamide

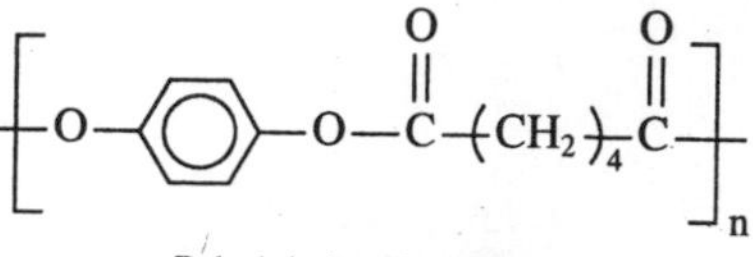

Poly 1,4-phenylene adipate

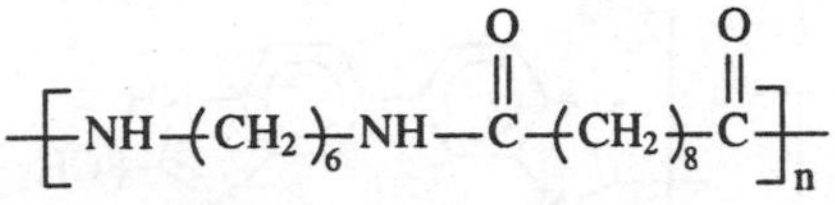

Polyhexamethylenesebacamide (nylon 6,10)

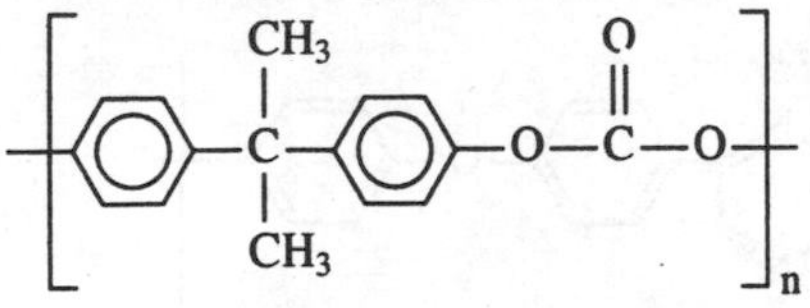

Poly 2,2-propanebis 4-phenyl carbonate

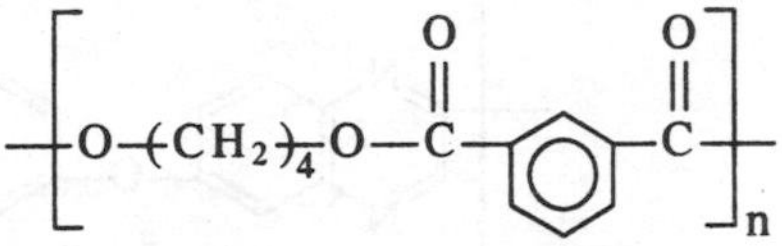

Polytetramethylene isophthalate

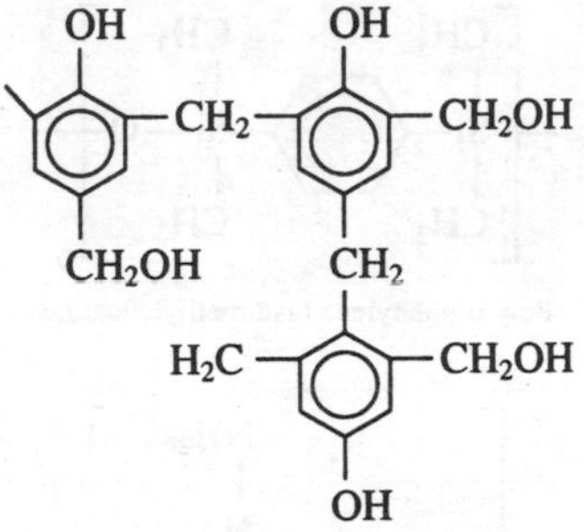

Phenol-formaldehyde resins

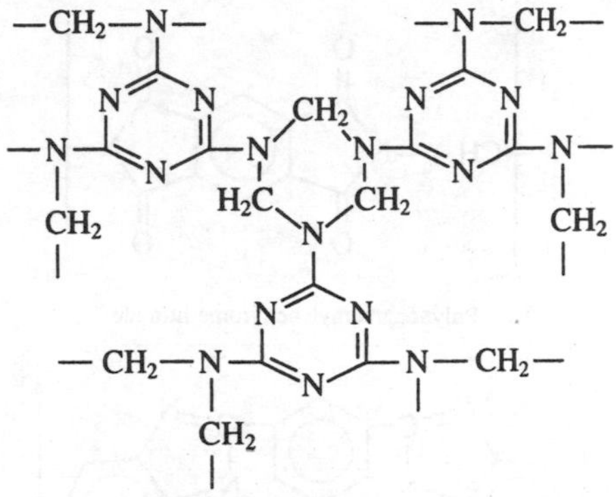

Melamine-formaldehyde resins

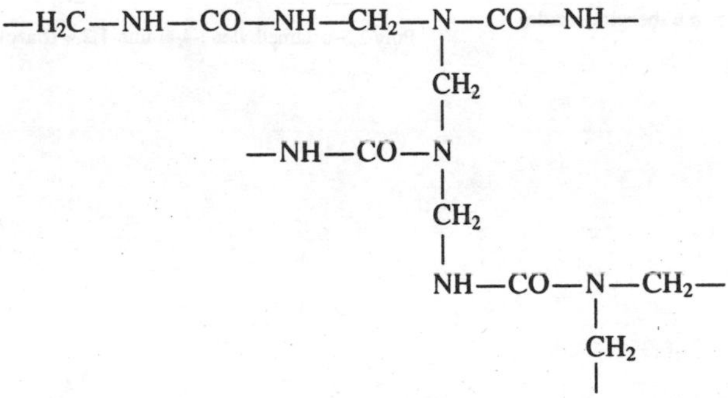

Urea-formaldehyde resins

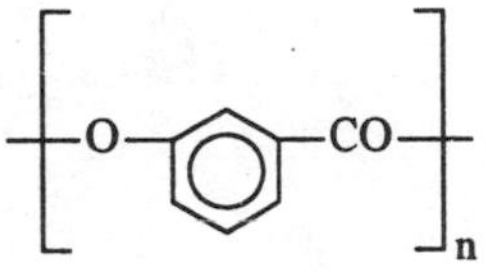

Poly m-phenyl carboxylate

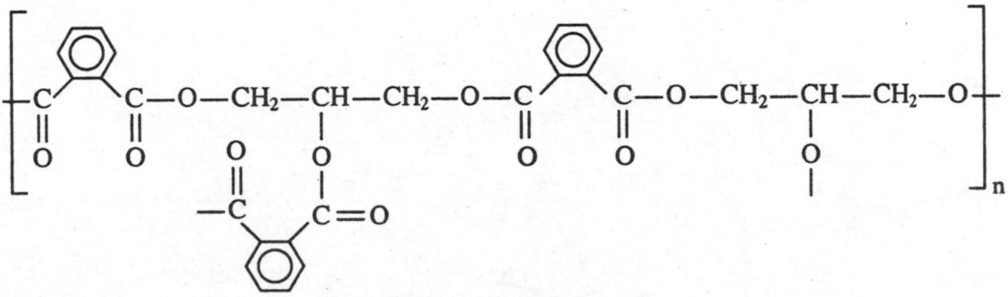

Polyglyceryl phthalate

Polybenzobenzimidazole

Poly 2,2'-octamethylene-5,5'-dibenzimidazole

Poly 2,2'-4,4' oxyphenylene-6,6'-oxydiquinoxaline

Polynonamethylenepyromellitimide

Poly p-phenylene bisdimethylsiloxane

Poly 2,2'-m-phenylene-5,5'(6,6'-benzimidazole)

Poly 3,5-octamethylene-4-amino-1,2,4-triazole

Appendix H

ASTM Methods

ASTM	Type of test or method
C 177	To measure the thermal conductivity of an insulating plastic
D 149	Measurement of dielectric strength of an insulating polymer material
D 150	Determination of dielectric constant and dissipation factor for a insulating polymer material
D 265	To measure the relative susceptibility of a standard test thick polymer specimen to pendulum type impact load (or Izod – Charpy impact test)
D 495	Estimation of arc resistance i.e. ability of plastic material to resist the action of high voltage arc
D 523	To measure the surface relative shininess on a surface of plastic material at a specified angle by using gloss meter
D 542	Measurement of refractive index of transparent plastic materials by Abbe refractometer and microscopical methods
D 543	Evaluation of resistance offered by plastics to chemical agents when they are immersed in the latter (immersion test)
D 568	To evaluate the flammability of flexible plastics
D 570	Determination of water absorption and hygroscopicity of plastic materials by oven drying and TVI drying test
D 630	To evaluate the flammability of semi flexible or self supporting plastics
D 638	Tensile test for estimation of tensile elongation and tensile modulus for semi-finished or finished plastic materials
D 648	To monitor the short term effects of temperature on a polymer material by determining heat deflection temperature (HDT)
D 671	To determine the resistance to deterioration of plastic specimen from cyclic stress (or flexural fatigue test)
D 695	Determination of compressive properties (i.e. behavior of a plastic material under applied compressive load) viz. modulus of elasticity, yield stress, extent of deformation, compressive strain and strength and slender ratio etc. of polymer plastic materials
D 696	To measure the coefficient of linear thermal expansion of plastic material
D 731	Molding index of thermosetting resin molding powder by cup flow test
D 732	To determine the shear strength of a molded plastic or a cut from a polymer sheet against an applied load
D 746	To measure the temperature at which plastics become rigid and brittle (brittleness temperature)

ASTM	Type of test or method
D 785	Determination of Rockwell hardness of a polymer, plastic material
D 790	To estimate flexural properties viz. flexural modulus or modulus of elasticity of plastic materials
D 792	Determination of specific gravity of a plastic material (in the sheet, rod, tube and moulded forms) and (in the form of moulding powder, flakes and pellets)
D 794	To monitor the long term heat resistance of polymeric materials
D 864	To measure the coefficient of cubical thermal expansion of plastics
D 1003	To measure the appearance of cloudiness (or haze) in the polymer when the light is scattered from the bulk or surface of the specimen
D 1044	To test the resistance of transparent plastic materials to abrasion
D 1238	Melt index test for polymers
D 1435	Evaluation of stability of plastic materials to out door exposures such as climatic changes polluted environment and moisture variation in atmosphere etc.
D 1499	Determination of resistance offered by plastic materials when they are exposed to radiation produced by carbon arc lamps
D 1505	Determination of density of plastic materials by density gradient method
D 1525	To monitor the short term effects of temperature on polymer material by determining vicat softening temperature
D 1692	Measurement of flammability of cellular plastics in a horizontal position
D 1693	Evaluating the widening of a brittle fracture made on the surface of polyethylene sample in the presence of external crack sensitizing agents such as detergents, lubricants, sunlight or oil etc. (environmental stress cracking resistance test)
D 1703	Measurement of apparent viscosity or melt index of polymer by capillary rheometer
D 1712	Sulfide staining test for measuring the resistance offered by plastics to hydrogen sulfide
D 1729	Matching or evaluation of color differences of plastic materials with reference to a standard by using calorimeter, or spectrophotometer
D 1822	To measure the tensile impact strength of a very thin or flexible polymer specimen
D 1824	Measurement of viscosity of concentrated polymer solutions, or polymer melts by Brookfield viscometer based on rotational rheology
D 1895	Determination of apparent (bulk) density, bulk factor and pourability of plastic material
D 1929	To determine the ignition temperature of a polymer material specimen
D 1939 (and D 2393)	Estimation of resistance of plastics especially ABS plastics to exposure to acetic acid by acetic acid immersion test
D 2152	Estimation of resistance of polyvinyl chloride to acetone by acetone immersion test
D 2236	To monitor the short term effects on a polymer sample at elevated temperatures by torsion pendulum test
D 2240	To determine the resistance of a soft plastic material to permanent deformation, indentation, or scratching by durometer (or Durometer hardness)
D 2290	Evaluation of resistance of plastics for the development of stains caused by staining agents present in food, cosmetic, detergents and pharmaceuticals etc.

D 2471	Estimation of gel time and peak exothermic temperature of thermosetting polymer resins
D 2565	Determination of resistance offered by plastic material when they are exposed to light produced by water cooled xenon arc
D 2583	To determine the Barcol hardness of both reinforced or non-reinforced rigid plastic materials
D 2843	Measurement of density of smoke generated from burning plastics (smoke burn test)
D 2857	Determination of dilute solution viscosity of polymers
D 3014	Comparison of relative extent, time of burning and loss of weight of rigid cellular plastics (against a standard), in vertical position
D 3123	Spiral flow of low pressure moulded thermosetting polymer resins
D 3417 (and D 3418)	Studying of temperature induced phase changes in polymers, degree of curing process and thermal degradation of polymers by differential scanning calorimetry (DSC), thermogravimetry analysis (TGA) and thermomechanical analysis (TMA)
D 3593	Determinaton of molecular weight distribution (MWD) in polymers by gel permeation chromatography (GPC)
E 162	To estimate the surface flammability of a plastic material by using a radiant heat energy source
G 21	To estimate the resistance of plastic materials for fungi
G 22	To estimate the resistance of plastic materials to bacteria
G 53	Monitoring the deterioration or degradation of polymers by the effects due to their exposure to fluorescent ultraviolet light and condensation

Appendix I

SI Prefixes

10^{18}	exa	E	10^{-18}	atto	a
10^{15}	peta	P	10^{-15}	femto	f
10^{12}	tera	T	10^{-12}	pico	p
10^{9}	giga	G	10^{-9}	nano	n
10^{6}	mega	M	10^{-6}	micro	μ
10^{3}	kilo	k	10^{-3}	milli	m
10^{2}	hecto	h	10^{-2}	centi	c
10^{1}	deca	da	10^{-1}	deci	d

Appendix J

Some Fundamental Physical Constants

	Symbol	Unit	Value
Avogadro constant	N_A	mol^{-1}	6.0222×10^{23}
Boltzmann constant	k	$J\ K^{-1}$	1.3807×10^{-23}
Planck constant	h	$J\ s$	6.6262×10^{-34}
Faraday constant	F	$C\ mol^{-1}$	9.6485×10^4
Gas constant	R	$J\ mol^{-1}\ K^{-1}$	8.3143
		$bar\ cm^3\ mol^{-1}\ K^{-1}$	83.144
		$atm\ cm^3\ mol^{-1}\ K^{-1}$	82.057
Elementary charge	e	C	1.6022×10^{-19}
Speed of light in vacuum	C_o	$m\ s^{-1}$	2.998×10^8
Permittivity of vacuum	ε_o	$F\ m^{-1}$	8.8542×10^{-12}
Permeability of vacuum	μ_o	$H\ m^{-1}$	$4\ \mu \times 10^{-7}$
Nuclear magneton	μ_N	$J\ T^{-1}$	5.0509×10^{-27}

Appendix K

Physical Quantities and SI Units

Physical quantity	Unit	Symbol	Value in SI units
time	minute	min	60 seconds
length	angstrom	Å	10^{-10} m = 0.1 nm
mass	unified atomic mass unit	$U = m\,a\,(^{12}C)/12$	1.6605×10^{-27} kg
area	barn	b	10^{-28} m^2
volume	liter	L	10^{-3} m^3
current	ampere	I	
energy	electron volt	eV	1.60218×10^{-19} J
pressure	bar	bar	10^5 Pa
plane angle	degree	°	$(\pi/180)$ rad
temperature	celsius temperature	°C	(T/K) − 273.15

Appendix L

Important Derived SI Units of Physical Quantities with their Symbols

Physical quantity		Physical unit			Unit
Name	Symbol	Name	Symbol	Derived	Base
frequency	ν	hertz	Hz	s^{-1}	
force	F	newton	N	$J\,m^{-1}$	$m\,kg\,s^{-1}$
energy/work/heat	E	joule	J	J	N m
power	P	watt	W	$J\,s^{-1}$	$m^{-1}\,kg\,s^{-2}$
pressure, stress	p, σ	pascal	Pa	$N\,m^{-2}$	$m^{-1}\,kg\,s^{-2}$
electric charge	Q	coulomb	C	As	
electric potential	U	volt	V	$J\,C^{-1}$	$m^2\,kg\,s^{-3}\,A^{-1}$
electric capacitance	C	farad	F	CV^{-1}	$m^{-2}\,kg^{-1}\,s^4\,A^2$
resistance/ impedance	R	ohm	Ω	VA^{-1}	$m^{-2}\,kg\,s^{-3}\,A^{-2}$
electric conductance	G	siemens	S	Ω^{-1}	$m^{-2}\,kg^{-1}\,s^3\,A^2$
electric inductance	L	henry	H	$VA^{-1}\,s$	$m^2\,kg\,s^{-2}\,A^{-2}$
electric dipole moment	μ	debye	D cm	3.335×10^{-30}	10^{-18} Fr cm
magnetic flux density	B	tesla	T.	$Vs\,m^{-2}$	$kg\,s^{-2}\,A^{-1}$
magnetic flux	ϕ	weber	Wb	Vs	$m^2\,kg\,s^{-2}\,A^{-1}$
luminous flux		lumen	lm	cd sr	
illuminance		lux	lx	$lm\,m^{-2}$	
absorbed radiation dose	D	gray	Gy	$J\,kg^{-1}$	$m^2\,s^{-2}$
radiation dose equivalent		sievert	Sv	$J\,kg^{-1}$	
radioactivity	A	becqueral	Bq	s^{-1}	
dynamic viscosity	η	poise	cP	1 mPa s	$m^{-1}\,kg\,s^{-1}$
kinematic viscosity	ν	stokes	St	$1.10^{-4}\,m^2\,s^{-1}$	
heat capacity	C_p or Cv	clausius	Cl	$4.184\,J\,K^{-1}$	

Appendix M

Other Units and Conversion

1 Å (1 A) = 10^{-10} m
1 light year (1 ly) = 9.461×10^{15} m
1 astronomical unit (1 au) = 1.496×10^{11} m
1 inch (1 in) = 2.540 cm = 0.0254 m
1 foot (1 ft) = 30.48 cm = 0.3048 m
Land mile (1 mile) = 1609.344 m = 1.609 km
1 year (1 y) = 3.156×10^7 s
1 liter (1l) = 1000 cm^3 = 10^{-3} m^3
1 pound (1 lb) = 453.592 g = 0.453592 kg
1 calorie (1 cal) = 4.1868 J
1 electron volt (1 eV) = 1.60218×10^{-19} J
1 horsepower (1 hp) = 745.7 W
1 atmosphere (1 atm) = 1.013×10^5 Pa = 0.1013 M Pa
1 mm of mercury (1 mm Hg) = 133.322 Pa = 0.000133 M Pa
degree celsius (1°C) = (T/K) − 273.15
degree fahrenheit (1 °F) = ((T/°C) × 9/5)
British thermal unit (1 BTU) = 1.05579 kJ
dyne (1 dyn) = 1×10^{-5} N

Appendix N

Definition of the Basic Units

Ampere: The ampere is that constant current which, if maintained in two straight parallel conductors of infinite length of negligible circular cross section and placed one meter apart in a vacuum, would produce between these conductors a force equal to 2×10^7 newton/metre of length.

Candela: The candela is the luminous intensity in the perpendicular direction of a surface of 1/600 000 square metre of a black body at the temperature of the freezing platinum under a pressure of 101 325 newton per square metre.

Kelvin: The kelvin unit of thermodynamic temperature is the fraction 1/273.16 of the thermodymamic temperature of the triple point of water.

Kilogramme: The kilogramme is the unit of mass and is equal to the mass of the international prototype of the kilogramme (kept in Sevres).

Mass fraction: It is the mass of a given component divided by total mass of all the i, components, i.e. $m_1 = m_1/\Sigma_i \, m_i$. Mass fraction is also called weight fraction, since same gravity field is applied every time under normal conditions.

Metre: The metre is the length equal to 1650763.73 wavelengths in vacuum of the radiation corresponding to the transition between the levels $2p_{10}$ and $5d_5$ of the Krypton–86 atom.

Mole: The mole is the amount of substance, which contains as many elementary units as there are atoms in 0.012 kg of C^{12}. The elementary units must be specified and may be an atom, a molecule, an ion, an electron, a photon etc. or a given group of such entities.

Mole Fraction: It is the number of moles, n_1 of a given component divided by total number of moles of i component, i.e. $x_1 = n_1/\Sigma_i \, n_i$. Mole i.e. unit representing the amount of substance. The amount of substance equal to one mole is invariably equal to its molecular weight. In other words mole is defined as weight of substance divided by its molecular weight.

Radian: The angle subtended at the centre of a circle by an arc of the circle equal in length to the radius of the circle.

Second: The second is the duration of 9 192 631 770 periods of the radiation corresponding to the transition between the two hyperfine levels of the ground state of the cesium–133 atom.

steradian: The solid angle subtended at the centre of a sphere by an area on the surface of the sphere equal in magnitude to the area of a square having sides equal in length to the radius of the sphere.

Volume fraction: It is the volume of a given component divided by the sum of volumes of all the i, components, i.e. $\phi_1 = V_1/\Sigma_i V_i$. The volumes refer to the volume of components before mixing.

General References

Encyclopedias, Dictionaries and Source Books
Alger, M.S.M., *Polymer Science Dictionary*, London, Chapman and Hall, 1994.
Ash, M., and I. Ash (eds), *Encyclopedia of Plastics, Polymers, and Resins*, 4 vols. New York: Chemical Publishing, 1987.
Brandrup, J., and E.H. Immergut (eds.), Polymer Handbook, 3d ed., New York, Wiley, 1989.
Carley, J.F. (ed.), Whiltington's Dictionary of Plastics, 3d ed., Lancester, Pa,: Technomic, 1993.
Chanda, M., and S.K. Roy, Plastics Technology Handbook, New York: Dekker, 1987.
Cheremisinoff, N.P. (ed.), Handbook of Polymer Science and Technology, multivolume, New York: Dekker, 1994.
Salamone, J.C. (ed.), Polymeric Materials Encyclopedia, Boca Raton: CRC Press, 1996.
Harper, C.A. (ed.), Handbook of Plastics, Elastomers, and Composites, 2nd ed., New York: McGraw-Hill, 1992.
Heath, R.J., and A.W. Birley, Dictionary of Plastics Technology, New York: Chapman and Hall, 1993.
Kroschwitz, J.I. (ed.), Polymers, an Encyclopedic Source Book of Engineering Properties, New York: Wiley, 1987.
Mark, H., C. Overberger, G. Menges, N.M. Bikales, ets., Encyclopedia of Polymer Science and Engineering, 2d ed., 18 vols. and 2 supplements, New York: Wiley, 1985.
Rubin, I.I., (ed.), Hand book of Plastic Materials and Technology, New York: Wiley, 1992.
Whelan, T., Polymer Technology Dictionary, London: Chapman and Hall, 1994.

Review Journals
Advances in Polymer Science
Polymer Science and Technology

Research Journals
Angewandte Makromolecular Chemie
British Polymer Journal
Colloid and Polymer Science
Die Makromolecular Chemie
European Polymer Journal
High Performance Polymer
J. Macromolecular Science
J. Polymer Science, Polymer Chemistry, Polymer Physics and Polymer Letters (Eds.)
J. Applied Polymer Science
International Journal of Biological Macromolecules
International Journal of Polymeric Materials

Macromolecules
Polymer
Polymeric Acta
Polymer Bulletin
Polymer Composites
Polymer Journal
Polymer J. USSR
Polymer Materials

Besides these journals a good number of papers on polymers are published in Nature, J. American Chemical Society, J. Physical Chemistry, Langmuir and J. Organic Chemistry.

Index